D0152709

Parallel Algorithms

CHAPMAN & HALL/CRC
Numerical Analysis and Scientific Computing

Aims and scope:
Scientific computing and numerical analysis provide invaluable tools for the sciences and engineering. This series aims to capture new developments and summarize state-of-the-art methods over the whole spectrum of these fields. It will include a broad range of textbooks, monographs and handbooks. Volumes in theory, including discretisation techniques, numerical algorithms, multiscale techniques, parallel and distributed algorithms, as well as applications of these methods in multi-disciplinary fields, are welcome. The inclusion of concrete real-world examples is highly encouraged. This series is meant to appeal to students and researchers in mathematics, engineering and computational science.

Proposals for the series should be submitted to one of the series editors above or directly to:
CRC Press, Taylor & Francis Group
4th, Floor, Albert House
1-4 Singer Street
London EC2A 4BQ
UK

Published Titles

Numerical Linear Approximation in C
Nabih N. Abdelmalek and William A. Malek

Parallel Algorithms
Henri Casanova, Arnaud Legrand, and Yves Robert

Parallel Iterative Algorithms
Jacques M. Bahi, Sylvain Contassot-Vivier, and Raphael Couturier

Parallel Algorithms

Henri Casanova
Arnaud Legrand
Yves Robert

CRC Press is an imprint of the
Taylor & Francis Group, an **informa** business
A CHAPMAN & HALL BOOK

Cover photo by Doug Young.

Chapman & Hall/CRC
Taylor & Francis Group
6000 Broken Sound Parkway NW, Suite 300
Boca Raton, FL 33487-2742

Printed in the United States of America on acid-free paper
10 9 8 7 6 5 4 3 2 1

International Standard Book Number-13: 978-1-58488-945-8 (Hardcover)

Library of Congress Cataloging-in-Publication Data

Casanova, Henri.
Parallel algorithms / Henri Casanova, Arnaud Legrand, and Yves Robert.
p. cm. -- (Numerical analysis and scientific computing ; 3)
Includes bibliographical references and index.
ISBN 978-1-58488-945-8 (alk. paper)
1. Parallel algorithms. 2. Parallel programming (Computer science) I. Legrand, Arnaud. II. Robert, Yves. III. Title. IV. Series.

QA76.642.C39 2008
005.2'75--dc22 2008019142

Visit the Taylor & Francis Web site at
http://www.taylorandfrancis.com

and the CRC Press Web site at
http://www.crcpress.com

Contents

Preface

Parallel computing has undergone a stunning evolution, with high points (e.g., being able to solve many of the grand-challenge computational problems outlined in the 80s) and low points (e.g., the demise of countless parallel computer vendors). Today, parallel computing is omnipresent across a large spectrum of computing platforms. At the "microscopic" level, processor cores have used multiple functional units in concurrent and pipelined fashions for years, and multiple-core chips are now commonplace with a trend toward rapidly increasing numbers of cores per chip. At the "macroscopic" level, one can now build clusters of hundreds to thousands of individual (multi-core) computers. Such distributed-memory systems have become mainstream and affordable in the form of commodity clusters. Furthermore, advances in network technology and infrastructures have made it possible to aggregate parallel computing platforms across wide-area networks in so-called "grids."

Objective

The aim of this book is to provide a rigorous yet accessible treatment of parallel algorithms, including theoretical models of parallel computation, parallel algorithm design for homogeneous and heterogeneous platforms, complexity and performance analysis, and fundamental notions of scheduling. The focus is on algorithms for distributed-memory parallel architectures in which computing elements communicate by exchanging messages. While such platforms have become mainstream, the design of efficient and sound parallel algorithms is still a challenging proposition. Fortunately, in spite of the "leaps and bounds" evolution of parallel computing technology, there exists a core of fundamental algorithmic principles. These principles are largely independent from the details of the underlying platform architecture and provide the basis for developing applications on current and future parallel platforms. This book identifies and synthesizes fundamental ideas and generally applicable algorithmic principles out of the mass of parallel algorithm expertise and practical implementations developed over the last decades.

Intended Audience and Use

The target audience for this book is graduate students and post-graduate researchers in computer science and related fields. Each chapter is organized in three parts: (i) lecture material with many proofs, examples, and case-studies;

(ii) a set of exercises; and (iii) solution sketches for exercises marked with a $\diamond$. This book should be ideally suited for teaching a course on parallel algorithms, or as a complementary text for teaching a course on high-performance computing. Importantly, although most of the content of the book is about algorithm design and analysis, it is nevertheless a sound basis for teaching applied parallel programming. Many of the included examples, case studies, and exercises are natural starting points for hands-on homework assignments.

Book Content and Organization

Only a very brief introduction to parallel computing is provided because this book is intended for a graduate-level audience. We refer the reader to, for instance, the book by El-Rewini and Lewis [52] for an introduction to the field if needed. Also, it is important to note that this is not a "high-performance computing" book. Indeed, this book provides very little content pertaining to parallel architectures, programming for the memory hierarchy, and the like. We refer the reader to the book by Culler and Singh [48] and to the more recent book by El-Rewini and Abd-El-Barr [51] for details on parallel architectures and high-performance computing programming tools and techniques.

This book is structured in three distinct parts: Models (Chapters 1–3), Parallel Algorithms (Chapters 4– 6), and Scheduling (Chapters 7–8).

Chapters 1 and 2 cover two classical theoretical models of parallel computation: PRAMs and sorting networks. Both models provide theoretical foundations for reasoning about parallel algorithms, but also point to particular techniques that can be used to implement efficient parallel algorithms in practice.

Chapter 3 discusses network models, both for topology and performance, and defines several classical communication primitives that will serve as the foundations for many of the parallel algorithms presented in the next two chapters.

Chapters 4 and 5 discuss parallel algorithms on ring and grid logical topologies, two useful and popular abstractions. Both chapters expose fundamental issues and trade-offs encountered when designing and implementing parallel algorithms in practice.

Chapter 6 deals with the issue of load balancing on heterogeneous computing platforms. This topic is highly relevant to emerging computing platforms, for instance, those that aggregate several clusters together.

Chapter 7 introduces the notion of scheduling, or the art of orchestrating communication and computation to achieve higher performance. This chapter presents fundamental results and approaches for common scheduling problems that arise when developing parallel algorithms. Chapter 6 describes advanced scheduling topics. These topics include divisible load scheduling and steady-state scheduling, which are crucial because they are directly applicable to the popular master-worker application model. The last section discusses the automatic parallelization of loop nests.

Although the content in the more theoretical chapters and that in the more applied chapters are complementary, it is possible to cover only a subset of the chapters. For a course focused on theoretical foundations of parallel algorithm design, one may opt for covering only Chapters 1, 2, 7, and 8. For a course focused on more practical algorithm design, one may opt for covering only Chapters 3, 4, 5, and 6.

Acknowledgments

The content of this book, or at least preliminary versions of it, has been used to teach courses at École Normale Supérieure de Lyon, École Polytechnique, the University of Tennessee, Knoxville, and the University of Hawai'i at Mānoa. We are grateful to the students for their feedback and suggestions.

We also wish to thank the following people, who have contributed to some of the content by their insightful suggestions, their own previously published work, or their help reviewing draft chapters: Olivier Beaumont, Mahdi Belcaid, Anne Benoit, Rémi Bertin, Vincent Boudet, Benjamin Depardon, Larry Carter, Alain Darte, Frédéric Desprez, Jack Dongarra, Jeanne Ferrante, Matthieu Gallet, Philip Johnson, Jean-Yves L'Excellent, Loris Marchal, Fabrice Rastello, Arnold Rosenberg, Veronika Rehn-Sonigo, Mark Stillwell, Marc Tchiboukdjian, Jean-Marc Vincent, Frédéric Vivien, and Joshua Wingstrom.

Finally, we wish to thank Doug Young for providing the art for the cover of this book.

Henri Casanova
Arnaud Legrand
Yves Robert

Part I

Models

Chapter 1

PRAM Model

A Turing machine is a simple abstract computational device intended to investigate the extent of what can be computed on a sequential computer. This model enables one to define the cost of an algorithm (e.g., in number of operations) *precisely* and is the basis for complexity results (e.g., polynomial vs. NP-complete, etc.).

In view of the wide variety of parallel architectures, defining a precise yet general model for parallel computers seems hopeless. Perhaps most daunting is the modeling of the costs associated to data communication within the parallel computer. It turns out that a reasonable option is to simply ignore communication costs altogether! One then obtains an imperfect model in which the cost of an algorithm only bears a vague relationship to the algorithm's execution times on real-world parallel computers. However, this model makes it possible to determine a precise *classification* of problems and algorithms and to obtain complexity results (e.g., to show that a given algorithm is optimal, to establish the minimal complexity of a problem).

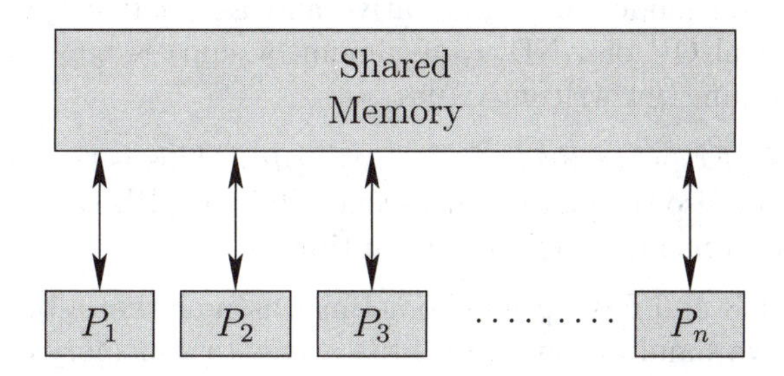

FIGURE 1.1: The PRAM model.

The PRAM model, which stands for Parallel RAM,[1] comprises a shared central memory that can be accessed by the various *processing units* (see Figure 1.1), or PUs. All PUs execute synchronously the same algorithm and work on distinct (or not) memory areas. In this model neither the number of PUs nor the size of the memory is bounded, which clearly does not hold

[1]A RAM, or random access machine, is an abstract model of a sequential processor.

for real-world parallel computers. Another simplifying assumption is that any PU can access any memory location in one unit of time. This assumption would also never hold in the real world. It is physically impossible to ensure that a large number of PUs can simultaneously access an arbitrarily large shared memory without incurring prohibitive overhead. In this model, the same memory location, or *cell*, can be accessed simultaneously by multiple PUs. For this reason, one typically considers three different PRAM models that specify which concurrent accesses to a memory cell are allowed:

- **CREW** (*Concurrent Read Exclusive Write*): This is the most common model. It is assumed that at a given time, i.e, during a given step of an algorithm, arbitrarily many PUs can read the value of a cell simultaneously while at most one PU can write a value to a cell.
- **CRCW** (*Concurrent Read Concurrent Write*): This is the most powerful model, with an unbounded number of simultaneous writes to a memory cell. Several possible "modes" are used to define the semantics of concurrent writes to a single memory cell:
  - Consistent mode: All PUs must write the same value.
  - Arbitrary mode: PUs can write different values but only *one of them* will be written. This mode does not specify which value is written, so the algorithm designer must be prepared for a non-deterministic execution.
  - Priority mode: The value written by the PU with the lowest index is chosen.
  - Fusion mode: A commutative and associative operation (e.g., a logical OR or AND, a maximum, a sum) is applied on-the-fly to the different written values.
- **EREW** (*Exclusive Read Exclusive Write*): This is the more restrictive model but also the more realistic one. Only one PU can have read/write access to a memory cell at a given time.

The above ER and EW options sometimes assume that a bounded number of PUs can have simultaneous read/write access to a memory cell, as opposed to just a single processor. In the EW case one must then specify the mode of concurrent writes, as in the CW case.

1.1 Pointer Jumping

In this section, we present three simple PRAM algorithms that all rely on a fundamental technique called *pointer jumping*. As we will see, this is a natural extension of the *divide and conquer* technique to a parallel setting.

1.1.1 List Ranking

Let us assume that we have a linked list L of n objects arbitrarily distributed throughout our PRAM's memory. We wish to determine the distance $d[i]$ to the end of the list for every element i:

$$d[i] = \begin{cases} 0 & \text{if } next[i] = \textbf{Nil} \\ d[next[i]] + 1 & \text{if } next[i] \neq \textbf{Nil} \end{cases}.$$

A straightforward sequential algorithm consists in propagating and incrementing the d value from the end of the list toward the beginning of the list in n steps (assuming that the list is doubly-linked). The question is whether there is a PRAM algorithm with a complexity lower than $O(n)$. Note that we assume that each algorithm step is executed in one time unit, so we use the number of steps, the execution time, and the complexity interchangeably. It turns out that one can design an algorithm with $O(\log n)$ complexity, by associating each element of the list, i, to a different processor, P_i. This is possible because the PRAM model assumes an unbounded number of PUs.

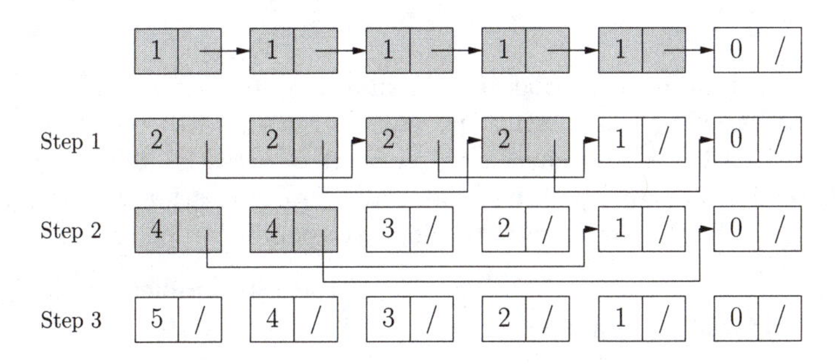

FIGURE 1.2: Typical execution of the list ranking algorithm. Gray cells indicate active values, i.e., values that are in the process of being computed.

An example execution of the list ranking algorithm is depicted in Figure 1.2 for $n = 6$. The principle is straightforward: At each step one divides each list into two sublists. For instance, in the first step, all elements with even indices are placed in a sublist, and all elements with odd indices are placed in another sublist. The size of each sublist is divided by two at each step, leading to a number of steps equal to $\lceil \log n \rceil$. [2]

We show the pseudo-code for our list ranking algorithm in Algorithm 1.1. The **forall** i **in parallel do** loop is executed in parallel across all the PUs that are responsible for the memory cells referenced in the body of the loop. Care must be taken, however, to ensure that all iterations of the loop can be

[2]As in most computer science books, all logarithms are calculated in base 2 and we use $\log n$ to denote $\log_2 n$.

```
1 RANK_COMPUTATION(L)
2   forall i in parallel do                                    { Initialization }
3     if next[i] = Nil then d[i] ← 0 else d[i] ← 1
4   while there exists a node i such that next[i] ≠ Nil do     { Main loop }
5     forall i in parallel do
6       if next[i] ≠ Nil then
7         d[i] ← d[i] + d[next[i]]
8         next[i] ← next[next[i]]
```

ALGORITHM 1.1: List ranking algorithm.

executed safely in parallel. For instance, consider the following loop (assuming that indices for array A start at 1 and that PU P_i is responsible for updating array element $A[i]$):

```
forall i in parallel do
  if i > 1 then
    A[i − 1] ← A[i]
```

The problem here is that processor P_2 may update $A[2]$ before P_1 can read it. The same, if not as glaring, problem occurs in statements 7 and 8 of Algorithm 1.1. To ensure that the loop executes correctly, it suffices to ensure that all read operations happen before all write operations. Therefore, the semantic of a **forall** i **in parallel do** loop is assumed to be as follows:

```
                                forall i in parallel do
forall i in parallel do           temp[i] ← B[i]
  A[i] ← B[i]            ⟺     forall i in parallel do
                                  A[i] ← temp[i]
```

Another problem with the algorithm as it is written in Algorithm 1.1 is that the test in Statement 4 can be done in constant time only on a CRCW PRAM. On a CRCW PRAM, the test could be implemented by having all PUs write their boolean value of $next[i] =$ **Nil** to a single memory cell, say *done*. Using a CRCW with a fusion mode for concurrent writes based on a logical AND, *done* is true only once the algorithm has completed. Unfortunately, there is no such constant time solution on a CREW PRAM. On a CREW PRAM, the best approach is to perform pair-wise logical AND operations in a binary tree pattern, leading to $O(\log n)$ steps. We will further discuss this distinction between a CRCW and a CREW PRAM in Section 1.3.1.

In the particular case of our list ranking algorithm, we know that the algorithm proceeds in $\lceil \log n \rceil$ steps. Therefore, we can rewrite the main loop:

while there exists a node i such that $next[i] \neq$ **Nil do**

as

for $step$=1 **to** $\lceil \log n \rceil$ **do**

```
1  PREFIX_COMPUTATION(L)
2    forall i in parallel do                              { Initialization }
3      y[i] ← x[i]
4    while there exists a node i such that next[i] ≠ Nil do   { Main loop }
5      forall i in parallel do
6        if next[i] ≠ Nil then
7          y[next[i]] ← y[i] ⊗ y[next[i]]
8          next[i] ← next[next[i]]
```

ALGORITHM 1.2: Prefix computation of a list.

One remaining question is whether this algorithm can be executed in $O(\log n)$ time by any type of PRAM. In terms of writes, each PU P_i writes the values of $d[i]$ and of $next[i]$ in exclusive mode. In terms of reads, PUs P_i and P_j such that $next[j] = i$ may read the value of $d[i]$ concurrently, thus seemingly precluding the use of an EREW PRAM. A simple solution to overcome this problem is to introduce a temporary array and to replace

$$d[i] \leftarrow d[i] + d[next[i]]$$

by

$$\begin{aligned} temp[i] &\leftarrow d[next[i]] \\ d[i] &\leftarrow d[i] + temp[i] \end{aligned}$$

With this last simple transformation, we now obtain a $O(\log n)$ algorithm on an EREW PRAM, the most restrictive PRAM model.

1.1.2 Prefix Computation

Given a sequence $(x_1, x_2, \ldots, x_n)$ and a binary associative operation $\otimes$, we wish to compute the sequence $(y_1, y_2, \ldots, y_n)$ defined as

$$\begin{cases} y_1 = x_1 \\ y_k = y_{k-1} \otimes x_k = x_1 \otimes x_2 \otimes \ldots \otimes x_k & , \text{for } 1 < k \leqslant n. \end{cases}$$

Let us use a PRAM with n PUs. The sequence $(x_1, x_2, \ldots, x_n)$ is stored as a linked list such that $x[i] = x_i$ for $i \leqslant n$, $next[i]$ points to the cell that contains $x[i+1]$ for $i < n$, and $next[n] =$ **Nil**. We can devise a simple algorithm that uses the same pointer jumping technique as the list ranking algorithm in the previous section. The corresponding pseudo-code in shown in Algorithm 1.2.

An example execution of this algorithm is depicted in Figure 1.3. Just as for the list ranking problem, there are exactly $\lceil \log n \rceil$ steps and the global condition of the **while do** loop can thus easily be replaced by a **for do** loop. Likewise, concurrent read access can easily be removed to obtain a $O(\log n)$ EREW algorithm.

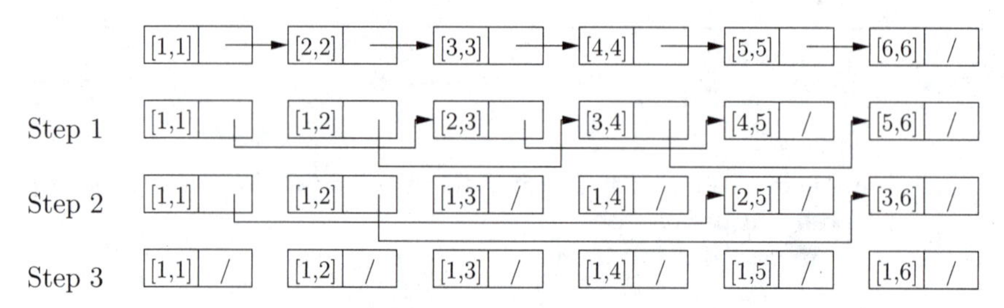

FIGURE 1.3: Example execution of the prefix computation algorithm. $[i, j]$ denotes $x_i \otimes x_{i+1} \otimes \ldots \otimes x_j$ for $i \leqslant j$.

1.1.3 Euler Tour

In this section, we present a clever application of the prefix computation in the previous section. Consider an arbitrary binary tree with n vertices. We want to compute the depth of each vertex, i.e., its distance to the root vertex. We use a simple data structure to store the tree in memory. Each vertex i is represented by three fields: *father*$[i]$, *left*$[i]$, and *right*$[i]$, which point to its father, its left child, and its right child, respectively, or **Nil** in case such parents and children do not exist.

A naive PRAM algorithm would consist of going through the tree in breadth-first fashion. The complexity is then $O(d)$, where d is the depth of the tree. This algorithm is satisfactory when the tree is balanced because in this case $d = O(\log n)$. However, when the tree is comb-shaped, $d = O(n)$. The Euler tour technique leads to a $O(\log n)$ algorithm regardless of the shape of the tree.

We associate three PUs to each vertex. We call these the A, B, and C PUs (one can think of them as three types of PUs, each with a different role in the algorithm). The principle of the algorithm is to transform the problem of computing the depth of all vertices into a single prefix computation. This is accomplished by having each PU store a value and a pointer to the value stored by another PU, thus creating a list (called the "Eulerian path"). We simply say that a PU points to another PU. We then do a prefix computation of this list. The trick consists in constructing the list so that the prefix computation leads to the desired depth values.

We depict the list for an example tree in Figure 1.4(a). The first element of the list is the value held by the A PU corresponding to the root vertex. The last element of the list is the value held by the C PU at the root also. We establish the links between the list elements as follows:

- The A PU of each vertex points to the A PU of the vertex's left child if any, or to the vertex's own B PU otherwise.
- The B PU of each vertex points to the A PU of the vertex's right child if any, or to the vertex's own C PU otherwise.

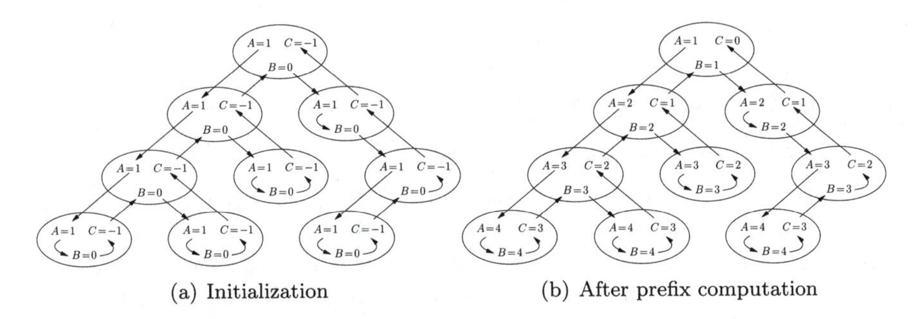

(a) Initialization (b) After prefix computation

FIGURE 1.4: Euler tour. (a) Creation of the Eulerian path and initialization. (b) Results after the prefix computation.

- If a vertex has a left child, then its C PU points to the B PU of the vertex's father; otherwise, its C PU points to the C PU of the vertex's father (the C PU of the root vertex points to **Nil**).

The above creates a depth-first path through the tree. We denote by $x[i]$ the value for which PU P_i is responsible. These values are initialized as follows:

$$x[i] = \begin{cases} 1 & \text{if } P_i \text{ is a PU of type } A, \\ 0 & \text{if } P_i \text{ is a PU of type } B, \\ -1 & \text{if } P_i \text{ is a PU of type } C. \end{cases}$$

Figure 1.4(a) shows these values, written as "PU type = value" for simplicity.

Once the above list is established and the values of its elements are initialized, which can be done in constant time on an EREW PRAM, we can perform a prefix computation for the addition operator. The reader can easily check that this computation leads to the values shown in Figure 1.4(b). The depth of a vertex is stored in the value store by the C PU associated to that vertex. In hindsight we now see the rationale for the initial values of the list elements:

- The A PU of vertex i adds 1 to the prefix sum, accounting for going down a level in the tree. More precisely, $d(\mathit{left}[i]) = d(i) + 1$ (where $d(i)$ is the depth of vertex i).

- The B PU's contribution to the sum is equal to 0 because the depth of the left child of a vertex is equal to the depth of the vertex's right child.

- The C PU's contribution to the sum is equal to -1, accounting for going up a level in the tree.

We have reduced our problem to a constant-time initialization phase and a prefix computation with $O(n)$ PUs. Therefore, using the algorithm in the previous section, we obtain an $O(\log n)$ EREW algorithm!

1.2 Performance Evaluation of PRAM Algorithms

1.2.1 Cost, Work, Speedup, and Efficiency

In this section, we give a few common definitions to evaluate the performance of a PRAM algorithm. Let P be a problem of size n that we want to solve (e.g., a computation over an n-element list). Let $T_{seq}(n)$ be the execution time of the best (known) sequential algorithm for solving P. Let us now consider a PRAM algorithm that solves P in time $T_{par}(p, n)$ with p PUs.

DEFINITION 1.1 (Cost and Work). The *cost* of a PRAM algorithm is defined as

$$C_p(n) = p.T_{par}(p, n) .$$

The *work* $W_p(n)$ of a PRAM algorithm is the sum over all PUs of the number of performed operations. The difference between cost and work is that the work does not account for PU idle time.

Intuitively, cost is a rectangle of area $T_{par}(p, n)$ (the execution time) times p (the number of PUs). Therefore, the cost is minimal if, at each step, all PUs are used to perform *useful* computations, i.e., computations that are part of the sequential algorithm. In this case, the cost is equal to the work.

The speedup of a PRAM algorithm is the factor by which the program's execution time is lower than that of the sequential program.

DEFINITION 1.2 (Speedup and Efficiency). The *speedup* of a PRAM algorithm is defined as

$$S_p(n) = \frac{T_{seq}(n)}{T_{par}(p, n)} .$$

The *efficiency* of the PRAM algorithm is defined as

$$e_p(n) = \frac{S_p(n)}{p} = \frac{T_{seq}(n)}{p.T_{par}(p, n)} .$$

Some authors use the following definition: $S_p(n) = \frac{T_{par}(1,n)}{T_{par}(p,n)}$, where $T_{par}(1, n)$ is the execution time of the algorithm with a single PU. This definition quantifies how the algorithm scales: If the speedup is close to p (i.e., is $\Omega(p)$), one says that the algorithm scales well. However, this definition does not reflect the quality of the parallelization (i.e., how much we can really gain by parallelization), and it is thus better to use $T_{seq}(n)$ than $T_{par}(1, n)$.

1.2.2 A Simple Simulation Result

In this section we present our first "simulation result." With this result one can reason about the performance of an algorithm for a given PRAM given a performance model of the algorithm for a PRAM with more PUs. The main idea is that a PRAM with fewer PUs can simulate a PRAM with more PUs by executing fewer operations concurrently (with some loss of performance).

PROPOSITION 1.1. *Let $\mathcal{A}$ be an algorithm whose execution time is t on a PRAM with p PUs. $\mathcal{A}$ can be simulated on a RAM of the same type with $p' \leqslant p$ PUs in time $O\left(\frac{t.p}{p'}\right)$. The cost of the algorithm on the smaller PRAM is at most twice the cost on the larger PRAM.*

Proof. Each step of $\mathcal{A}$ can be simulated in at most $\left\lceil \frac{p}{p'} \right\rceil$ time units with $p' \leqslant p$ PUs by simply reducing concurrency and having the p' PUs perform sequences of operations. Since there are at most t steps, the execution time t' of the simulated algorithm is at most $\left\lceil \frac{p}{p'} \right\rceil t$. Therefore, $t' = O\left(\frac{p}{p'}.t\right) = O\left(\frac{t.p}{p'}\right)$.

We also have $C_{p'} = t'.p' \leqslant \left\lceil \frac{p}{p'} \right\rceil p'.t \leqslant \left(\frac{p}{p'} + 1\right) p't = p.t\left(1 + \frac{1}{p'}\right) = C_p.(1 + 1/p') \leqslant 2C_p$. □

A direct consequence of this result is that the cost of an algorithm is at least of the same order of magnitude as the sequential complexity (otherwise, the simulation of this algorithm with a single PU would lead to a lower complexity). A PRAM algorithm is said to be *efficient* when its cost has the same order of magnitude as the complexity of the corresponding sequential algorithm (i.e., its efficiency is $\Omega(1)$). Consequently, the work of an efficient algorithm cannot be reduced by more than a constant factor.

When reducing the number of processors in Proposition 1.1, we have grouped operations from a parallel algorithm and simulated them on a single processor. In practice, it is often the case that these operations could be executed faster using a sequential algorithm directly. It is, for example, easy to design a mixed version of the prefix computation algorithm whose execution time with p PUs is $O(n/p + \log(p))$ instead of $O(n.\log(n)/p)$. Assuming the list is split into p sub-lists of n/p consecutive elements, each PU can sequentially compute the prefix for its own sub-list in time $O(n/p)$. Then, one can propagate sub-prefixes of the sub-lists in time $O(\log p)$. Lastly, one simply updates each sub-list with the new prefixes in time $O(n/p)$. Such an approach is referred to as "coarsening the grain of the algorithm" and is generally quite efficient. The same technique is used in Chapter 2 to sort an array on a one-dimensional network of processors efficiently (Section 2.2.2.)

The study of the efficiency of algorithms is extremely important in practical contexts. Indeed, in the real world, resources are not free and the efficiency can decrease sharply as p increases beyond some reasonable threshold. The typical reason for this decrease in efficiency is that when p becomes too large relative to n, communication costs become large. Further increases of p can even increase overall execution time.

There are scenarios in which it is possible to solve a problem more than p times faster using p processors than when using a single processor of the same type. In this case efficiency is greater than 1 and one speaks of *super-linear speedup*. One cause of super-linear speedup is that p processors typically have p times as much cache and memory as a single processor. Therefore, when using sufficiently many processors, the entire data for the problem at hand may suddenly fit entirely in memory (or in cache), thus avoiding costly swapping (or cache misses). It is important to understand that, although highly desirable in practice, achieving super-linear speedup does not mean that the parallel algorithm is particularly efficient. From a strictly parallel algorithms perspective, comparing a parallel execution of an algorithm that achieves super-linear speedup to a sequential execution that runs out of memory or cache is not a fair comparison. One option would be to compute a speedup relative to the execution time when using the smallest number of processors that leads to a super-linear speedup.

1.2.3 Brent's Theorem

THEOREM 1.1 (Brent). *Let $\mathcal{A}$ be an algorithm that executes a total number of m operations and that runs in time t on a PRAM (with some unspecified number of PUs). $\mathcal{A}$ can be simulated in time $O\left(\frac{m}{p}+t\right)$ PRAM of the same type that contains p PUs.*

Proof. Say that at step i $\mathcal{A}$ performs $m(i)$ operations (implying that $\sum_{i=1}^{t} m(i) = m$). Step i can be simulated with p PUs in time $\left\lceil\frac{m(i)}{p}\right\rceil \leqslant \frac{m(i)}{p}+1$. One can simply sum these upper bounds to prove the theorem. □

Brent's theorem makes it possible to quantify the performances of an algorithm when the number of PUs is reduced. Let us consider for example the computation of the maximum of a list of integers on an EREW PRAM. This computation can be done in $O(\log n)$ time by structuring the computation as a binary tree, and computing pair-wise maxima at each step until there is only one value left. The first step requires $O(n)$ PUs. But what if we had fewer than $O(n)$ PUs? Brent's theorem tells us that we can simulate the algorithm on p PUs in $O(\frac{n}{p}+\log n)$ time (we perform $m=n-1=O(n)$ comparisons). Therefore, if we choose $p=\frac{n}{\log n}$, we achieve the same execution time but with fewer PUs!

1.3 Comparison of PRAM Models

In this section, we discuss the comparative *power* of the EREW, CREW, and CRCW models.

```
1 COMPUTE_MAXIMUM(A, n)
2   forall i ∈ {1,...,n} in parallel do
3     m[i] ←True
4   forall i, j ∈ {1,...,n}^2, i ≠ j in parallel do
5     if A[i] < A[j] then m[i] ← False
6   forall i ∈ {1,...,n} in parallel do
7     if m[i] = True then max ← A[i]
8   return max
```

ALGORITHM 1.3: CRCW algorithm to compute the largest value of an array.

1.3.1 Model Separation

One may wonder whether a CRCW PRAM is more powerful than a CREW PRAM. The problem of model comparison can be stated as such: Is there a problem such that, using the same number of PUs, we can solve this problem with a CRCW PRAM strictly faster than with the best algorithm for solving this problem with a CREW PRAM?

The problem of finding the maximum value of an array gives us the answer to the previous question. With Algorithm 1.3, we can compute the largest value in array A in $O(1)$ time on a CRCW PRAM with $O(n^2)$ PUs. In this algorithm all PUs write the same value concurrently, **False**, and thus the PRAM runs in consistent mode. The first and the third loops use only n PUs whereas the second loop uses $n(n-1)$ PUs, with $P_{i,j}, i \neq j$, being responsible for the comparison $A[i] < A[j]$.

More generally, with a CRCW PRAM with n PUs one can compute the fusion $\bigotimes_{1 \leqslant i \leqslant n} e_i$ of n elements $(e_1, \ldots, e_n)$, where $\otimes$ is an associative operator, in $O(1)$ time. Such a *reduction* operation cannot be done in fewer than $\Omega(\log n)$ steps with a CREW PRAM. Indeed, with this model, at most two values can be merged into a single one in one step. Therefore, the number of values that need to be merged is halved at each step for a total of $\Omega(\log n)$ steps.

What about the CREW and EREW models then? Here again, there is a simple problem that separates these models: Determine whether a given element e belongs to a set $(e_1, \ldots, e_n)$ of n distinct elements. This problem can be easily solved in $O(1)$ steps using a CREW PRAM with n PUs. First, a boolean variable, *res*, is initialized to **False**. Then, in parallel, each PU P_i compares e with e_i (implying concurrent read access to e) and set *res* to **True** if both elements are equal. Since the e_1 are pair-wise distinct, at most one PU will modify *res* (ensuring exclusive writes).

On an EREW PRAM, PUs cannot read e simultaneously. Comparisons can be done in $O(1)$ units of time only if $\Omega(n)$ copies of e have been created, which cannot be done in less than $\Omega(\log n)$ time. More generally, broadcasting information to n PUs on an EREW PRAM requires $\Omega(\log n)$ steps, with double the number of copies of the information at each step.

1.3.2 Simulation Theorem

In the previous section, for each model, we have exhibited algorithms for which the separation factor between the different PRAM models was $\Omega(\log p)$. One may, however, wonder whether larger factors could be found. The following theorem proves that these factors are in fact maximal.

THEOREM 1.2. *Any CRCW algorithm with p PUs has an execution time of at most $O(\log p)$ times lower than the best EREW algorithm with p PUs for solving the same problem.*

Proof. Let us assume that the CRCW PRAM uses a consistent mode (only concurrent writes of the same values are authorized). We show a method to simulate concurrent writes with only exclusive writes (concurrent reads can be handled in a similar fashion).

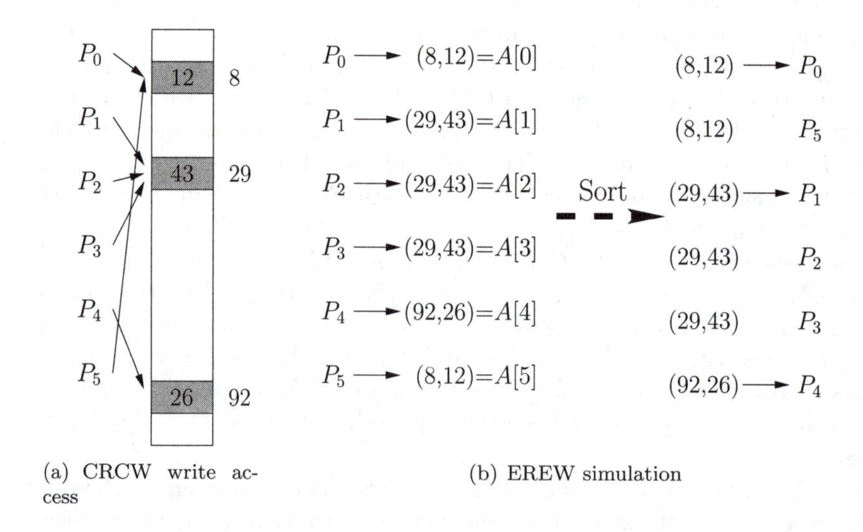

FIGURE 1.5: Simulation of concurrent writes with exclusive writes.

Let us consider a given step of the CRCW algorithm and simulate it with $O(\log p)$ steps of an EREW algorithm. Both PRAMs have the same computing power (i.e., the same number of processors). Therefore, we only need to focus on memory accesses. Our EREW algorithm requires a temporary array A of size p. When PU P_i of the CRCW algorithm writes a value x_i to address

l_i, the corresponding PU P_i of the EREW algorithm writes $A[i] = (l_i, x_i)$. We sort A in $O(\log p)$ time (let us assume for now that we have an EREW algorithm for sorting p values with $O(p)$ PUs and $O(\log p)$ steps). Then, each PU P_i reads the consecutive cells: $A[i] = (l_j, x_j)$ and $A[i-1] = (l_k, x_k)$ (where $0 \leqslant j, k \leqslant p-1$). If $i = 0$ or $l_j \neq l_k$, then P_i writes the value x_j at the address l_j. It is easy to check that these reads and writes are exclusive. This process is depicted in Figure 1.5 for an example.

All we need to do to complete this proof is to design an EREW algorithm that sorts p values with $O(p)$ PUs in $O(\log p)$ steps. This is done in the following section for the CREW model. One can then use known techniques to transform this CREW algorithm into an EREW algorithm [43, 100]. □

1.4 Sorting Machine

In 1986, Cole [43] proposed a very elegant CREW PRAM algorithm to sort n numbers in $O(\log n)$ time with $O(n)$ PUs. This algorithm is optimal as the sequential complexity of sorting (using only comparisons) is $O(n \log n)$. Cole's algorithm is based on the classical merge sort algorithm whose representation as a binary tree (see Figure 1.6) reveals potential parallelism: All merging steps of a given level of the tree can be done in parallel. Let us assume that we process each level in sequence. Since there are $\log n$ levels, we need to be able to process each level in $O(1)$ time to get a $O(\log n)$ total execution time. How can two arbitrary-sized lists be merged in $O(1)$ time? This is where the beauty of Cole's algorithm resides. Partial information from previous merges is used to compute current merges in only a constant number of steps, a very clever but by no means straightforward algorithm. Some readers may thus wish to skip this section due to its rather intricate mathematical content.

1.4.1 Merge

In all that follows we assume that we sort/merge arrays of integers. The merge of two sorted sequences J and K is denoted $J|K$. We say an integer x is *between* a and b if and only if $a < x \leqslant b$.

DEFINITION 1.3 (Rank). The *rank* of an element x in a sequence J is defined as the number of elements of J that are smaller than x:

$$rank(x, J) = \text{card}\{j \in J, j < x\} .$$

Likewise, the *cross-rank* of A in B is the function $R[A, B] : \left| \begin{array}{l} A \to \mathbb{N} \\ e \mapsto rank(e, B) \end{array} \right.$.

This function can be represented as an array of size $|A|$ whose i-th entry is the rank of the i-th element of A in B.

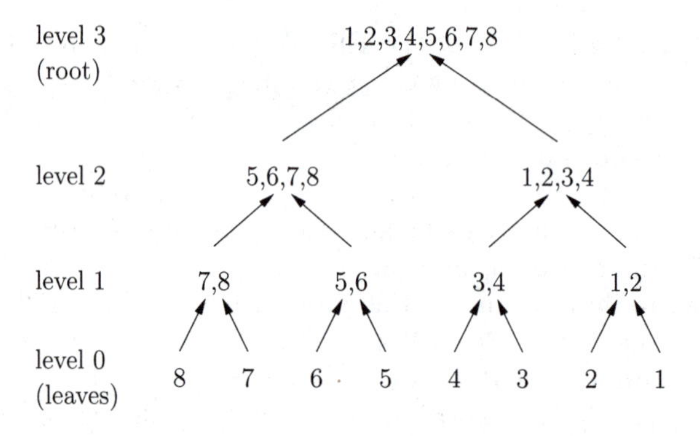

FIGURE 1.6: Example binary tree for Cole's parallel merge sort.

DEFINITION 1.4 (Good Sampler). A sequence L is said to be a *good sampler (GS)* of a sequence J if, for any $k \geqslant 1$, there are at most $2k+1$ elements of J between $k+1$ (arbitrary) consecutive elements of $\{-\infty\} \cup L \cup \{+\infty\}$.

Intuitively, L is a GS of J if elements of L are (almost) uniformly distributed among the elements of J. For $k = 1$, the definition imposes that there is no more than three elements of J between two consecutive elements of $\{-\infty\} \cup L \cup \{+\infty\}$. For example, the set Even(J) (resp. Odd(J)) of elements of even (resp. odd) indices in J is a GS of J. Let us consider $J = \{j_1, \ldots, j_n\}$ and check that Even(J) is a GS of J. Consider $k+1$ consecutive elements $\{j_{2i}, j_{2(i+1)}, \ldots, j_{2(i+k)}\}$ of $\{-\infty\} \cup \text{Even}(J) \cup \{+\infty\}$ (we use the convention $j_0 = -\infty$ and $j_{2(i+k)} = +\infty$ if $2(i+k) > n$). We can easily check that there are either $2k-1$ or $2k$ elements of J between j_{2i} and $j_{2(i+k)}$.

Using good samplers, two sorted sequences can be quickly merged: Let us consider two sorted sequences J and K, and let us assume that L is a GS of J and K (we set $l_0 = -\infty$ and $l_{|L|+1} = +\infty$). Figure 1.7 illustrates the partitioning of J and K.

We can now define a function MERGE_WITH_HELP(), which is shown in Algorithm 1.4. Let us explain how this function works with an example:

Let $J = [2, 3, 7, 8, 10, 14, 15, 17, 18, 21]$ and $K = [1, 4, 6, 9, 11, 12, 13, 16, 19, 20]$.

$L = [5, 10, 12, 17]$ is a GS of both J and K, which can be checked exhaustively for $k = 1$ (5 checks for $(-\infty, 5)$, $(5, 10)$, $(10, 12)$, $(12, 17)$, and $(17, +\infty)$), for $k = 2$ (4 checks for $(-8, 10)$, $(4, 12)$, $(10, 17)$, and $(12, +\infty)$), and so on. We obtain

- $J(1) = [2, 3]$, $J(2) = [7, 8, 10]$, $J(3) = \emptyset$, $J(4) = [14, 15, 17]$, and $J(5) = [18, 21]$.
- $K(1) = [1, 4]$, $K(2) = [6, 9]$, $K(3) = [11, 12]$, $K(4) = [13, 16]$, and $K(5) = [19, 20]$.

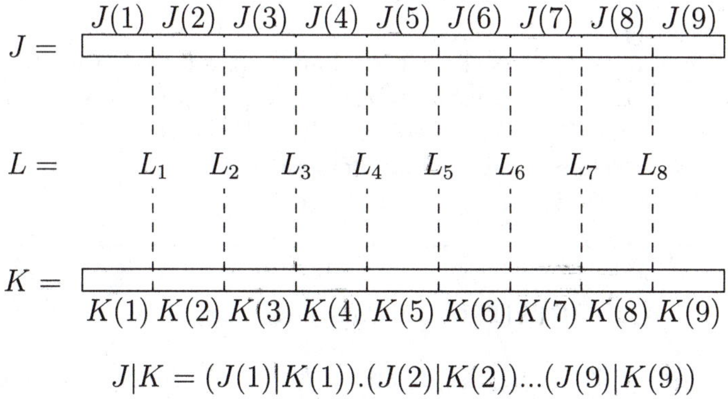

$$
\begin{array}{l}
J = \; J(1)\; J(2)\; J(3)\; J(4)\; J(5)\; J(6)\; J(7)\; J(8)\; J(9) \\
L = \; L_1 \; L_2 \; L_3 \; L_4 \; L_5 \; L_6 \; L_7 \; L_8 \\
K = \; K(1)\; K(2)\; K(3)\; K(4)\; K(5)\; K(6)\; K(7)\; K(8)\; K(9)
\end{array}
$$

$$J|K = (J(1)|K(1)).(J(2)|K(2))...(J(9)|K(9))$$

FIGURE 1.7: Merging J and K with the help of L.

```
MERGE_WITH_HELP(J, K, L)
1   J and K are partitioned in |L| + 1 subsets
    J(i) = {j ∈ J, l_{i-1} < j ⩽ l_i} and K(i) = {k ∈ K, l_{i-1} < k ⩽ l_i}, for
    1 ⩽ i ⩽ |L| + 1
       { As L is a GS of J and K, each subset has at most three elements }
2   forall i, 1 ⩽ i ⩽ |L| + 1 in parallel do { Subsets are merged in parallel }
      res_i ← MERGE(J(i), K(i))
3   J|K ← res_1 res_2 ... res_{|L|+1}        { Partial results are concatenated: J|K
    denotes the fusion of J and K }
    return J|K
```

ALGORITHM 1.4: Merge with help algorithm.

- $\begin{cases} res_1 = \text{MERGE}([2,3],[1,4]) = [1,2,3,4] \\ res_2 = \text{MERGE}([7,8,10],[6,9]) = [6,7,8,9,10] \\ res_3 = \text{MERGE}(\emptyset,[11,12]) = [11,12] \\ res_4 = \text{MERGE}([14,15,17],[13,16]) = [13,14,15,16,17] \\ res_5 = \text{MERGE}([18,21],[19,20]) = [18,19,20,21]. \end{cases}$

LEMMA 1.1. *If L is a GS of the sorted sequences J and K, and if cross-ranks $R[L,J]$, $R[L,K]$, $R[J,L]$, and $R[K,L]$ are known, then* MERGE_WITH_HELP(J,K,L) *runs in $O(1)$ time with $|J|+|K|$ PUs on a CREW PRAM.*

Proof. Each of the three steps can be done in $O(1)$ time using $R[L,J]$, $R[L,K]$, $R[J,L]$ and $R[K,L]$:

- Step 1: J is partitioned using $|J|$ PUs. Each P_j, $j \in J$, reads $rank(j,L) = r$ and inserts j in $J(r)$. As there are at most three PUs writing into $J(r)$, we can sort write accesses in exclusive mode. We proceed likewise to

```
COLE_MERGE()
1  Receive X(t+1) from the left child and Y(t+1) from the right child
2  Merge: val(t+1) ← MERGE_WITH_HELP(X(t+1), Y(t+1), val(t))
3  Reduce: Send Z(t+1) = REDUCE(val(t+1)) to the father.
     { REDUCE() keeps one value out of every four:
       REDUCE({z_1, z_2, ..., z_n}) = {z_4, z_8, z_12, ...}. }
```

ALGORITHM 1.5: Cole merge algorithm.

partition K. Knowing $R[J, L]$ and $R[K, L]$, it is thus possible to do the first step in $O(1)$ time with $|J| + |K|$ PUs.

- Step 2: Since $J(i)$ and $K(i)$ have at most three elements, parallel fusion can clearly be done in $O(1)$ time with $|L| + 1$ PUs.
- Step 3: Knowing $R[L, J]$ and $R[L, K]$, we can compute $R[L, J|K]$. Indeed, for each element $l \in L$ we have

$$rank(l, J|K) = rank(l, J) + rank(l, K) .$$

The PU P_i responsible for res_i computes the rank r of l_{i-1} in $J|K$ and stores the elements of res_i (at most six elements) starting from position $r + 1$. Step 3 can thus be performed with $|L| + 1$ PUs. □

1.4.2 Sorting Trees

We assume that $n = 2^m$ and use the sorting tree depicted in Figure 1.6. The key idea here is to use this tree in a *pipelined* fashion. Each node of the tree stores a sorted sequence $val(t)$ that grows step after step. Each step is done in $O(1)$ time because we ensure that $val(t)$ is a GS of the next step's inputs.

Mode of Operation

At step $t = 0$, all nodes store an empty sequence except for the leaves, which store one element of the array to be sorted. At step $t + 1$, each node runs an algorithm that uses a simple REDUCE() operation, as shown in Algorithm 1.5. We will prove in Section 1.4.3 that at any step t we have

$$val(t) = X(t)|Y(t) .$$

A node of the tree is said to be *complete* when it has *received* all its inputs, i.e., a sorted sequence of 2^k elements for a node at level k. As soon as this sequence is non-empty, its size doubles at each step from 1 to 2^k. If a node is complete at step t, then the REDUCE() operation is changed in the following way:

- At step $t+1$, it sends one element out of every four from $val(t+1)$ to its father.
- At step $t+2$, it sends one element out of every two (the second one, the fourth one, etc.) from $val(t+1)$ to its father.
- At step $t+3$, it sends all the elements of $val(t+1)$ to its father.
- From step $t+4$ on, it stops working and does not send anything to its father any longer.

The proof of correctness of Cole's algorithm is based on three main invariants:

- The size of the input of nodes at level k in the tree doubles at each step (from 1 to 2^k) until the node is complete.
- If $X(t)$ is a GS of $X(t+1)$ and $Y(t)$ is a GS of $Y(t+1)$, then $Z(t)$ is a GS of $Z(t+1)$.
- We can compute $R[S(t+1), S(t)]$ for any sequence of input or output of a node.

Study of the Example

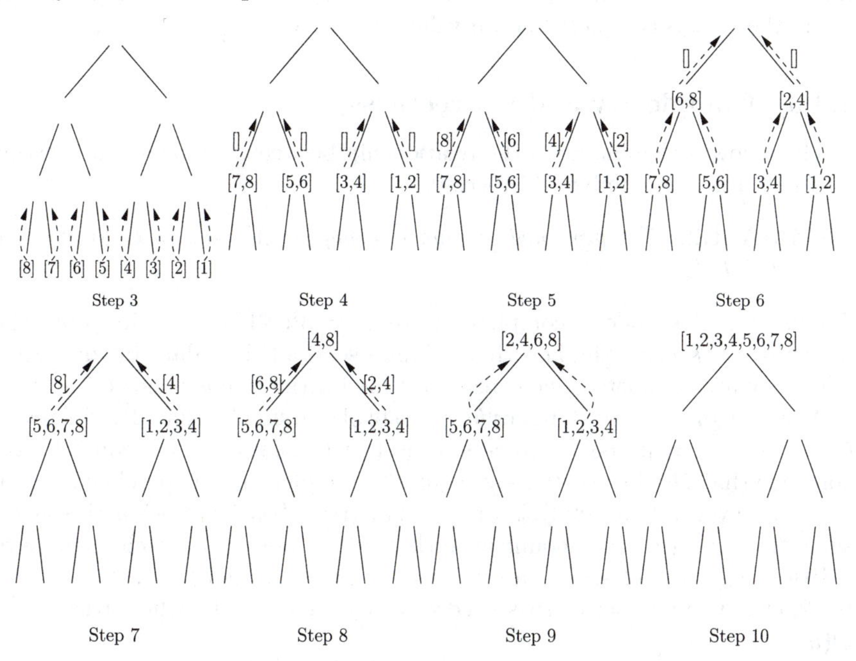

FIGURE 1.8: Sorting an array of size 8 with Cole's parallel merge sort.

In this section, we describe in detail how the algorithm works with an input of size 8 (see Figure 1.8). At step $t = 0$, leaves have sequences of size 2^0 and are hence complete. At steps $t = 1$ and $t = 2$, they send one element out of four, then one out of two, i.e., no element at all. At step $t = 3$, they send their unique value to their father and stop working.

Let us focus on the father [8, 7] of leaves 8 and 7. At step $t = 3$, it computes $val(3)$, i.e., MERGE_WITH_HELP($\{8\}, \{7\}, \emptyset$), and becomes complete (it is a level 1 node). At step $t = 4$, it does not send anything to its father. At step $t = 5$, it sends one element out of every two, i.e., $\{8\}$. At step $t = 6$, it sends its two elements, and then stops working.

Let us now focus on the root node of the subtree [8, 7, 6, 5]. At step $t = 5$, it receives $\{8\}$ and $\{6\}$ and computes $val(5) =$ MERGE_WITH_HELP($\{8\}, \{6\}, \emptyset$). At step $t = 6$, it receives $X(6) = \{7, 8\}$ and $Y(6) = \{5, 6\}$ and computes $val(6) =$ MERGE_WITH_HELP($\{7, 8\}, \{5, 6\}, val(5)$). We can check that $val(5)$ is a GS of $X(6)$ and $Y(6)$. At step $t = 6$, the node is complete (it is a level 2 node). At step $t = 7$, it sends $\{8\}$, and at step $t = 8$, it sends $\{6, 8\}$. At step $t = 9$, it sends its four elements and stops working.

Lastly, let us look at the root of the tree. At step $t = 7$, it receives $\{8\}$ and $\{4\}$ and computes $val(7) =$ MERGE_WITH_HELP($\{8\}, \{4\}, \emptyset$). At step $t = 8$, it receives $\{6, 8\}$ and $\{2, 4\}$ and computes $val(8) =$ MERGE_WITH_HELP($\{6, 8\}, \{2, 4\}, val(7)$). At step $t = 9$, it receives $\{5, 6, 7, 8\}$ and $\{1, 2, 3, 4\}$ and computes $val(9) =$ MERGE_WITH_HELP($\{5, 6, 7, 8\}, \{1, 2, 3, 4\}, val(8)$). On can easily check that $val(t)$ is a GS of $X(t + 1)$ and $Y(t + 1)$ for $t = 6, 7, 8$. At the end of step $t = 9$, the root is complete and all values are sorted.

1.4.3 Complexity and Correctness

The following lemma gives the relationship between the algorithm's execution time and the number of processors.

LEMMA 1.2. *The pipelined sorting tree algorithm runs in $O(\log n)$ time with $O(n)$ PUs.*

Proof. A level k node is complete at step $t = 3k$. This is easily proved by induction on k using the fact that children send all their data in three steps once they are complete. The execution time for the tree is hence $O(\log n)$.

At level k of the tree, $\frac{n}{2^k}$ nodes merge lists of size smaller than 2^k. Therefore, $O(n)$ PUs are required to process a level in $O(1)$ time. One could hastily conclude that $O(n \log n)$ PUs are required to implement the pipelined sorting tree. However, a finer analysis reveals that (i) all levels are not active at the same time and (ii) the amount of work per level doubles at each step. Level k PUs merge lists of size 2^k at step $t = 3k$, 2^{k-1} at step $t - 1$, 2^{k-2} at step $t - 2$, etc. The number of PUs needed for each step of the whole tree is thus $O(n)$. □

Let us now prove the following good sampler invariant.

LEMMA 1.3. *Let X, X', Y, and Y' be four sorted sequences. If X is a GS of X' and if Y is a GS of Y',* REDUCE*$(X|Y)$ is a GS of* REDUCE*$(X'|Y')$.*

Proof. We still use the notation $X|Y$ to denote the fusion of X and Y. If X is a GS of X', then $X|W$ is clearly still a GS of X' for any set W. However, if X is a GS of X' and Y is a GS of Y', there $X|Y$ is not necessarily a GS of $X'|Y'$.

Indeed, let us consider for example $X = [2,7]$, $X' = [2,5,6,7]$, $Y = [1,8]$, and $Y' = [1,3,4,8]$. Then we have $X|Y = [1,2,7,8]$, and $X'|Y' = [1,2,3,4,5,6,7,8]$ but there are five elements of $X'|Y'$ between 2 and 7 (that are yet consecutive elements of $X|Y$). This is the reason why we resort to the reduce operator.

Let us prove the following property: There are at most $2r+2$ elements of $X'|Y'$ between r consecutive elements of $X|Y$ (we assume that $-\infty$ and $+\infty$ are in X and Y).

Proof. Let us consider a sequence $e_1, e_2, \ldots, e_r$ of r consecutive elements of $X|Y$. h_X elements among these r elements come from X and h_Y come from Y (with $h_X + h_Y = r$). Without loss of generality, we can assume that $e_1 \in X$. We consider two cases:

Case 1 ($e_r \in X$)**:** As X is a GS of X', there are at most $2(h_X - 1) + 1$ elements of X' between e_1 and e_r. There are also at most $2(h_Y + 1) + 1$ elements of Y' as these elements are between $h_Y + 2$ elements of Y, which is a GS of Y'. Hence, there are at most $2(h_X - 1) + 1 + 2(h_Y + 1) + 1 = 2r + 2$ elements of $X'|Y'$ between $e_1, e_2, \ldots, e_r$.

Case 2 ($e_r \in Y$)**:** Let us add an element $e_0 \in Y$ preceding e_1 and an element $e_{r+1} \in X$ following e_r. Then, elements from X' and Y' lying between $e_1, e_2, \ldots e_r$ come from elements for X' lying between $h_X + 1$ elements of X and from elements of Y' lying between $h_Y + 1$ elements of Y. Therefore, we have $(2h_X + 1) + (2h_Y + 1) = 2r + 2$ elements of $X'|Y'$. □

Coming back to the proof of the lemma, let us define $Z =$ REDUCE$(X|Y)$ and $Z' =$ REDUCE$(X'|Y')$. Let us consider $k+1$ consecutive elements $z_1, z_2, \ldots,$ z_{k+1} of Z. Since the reduce operator keeps one element out of every four, we have $z_1 = e_{4h}$, $z_2 = e_{4(h+1)}, \ldots, z_{k+1} = e_{4(h+k)}$, where $X|Y = \{e_1, e_2, \ldots, e_p\}$. Thus, there are $4k+1$ elements of $X|Y$ between $z_1, z_2, \ldots, z_{k+1}$. Using the previous property with $r = 4k+1$, we know that there are at most $8k+4$ elements of $X'|Y'$ between these $4k+1$ elements. Since the reduce operator keeps one element out of every four, there are at most $\frac{8k+4}{4} = 2k+1$ elements of Z' between the $k+1$ consecutive elements of Z, proving that Z' is a GS of Z. □

In steady-state, a node receives a sorted sequence $X(t+1)$ from its left child and a sorted sequence $Y(t+1)$ from its right child. It computes $val(t+1) =$ MERGE_WITH_HELP$(X(t+1), Y(t+1), val(t))$ and sends $Z(t+1)$=REDUCE$(val(t+1))$

to its father. Since Lemma 1.3 shows that $Z(t)$ is a GS of $Z(t+1)$, we have the following invariants:

1. $val(t) = X(t)|Y(t)$.
2. $X(t)$ is a GS of $X(t+1)$.
3. $Y(t)$ is a GS of $Y(t+1)$.

Note that the property still holds true for the last two communications (one element out of every two is sent, then finally all the elements).

Lastly, we have to ensure that the requirements of Lemma 1.1 hold true: We need to know the cross-ranks for MERGE_WITH_HELP() to run in $O(1)$ time. At a given step, X is a GS of X' and Y is a GS of Y', $U = X|Y$, and $Z =$ REDUCE(U). We can assume that cross-ranks $R[X', X]$ and $R[Y', Y]$ are known by the induction hypothesis. To compute $U' = X'|Y'$ with MERGE_WITH_HELP(X', Y', U), we need to know the cross-ranks $R[X', U]$, $R[Y', U]$, $R[U, X']$, and $R[U, Y']$. Finally we can compute $Z' =$ REDUCE(U') and $R[Z', Z]$ to get our invariant. Of course, we assume that for each sorted sequence S we know the cross-rank $R[S, S]$. In other words, we know the index of each element in the sorted sequence. This is computed as part of the internal representation of S.

LEMMA 1.4. *If $S = [b_1, b_2, \ldots, b_k]$ is a sorted sequence, then the rank of a given element a in S can be computed in $O(1)$ time with $O(k)$ PUs on a CREW PRAM.*

Proof. Take $b_0 = -\infty$ and $b_{k+1} = +\infty$. The rank of a is then computed with the following loop:

```
forall i, 0 ⩽ i ⩽ k in parallel do
  if b_i < a ⩽ b_{i+1} then
    rank ← i
```

There are no write conflicts because only one processor stores the value of *rank*. □

LEMMA 1.5. *If we have three sorted sequences S_1, S_2, S such that $S = S_1|S_2$ and $S_1 \cap S_2 = \emptyset$, then we can compute the cross-ranks $R[S_1, S_2]$ and $R[S_2, S_1]$ in $O(|S|)$ time with $O(|S|)$ PUs.*

Proof. We assume that whenever a sequence S is sorted, we also know $R[S, S]$. For any $a \in S_1 \subset S$ we have

$$rank(a, S_2) = rank(a, S) - rank(a, S_1)\ .$$

Hence the results. □

We can now prove our invariant on cross-ranks for MERGE_WITH_HELP().

LEMMA 1.6. *Suppose we have sorted sequences X, Y, $U = X|Y$, X', and Y' such that X is a GS of X', Y is a GS of Y', and we know the cross-ranks $R[X', X]$ and $R[Y', Y]$. Then, we can compute the cross-ranks $R[X', U]$, $R[Y', U]$, $R[U, X']$, and $R[U, Y']$ in $O(1)$ time with $O(|X| + |Y|)$ PUs.*

Proof. We first show how to compute $R[X', U]$. Let $X = [a_1, a_2, \ldots, a_k]$ and take $a_0 = -\infty$ and $a_{k+1} = +\infty$. We partition the sequence X' with X:

$$X'(i) = \{x' \in X',\ a_{i-1} < x' \leqslant a_i\} \text{ for } 1 \leqslant i \leqslant k+1.$$

This partition is computed in $O(1)$ time with $O(|X|)$ PUs because we know $R[X', X]$. We also partition U with X (which is the same as partitioning Y with X because $U = X|Y$):

$$U(i) = \{y \in Y, a_{i-1} < y \leqslant a_i\} \text{ for } 1 \leqslant i \leqslant k+1.$$

We compute the rank of x in U as follows:

```
forall i, 1 ⩽ i ⩽ k + 1 in parallel do
    forall x' ∈ X'(i) do
        Compute rank(x', U(i))                    { with Lemma 1.4 }
        rank(x', U) ← rank(a_{i-1}, U) + rank(x', U(i))
```

For each i we use $|U(i)|$ PUs. Altogether we thus require $O(|U|)$ PUs. As X is a GS of X', each $X'(i)$ consists of at most three elements and thus the computation runs in $O(1)$ time. We have therefore computed $R[X', U]$. $R[Y', U]$ is of course computed in a similar fashion.

To compute $R[U, X']$ we need $R[X, X']$ and $R[Y, X']$. Let us see how to compute $R[X, X']$. Consider an element a_i from $X \setminus X'$ and search for the minimal element a' of $X'(i+1)$. The rank of a_i in X' is the same as the rank of a' in X'. This rank is already computed as part of the internal representation of the sorted sequence X'. Thus, we can compute $rank(a_i, X')$ in $O(1)$ time with a single processor. To compute $R[Y, X']$ consider $y \in Y$. We compute $rank(y, X)$ using Lemma 1.5, because $U = X|Y$ is already computed. Then, we compute $rank(y, X')$ using $rank(y, X)$ and $R[X, X']$. This way, we compute $R[U, X']$ in $O(1)$ time with $O(|U|)$ PUs. We can compute $R[U, Y']$ in a similar way. □

LEMMA 1.7. *Using Lemma 1.6's notation, let*

$$Z = \text{Réduction}(U) \text{ and } Z' = \text{Réduction}(U').$$

We can compute $R[Z', Z]$ in $O(1)$ time with $O(|X| + |Y|)$ PUs.

Proof. The proof is straightforward. $R[X', Z]$ is computed from $R[X', U]$. Likewise, $R[Y', Z]$ is computed from $R[Y', U]$ and we obtain $R[U', Z]$. Z' is a subset of Z, hence the result. □

We have at last proved the following theorem:

THEOREM 1.3. *We can sort a sequence of n values in time $O(\log n)$ on a CREW PRAM with $O(n)$ PUs.*

The cost of Cole's algorithm is thus $O(n \log n)$. This is optimal because the sequential complexity of sorting is $O(n \log n)$.

1.5 Relevance of the PRAM Model

The central question in complexity theory for sequential algorithms is the P = NP question. The corresponding question in the complexity theory of parallel algorithms is the P = NC question. NC is the class of all problems that, with a polynomial number of PUs, can be solved in polylogarithmic time. An algorithm of size n is polylogarithmic if it can be solved in $O(\log(n)^c)$ time with $O(n^k)$ PUs, where c and k are constants. Problems in NC can be solved efficiently on a parallel computer. All the problems considered in this chapter belong to NC. "NC" is an abbreviation of "Nick (Pippenger)'s Class."

Besides these theoretical complexity considerations, skeptical readers may question the relevance of the PRAM model for practical implementation purposes. A common criticism of the PRAM model is the unrealistic assumption of an immediately addressable, unbounded shared parallel memory. This criticism, similar to the criticism of $O(n^{17})$ "polynomial" time algorithms, often comes from a misunderstanding of the role of theory:

- Theory is not everything: Theoretical results are not to be taken as is and implemented by engineers.
- Theory is also not nothing: The fact that an algorithm cannot in general be implemented as is does not mean it is meaningless.

When a $O(n^{17})$ algorithm is designed for a problem, it does not lead to a practical way to solve that problem. However, it proves something inherent about the problem, namely, that it is in P, and thus that its hardness does not grow exponentially with the input size. Hopefully, in the process of proving this result, key insights may be developed that can later be used in practical algorithms for this or other problems, or for other theoretical results. Similarly, the design of PRAM algorithms for a problem proves something inherent to the problem (namely, that it is parallelizable) and can in turn lead to new ideas.

Lastly, even if communications are not taken into account in the performance evaluation of PRAM algorithms (a potentially considerable discrepancy between theoretical complexity and practical execution time), trying to design fast PRAM algorithms is not a useless pursuit. Indeed, PRAM algorithms can be simulated on other models and do not necessarily incur prohibitive communication overheads. It is commonly admitted that only cost-optimal PRAM algorithms have potential practical relevance and that the most promising are those cost-optimal algorithms with minimal execution time.

Bibliographical Notes

Some of the introduction material in this chapter is inspired by the books by Cormen, Leiserson, and Rivest [44] and by Gengler, Ubéda, and Desprez [59]. The presentation of Cole's parallel merge sort algorithm is taken from the book by Gibbons and Rytter [61]. The original article by Cole [43] also presents an EREW version of the sorting machine with the same performance. For additional information on the PRAM model, we refer the reader to the book by Reif [100].

1.6 Exercises

We start with six classical exercises. The first three are based on the pointer jumping technique, which the reader should have mastered by now. The fourth one comes back to model separation. The fifth one is a nice application of the $O(1)$ CRCW algorithm for finding the largest value of an array. We do not spoil the surprise about the sixth one and let the reader discover it. We end with a seventh exercise on connected components [61], which represents a nice example of the idea contained within Brent's theorem.

Exercise 1.1 : Matrix Multiplication

Describe a $O(\log(n))$ EREW PRAM algorithm that uses n^3 PUs to multiply two $n \times n$ matrices.

◇ Exercise 1.2 : Selection in a List

Let L be a list with n objects that are either red or blue. Design an efficient EREW algorithm that selects blue elements.

Exercise 1.3 : Splitting an Array

Let A be an array of length n whose elements are either 0 or 1. Design a $O(\log n)$ EREW algorithm using $O(n)$ PUs to move all the non-zero elements to the right side of the array while maintaining their original order.
Hint: Perform a prefix computation to find out what the index of each element should be.

◇ Exercise 1.4 : Looking for Roots in a Forest

In this exercise we see another example of a problem that separates EREW and CREW PRAMs. Let $\mathcal{F}$ be a forest of binary trees. Each node i of a tree is associated with a processor $P(i)$ and has a pointer to its father $\mathit{father}(i)$. In this exercise, we design an EREW and a CREW algorithm that make each node know the root $\mathit{root}(i)$ of the tree it belongs to and thus prove the advantage of concurrent reads.

1. Propose a CREW algorithm so that each node determines $\mathit{root}(i)$. Prove that your algorithm only uses concurrent reads and give its complexity.

2. What is the best execution time that can be expected on an EREW PRAM?

Exercise 1.5 : First Non-Zero Element

Let A be an array of length n whose elements are either 0 or 1. Design a $O(1)$ CRCW algorithm using $O(n)$ PUs to find the first element k such that $A[k] = 1$.
Hint: Partition A into $\sqrt{n}$ equal parts and find the first part that contains a 1 using Algorithm 1.3.

◇ Exercise 1.6 : Mystery Function

We define the following two operators on an array $A = [a_0, a_1, \ldots, a_{n-1}]$ of integers:

- PRESCAN(A) returns $[0, a_0, a_0 + a_1, a_0 + a_1 + a_2, \ldots, a_0 + a_1 + \ldots + a_{n-2}]$,
- SCAN(A) returns $[a_0, a_0 + a_1, a_0 + a_1 + a_2, \ldots, a_0 + a_1 + \ldots + a_{n-1}]$.

We saw in Section 1.1 how to implement these two operators in time $O(\log n)$ on an EREW PRAM. Consider the SPLIT function below:

```
1 SPLIT(A, Flags)
2     Idown ←PRESCAN(not(Flags))
3     Iup ←n - REVERSE(SCAN(REVERSE(Flags)))
4     forall i ∈ {1, ..., n} in parallel do
5         if Flags(i) then Index[i] ←Iup[i]
6         else Index[i] ←Idown[i]
7     Result ←PERMUTE(A,Index)
8     return Result
```

This function uses two functions: REVERSE(A) and PERMUTE(A, *Index*). The former reverses array A, and the latter reorders array A according to a permutation specified as an array of indices, *Index*. The slightly cumbersome REVERSE(SCAN(REVERSE(*Flags*))) simply scans from the end of the array *Flags*, considering its elements as integers.

1. Given an array *Flags* of booleans, what does the SPLIT function return? What is its execution time?

2. Consider the following MYSTERY function:

```
1 MYSTERY(A, Number_Of_Bits)
2     for i = 0 to Number_Of_bits − 1 do
3         bit(i) ←array indicating whether the i−th bit of
              elements of A is equal to 1 or not
4         A ←SPLIT(A, bit(i))
```

(a) What is the result of the MYSTERY function when applied to $A = [5, 7, 3, 1, 4, 2, 7, 2]$ and $Number\_Of\_Bits = 3$?

(b) What does the MYSTERY function compute?

(c) Assuming the size of integers is $O(\log n)$ bits, what is the execution time of MYSTERY with n PUs? What if only p PUs are used? What are the values of p that lead to an optimal value of the algorithm's work?

◇ Exercise 1.7 : Connected Components

We want to design a CREW algorithm to compute the connected components of an undirected graph $G = (V, E)$ where $V = \{1, \ldots, n\}$. More precisely, the output from the algorithm should be an array C of size n such that $C(i) = C(j) = k$ if and only if vertex i is in same component as vertex j. Moreover, k is the minimum index of the vertices in the connected component that contains vertices i and j. Let us introduce a number of useful definitions.

DEFINITION 1.5. At any step of the algorithm, the *pseudo-vertex labeled by* i is the set of vertices $j, k, l, \cdots \in V$ such that $C(j) = C(k) = C(l) = \cdots = i$. If i labels a pseudo-vertex, then i is called a *p-vertex.* For any vertex j that is not a p-vertex, $C(j)$ denotes the p-vertex of j.

An invariant of the algorithm is that a p-vertex i is the smallest index of vertices labeled by i and that all these vertices belong to the same connected component. This holds true when initializing $C(i) = i$ for all $i \in V = \{1, \ldots, n\}$. In other words, at the beginning of the algorithm, each PU considers itself as the reference vertex of its component. The algorithm then iteratively changes this rather egocentric perspective.

DEFINITION 1.6. A k-tree-loop ($k \geqslant 0$) is a weakly connected directed graph (i.e., its underlying undirected graph is connected) such that:

- Every vertex has out-degree 1.
- There exists exactly one circuit whose length is equal to $k + 1$.

Removing any one edge of this circuit results in an in-tree. A star is a 0-tree-loop.

The previous invariant can thus be read as: "The directed graph $(V, \{(i, C(i)) | i \in V\})$ consists of stars." We can freely identify pseudo-vertices and stars: The center of a star is the label of the corresponding pseudo-vertex. Connected components are computed by applying the following two functions in sequence as many times as needed:

```
1  GATHER()
2    forall i ∈ S in parallel do
3      T(i) ← min{C(j) | {i,j} ∈ E, C(j) ≠ C(i)}
4      { if this set is empty, then C(i) is associated to T(i) }
5    forall i ∈ S in parallel do
6      T(i) ← min{T(j) | C(j) = i, T(j) ≠ i}
7      { if this set is empty, then T(i) is associated to T(i) }

8  JUMP()
9    forall i ∈ S in parallel do  B(i) ← T(i)
10   repeat log n times
11     forall i ∈ S in parallel do  T(i) ← T(T(i))
12   forall i ∈ S in parallel do
       C(i) ← min{B(T(i)), T(i)}
```

1. Consider the following graph:

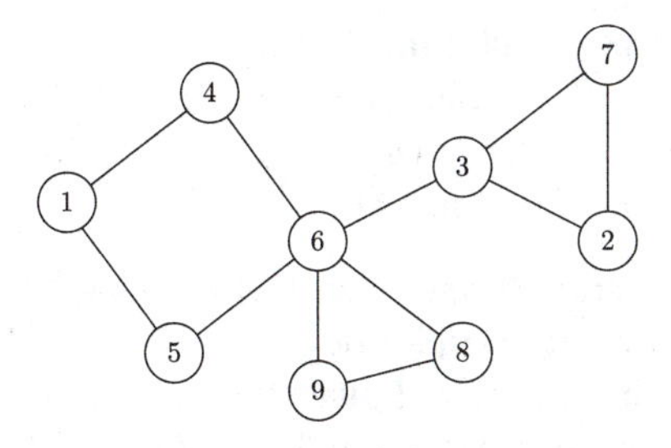

Apply the GATHER function to this graph, then the JUMP function, then the GATHER function, and so on. Follow the effect of these steps on the directed graphs $(V, \{(i, T(i)) \mid i \in V\})$ and $(V, \{(i, C(i)) \mid i \in V\})$.

2. Prove that after applying GATHER, connected components comprising more than one pseudo-vertex lead to 1-tree-loops in $(V, \{(i, T(i)) \mid i \in V\})$. Prove that the smallest pseudo-vertex of a 1-tree-loop always belongs to the cycle.

3. Prove that JUMP turns 1-tree-loops into stars.

4. Prove that applying the GATHER and JUMP functions $\lceil \log n \rceil$ times enables pseudo-vertices induced by C to correspond exactly to the connected components of the original graph.

5. What is the complexity of this algorithm? How many PUs are used?

1.7 Answers

Exercise 1.2 (Selection in a List)

We simply use a version of the pointer jumping technique to obtain Algorithm 1.6. Each PU determines the location of the first blue object after it in the list in $O(\log n)$ time.

This EREW algorithm requires a bit of explanation. It works like classical pointer jumping but in parallel on many lists. The last object in each of these lists is either blue or **Nil**. Each PU ends up with a pointer to the end of its list, i.e., to the next blue element. All pointer jumps are made on independent lists and thus do not interfere with each other. At the end of the algorithm the list of blue elements starts either with the first element of the initial list if it was blue, or with its first blue successor.

```
1  SELECT-BLUE()
2    forall i in parallel do
3      if next(i) = Nil Or color(next(i)) = blue then
4        done(i) ←True
5        blue(i) ←next(i)
6    while there is a node i such that done(i) = False do
7      forall i in parallel do
8        if done(i) = False then
9          done(i) ← done(next(i))
10         if done(i) = True then
11           blue(i) ← blue(next(i))
12         next(i) ← next(next(i))
```

ALGORITHM 1.6: Algorithm to select blue elements in a list.

Exercise 1.4 (Looking for Roots in a Forest)

▷ **Question 1.** A natural algorithm to find roots is based on the pointer jumping technique, exploiting the fact that a node and its ancestors share a path to the root of a tree. A simple transformation of the algorithm developed in Section 1.1 leads to Algorithm 1.7. This is a CREW algorithm since all writes performed by processor i are for its own data ($root(i)$ and $father(i)$). This is, however, not an EREW algorithm because many PUs have the same

```
1  Find_Root()
2      forall i in parallel do
3          if father(i) = Nil then
4              root(i) ← i
5      while there exists a node i such that father(i) ≠ Nil do
6          forall i in parallel do
7              if father(i) ≠ Nil then
8                  if father(father(i)) = Nil then
9                      root(i) ← root(father(i))
10                 father(i) ← father(father(i)).
```

ALGORITHM 1.7: CREW algorithm for finding roots in a forest.

father (especially in the end!) and thus will need to access simultaneously the same data *father*(i).

The same analysis as the one for list ranking (see Section 1.1) can be applied. Each node in the forest finds its root in at most time $O(\log d)$, where d is the maximal depth of the trees, and all nodes find their roots in parallel. Therefore, the algorithm runs in time $O(\log d)$.

▷ **Question 2.** Let us consider the worst case for the EREW model, i.e., the case where the forest contains only one tree. Let us count the number of PUs that may know the root at each step. In the EREW model, the number of PUs that know an information can at most double at each step. In our case, exactly one PU knows the root at the beginning and therefore at least $\Omega(\log n)$ steps are required to propagate this information to all PUs.

Exercise 1.6 (Mystery Function)

▷ **Question 1.** Let us look at an example:

$$\begin{aligned}
A &= [\ 5\ \ 7\ \ 3\ \ 1\ \ 4\ \ 2\ \ 7\ \ 2\ ] \\
\mathit{Flags} &= [\ 1\ \ 1\ \ 1\ \ 1\ \ 0\ \ 0\ \ 1\ \ 0\ ] \\
\mathit{Idown} &= [\ 0\ \ 0\ \ 0\ \ 0\ \ \boxed{0}\ \ \boxed{1}\ \ 2\ \ \boxed{2}\ ] \\
\mathit{Iup} &= [\ \boxed{3}\ \ \boxed{4}\ \ \boxed{5}\ \ \boxed{6}\ \ 7\ \ 7\ \ \boxed{7}\ \ 8\ ] \\
\mathit{Index} &= [\ 3\ \ 4\ \ 5\ \ 6\ \ 0\ \ 1\ \ 7\ \ 2\ ] \\
\mathit{Result} &= [\ 4\ \ 2\ \ 2\ \ 5\ \ 7\ \ 3\ \ 1\ \ 7\ ]
\end{aligned}.$$

With this example it is easy to see what function Split does: Elements of A whose corresponding elements in *Flags* are equal to 0 are moved to the beginning of *Result*, remaining in the same order. Likewise, elements of A whose corresponding elements in Flags are equal to 1 are grouped at the end of *Result*, remaining in the same order.

Using $O(n)$ PUs, SCAN and PRESCAN can be done in time $O(\log n)$. As other operations only require constant time, SPLIT runs in time $O(\log n)$.

▷ **Question 2.**

(a)
$$\begin{array}{lcl} A & = & [\ 5\ 7\ 3\ 1\ 4\ 2\ 7\ 2\ ] \\ bit(0) & = & [\ 1\ 1\ 1\ 1\ 0\ 0\ 1\ 0\ ] \\ A \leftarrow \textsf{SPLIT}(A, bit(0)) & = & [\ 4\ 2\ 2\ 5\ 7\ 3\ 1\ 7\ ] \\ bit(1) & = & [\ 0\ 1\ 1\ 0\ 1\ 1\ 0\ 1\ ] \\ A \leftarrow \textsf{SPLIT}(A, bit(1)) & = & [\ 4\ 5\ 1\ 2\ 2\ 7\ 3\ 7\ ] \\ bit(2) & = & [\ 1\ 1\ 0\ 0\ 0\ 1\ 0\ 1\ ] \\ A \leftarrow \textsf{SPLIT}(A, bit(2)) & = & [\ 1\ 2\ 2\ 3\ 4\ 5\ 7\ 7\ ] \end{array}$$

(b) Based on the example, it looks like MYSTERY sorts its input array. In fact, it is a parallel implementation of the well-known *radix-sort* algorithm: Starting with the least-significant bit, the SPLIT function splits the array into two parts, depending on the value of this bit. Each call to SPLIT sorts elements according to the current bit value while maintaining the order obtained with previous bits. This is why the algorithm goes from the least-significant bit to the most-significant bit.

(c) There are $O(\log n)$ iterations of the main loop. The execution time of the MYSTERY function is thus $O(\log^2 n)$ with $O(n)$ PUs. When using only p PUs, the execution time of SPLIT becomes $O(\frac{n}{p} + \log p)$ and the execution time of the parallel radix-sort becomes $O((\frac{n}{p} + \log p)\log n) = O(\frac{n}{p}\log n + \log n \log p)$. The work is optimal (i.e., equal to $O(n \log n)$) for p such that $p \log p \leqslant n$, e.g., for $p = n^q$ with $0 < q < 1$.

Exercise 1.7 (Connected Components)

▷ **Question 1.** At the beginning of the algorithm, C is initialized as follow:

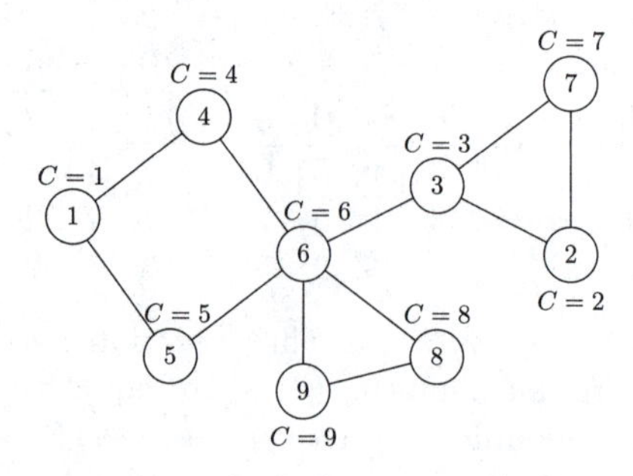

C remains unchanged after applying GATHER, but T is as follows:

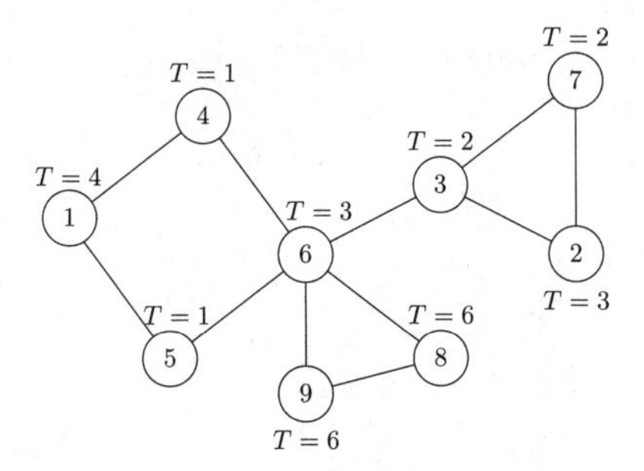

First, we can notice that the directed graph $(V, \{(i, T(i)) \mid i \in V\})$ consists of 1-tree-loops:

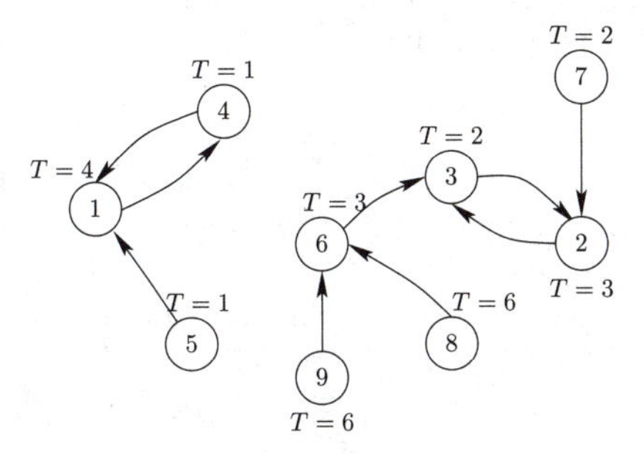

After pointer jumping in the Jump function, all these 1-tree-loops have been transformed into stars:

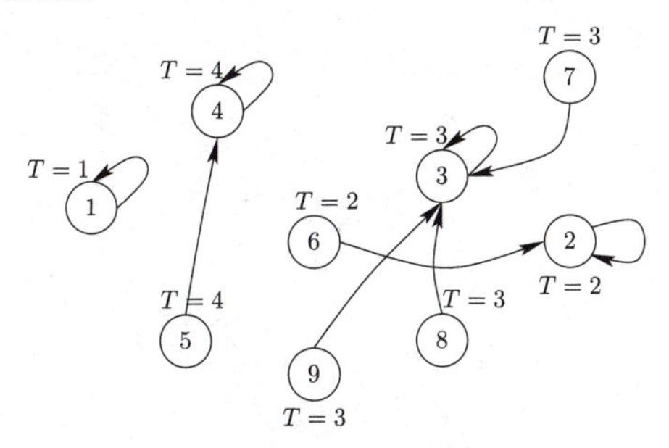

These stars are merged in $(V, \{(i, C(i)) \mid i \in V\})$ in the last part of the Jump function:

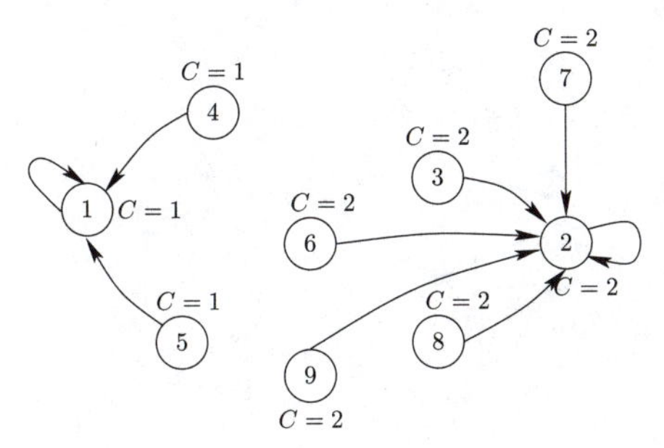

There are only two remaining pseudo-vertices and C is thus as follows before applying GATHER again:

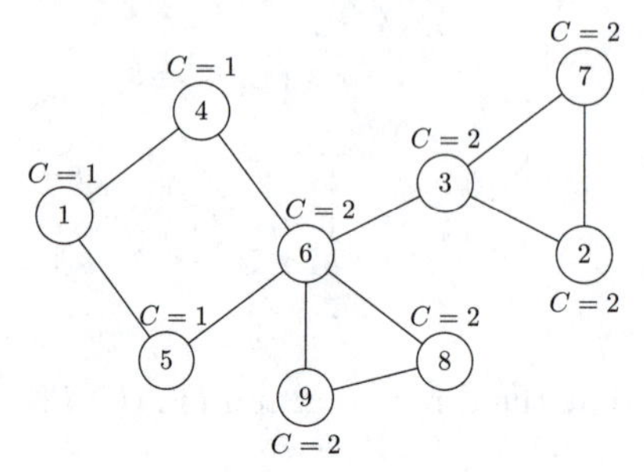

After the first step, T is updated:

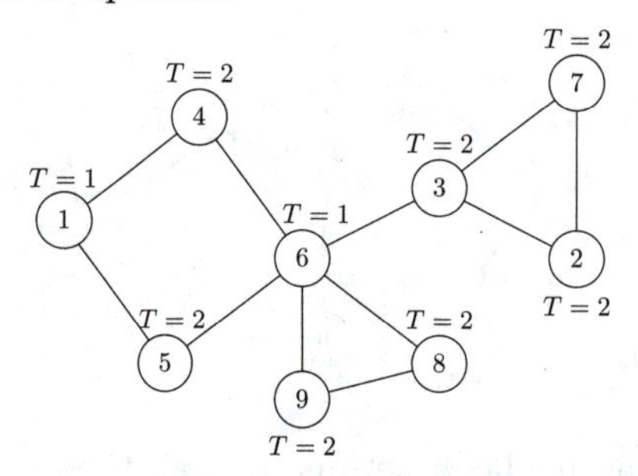

and finally

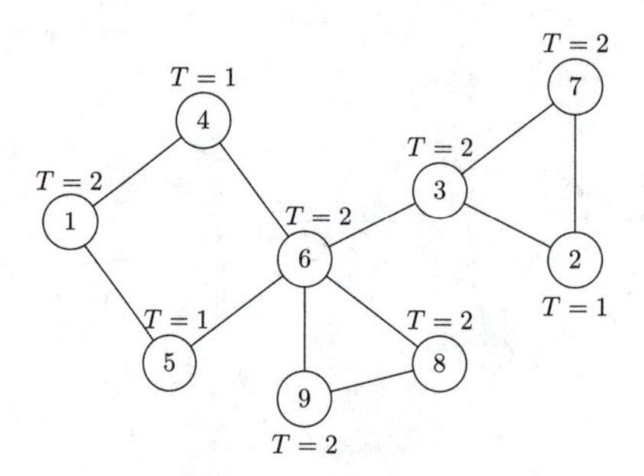

We end up with a directed graph $(V, \{(i, T(i)) \mid i \in V\})$ that consists solely of 1-tree-loops:

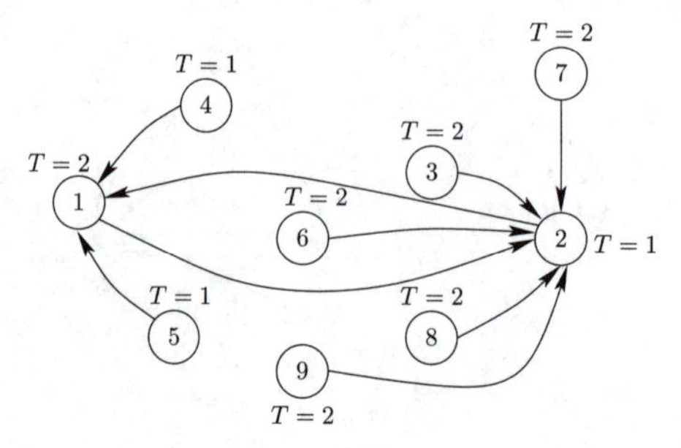

Pseudo-vertices 1 and 2 are thus merged after applying JUMP:

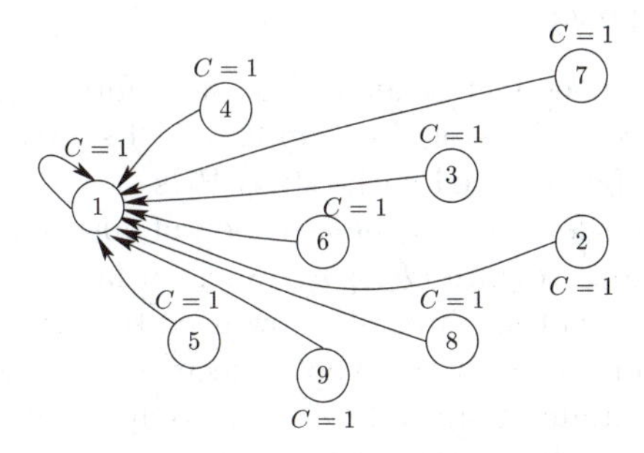

Each vertex is now aware of its p-vertex, and hence connected components have been computed.

▷ **Question 2.** First of all, any star associated with a graph component containing a single pseudo-vertex is transferred to T without any change.

When a connected component has more than one pseudo-vertex, T describes the set of 1-tree-loops for this component. Indeed, let us consider a pseudo-vertex of such a component. It contains at least one vertex adjacent to a vertex of another pseudo-vertex. GATHER causes each p-vertex to point – *via* T – to one another, while other vertices of the pseudo-vertex remain unchanged and still point to their original p-vertex. In short, when two groups are in contact, the corresponding p-vertices get linked in the directed graph induced by T. Each component in this directed graph has at least one loop since the out-degree of each vertex is exactly 1. There is at most one loop; otherwise, a p-vertex would have two values for T. Lastly, this loop can only have length 2. Otherwise, if its length were equal to 1, i and $T(i)$ would be the same or (if its length were larger than 2) there would be a smaller vertex i on the loop such that $T(i)$ is not the smallest p-vertex of vertices adjacent to vertices in pseudo-vertex i.

▷ **Question 3.** Using pointer jumping, JUMP merges all vertices of a 1-tree-loop into a star labeled by the smallest vertex. Indeed, after the pointer jumps, the T value of each vertex is either one or the other value of the vertices of the loop. In the last step, the T value of all vertices is set to the smallest value of this tree-loop.

▷ **Question 4.** We just need to prove that each step halves the number of connected components. Let us focus on p-vertices and on the graph induced by T on these vertices. In this graph, two pseudo-vertices i and j are connected if and only if there are two vertices k and l that are connected in the original graph and such that $C(k) = i$ and $C(l) = j$. The function JUMP merges all these pseudo-vertices into a single pseudo-vertex. Therefore, the number of pseudo-vertices is at least halved at each step. As there are originally n

pseudo-vertices, $\lceil \log n \rceil$ calls to GATHER and JUMP are sufficient to compute the connected components.

▷ **Question 5.** The sequential loop of JUMP enforces an execution time of at best $O(\log^2 n)$, regardless of the number of PUs. We first show that this execution time can be achieved with $O(n^2)$ PUs.

With $O(n^2)$ PUs, the first and the last loop of JUMP only require $O(1)$ time and the pointer jumps require $O(\log n)$ time. In fact, $O(n)$ PUs are enough to achieve such an execution time. We now need to show that GATHER can run in $O(\log n)$ time to prove that the whole execution time is $O(\log^2 n)$.

Computing the minimum of n values can be done in time $O(1)$ with n^2 PUs. However, GATHER actually computes the minima of many sets, which can be written abstractly as in Algorithm 1.8.

```
1 forall i ∈ S in parallel do
2     T(i) ← min {C(j) | {i,j} ∈ E, C(j) ≠ C(i)}
3                { if this set is empty, then C(i) is associated to T(i) }
```

ALGORITHM 1.8: Abstract algorithm to compute minima.

```
1 forall i, j ∈ S in parallel do
2     if {i,j} ∈ E And C(i) ≠ C(j) then Temp(i,j) ← C(j)
3     else Temp(i,j) ← ∞
4 forall i ∈ S in parallel do
5     Temp(i,1) ← min {Temp(i,j) | j ∈ S}
6 forall i ∈ S in parallel do
7     if Temp(i,1) = ∞ then T(i) ← C(i)
8     else T(i) ← Temp(i,1)
```

ALGORITHM 1.9: A clever algorithm to compute minima.

Algorithm 1.8 can be transformed into Algorithm 1.9. The first loop of Algorithm 1.9 clearly runs in time $O(1)$ with $O(n^2)$ PUs (more precisely, with $O(|E|)$ PUs) on a CREW. The next two loops run in time $O(\log n)$ with $O(n^2)$ PUs using classical pointer jumping (n PUs are assigned to each row and each PU computes its minimum independently of others).

Computing connected components can thus be done in time $O(\log^2 |V|)$ with $O(|V| + |E|)$ PUs. However, the JUMP function wastes resources in the minima computation. With Brent's theorem we can lower the number of PUs down to $O\left(\frac{n^2}{\log n}\right)$ (actually $O\left(\frac{|E|}{\log |V|} + |V|\right)$ PUs) without changing the execution time.

Chapter 2

Sorting Networks

A sorting network is a more realistic, but less general, model of a parallel computer than a PRAM. A sorting network is built by organizing and connecting comparator modules, or *comparators* for short, to sort a sequence of numbers. Each comparator has two input and two output "wires," with the wires holding numerical values. One of the output wires always outputs the smallest input value and the other outputs the largest input value, as depicted in Figure 2.1. The objective is to find a sorting network architecture that only depends on the length of the input sequence and that is independent of the values of the sequence. Therefore, the main difference between sorting networks and traditional comparison-based sorting algorithms is that the sequence of comparisons is set in advance, regardless of the outcome of previous comparisons.

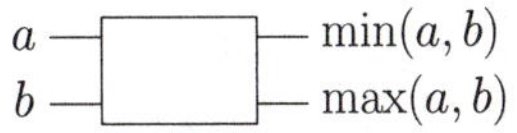

a — min(a,b)
b — max(a,b)

FIGURE 2.1: A comparator.

In this chapter, we present two sorting networks. The first network implements a merge sort, just like Cole's PRAM algorithm presented in Section 1.4. The second network uses an odd-even transposition scheme that can easily be mapped to a one-dimensional (1-D) network of processors.

2.1 Odd-Even Merge Sort

Batcher [15] proposed a merge sorting network constructed recursively from a network for merging two shorter sequences. We first describe the merging network. Throughout this section we assume that the length of the input sequence is a power of two.

2.1.1 Odd-Even Merging Network

Let us define a few notations:

- For an arbitrary sequence $\langle c_1, c_2, \ldots, c_n \rangle$, $\textsf{Sort}(\langle c_1, c_2 \ldots, c_n \rangle)$ denotes the sorted sequence of the c_i's.
- Whenever a sequence c is sorted, i.e., if $c_1 \leqslant c_2 \leqslant \cdots \leqslant c_n$, we write $\textsf{Sorted}(\langle c_1, c_2 \ldots, c_n \rangle)$.
- We define $\textsf{Merge}()$, the merging operator of two sorted sequences, as

$$\text{if } \textsf{Sorted}(\langle a_1, \ldots, a_n \rangle) \text{ and } \textsf{Sorted}(\langle b_1, \ldots, b_n \rangle) \text{ then}$$

$$\textsf{Merge}(\langle a_1, \ldots, a_n \rangle), \langle b_1, \ldots, b_n \rangle) = \textsf{Sort}(\langle a_1, \ldots, a_n, b_1, \ldots, b_n \rangle).$$

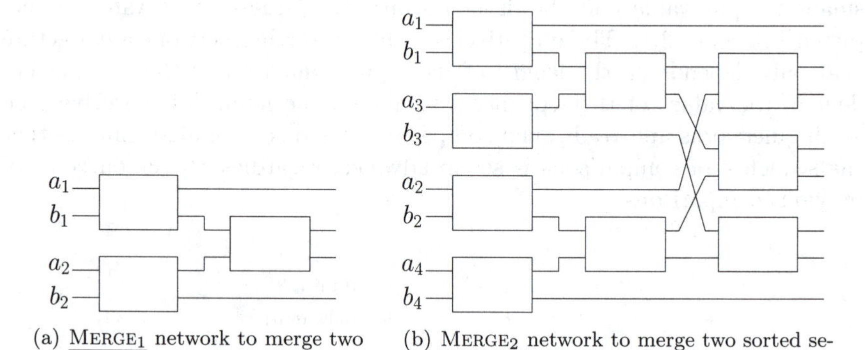

(a) $\underline{\text{MERGE}_1}$ network to merge two sorted sequences of length 2.

(b) $\underline{\text{MERGE}_2}$ network to merge two sorted sequences of length 4.

FIGURE 2.2: Merge networks for sequences of lengths 2 and 4.

Let us build a $\underline{\text{MERGE}_m}$ network that merges two sorted sequences of length 2^m. For $m = 0$, we only need a single comparator. For $m = 1$, assuming $\textsf{Sorted}(\langle a_1, a_2 \rangle)$ and $\textsf{Sorted}(\langle b_1, b_2 \rangle)$, we can use three comparators as depicted in Figure 2.2(a). It is not hard to see that this $\underline{\text{MERGE}_1}$ works as expected. We know that $a_1 \leqslant a_2$ and $b_1 \leqslant b_2$. The upper output is $\min(a_1, b_1)$ and the lower output is $\max(a_2, b_2)$. An additional comparator is added to sort the two outputs from the middle, i.e., $\max(a_1, b_1)$ and $\min(a_2, b_2)$.

Determining that $\underline{\text{MERGE}_2}$ (depicted in Figure 2.2(b)) works is more difficult. So instead let us prove the result in the general case, by induction. The $\underline{\text{MERGE}_m}$ network is built with two copies of the $\underline{\text{MERGE}_{m-1}}$ network followed with a column of $2^m - 1$ comparators. The first copy of $\underline{\text{MERGE}_{m-1}}$ merges odd elements from the input sequences and the second copy merges even elements. Surprisingly, a simple column of comparators suffices to complete the merging of the two input sequences.

PROPOSITION 2.1. *Consider two sequences* $A = \langle a_1, \ldots, a_{2n} \rangle$ *and* $B = \langle b_1, \ldots, b_{2n} \rangle$ *such that* $\textsf{SORTED}(A)$ *and* $\textsf{SORTED}(B)$*. Let us denote*

$$\begin{aligned} \langle d_1, \ldots, d_{2n} \rangle &= \textsf{MERGE}(\langle a_1, a_3, \ldots, a_{2n-1} \rangle, \langle b_1, b_3, \ldots, b_{2n-1} \rangle) \\ \langle e_1, \ldots, e_{2n} \rangle &= \textsf{MERGE}(\langle a_2, a_4, \ldots, a_{2n} \rangle, \langle b_2, b_4, \ldots, b_{2n} \rangle). \end{aligned}$$

Then, we have

$$\begin{aligned} &\textsf{SORTED}(\langle d_1, \min(d_2, e_1), \max(d_2, e_1), \ldots, \\ &\qquad \ldots, \min(d_{2n}, e_{2n-1}), \max(d_{2n}, e_{2n-1}), e_{2n} \rangle). \end{aligned}$$

Proof. Without loss of generality, we can assume that the elements are distinct. In the resulting sequence d_1 is in first position, which is correct as it is the smallest element of the whole sequence. Likewise, e_{2n}'s position is correct. In the general case, d_i and e_{i-1} (for $i \geqslant 2$ and $i \leqslant 2n$) are in position $2i-2$ or $2i-1$ in the resulting sequence. We show that their positions are correct by showing that they both dominate $2i-3$ elements of the complete sequence and are dominated by $4n-2i+1$ elements of the complete sequence. Therefore, their position in the whole sequence is necessarily $2i-2$ or $2i-1$ and the final comparison finally sets them in their correct positions.

We have four items to prove for $2 \leqslant i \leqslant 2n$:

(i) d_i dominates $2i-3$ elements.

(ii) e_{i-1} dominates $2i-3$ elements.

(iii) d_i is dominated by $4n-2i+1$ elements.

(iv) e_{i-1} is dominated by $4n-2i+1$ elements.

Let us start by proving (i): Assume for example that d_i is in sequence A (the case where d_i is in B is similar). Let k be the number of elements from $\langle d_1, d_2, \ldots, d_i \rangle$ that come from sequence A. Then, we have $d_i = a_{2k-1}$ and d_i dominates $2k-2$ elements from A. There are $i-k$ elements from B in $\langle d_1, d_2, \ldots, d_{i-1} \rangle$. The largest one is thus $b_{2(i-k)-1}$ and d_i dominates $2(i-k)-1$ elements from B. Therefore, d_i dominates $(2k-2)+(2(i-k)-1) = 2i-3$ elements. (ii) is proved in a similar way.

Let us now prove (iv): Assume that e_{i-1} is in sequence B. Let k be the number of elements from $\langle e_1, e_2, \ldots, e_{i-1} \rangle$ that come from sequence B. Then, we have $e_{i-1} = b_{2k}$ and e_{i-1} is dominated by $2n-2k$ elements from B. There are also $i-1-k$ elements from sequence A in sequence $\langle e_1, \ldots, e_{i-1} \rangle$. e_{i-1} is thus dominated by $a_{2(i-k)}$ and its successors, that is, $2n-2(i-k)+1$ elements from A. Therefore, e_{i-1} is dominated by $(2n-2(i-k)+1)+(2n-2k) = 4n-2i+1$ elements. The proof of (iii) is similar and is left as an exercise for the reader.

The positions of d_i and e_{i-1} in the resulting sequence are thus correct. □

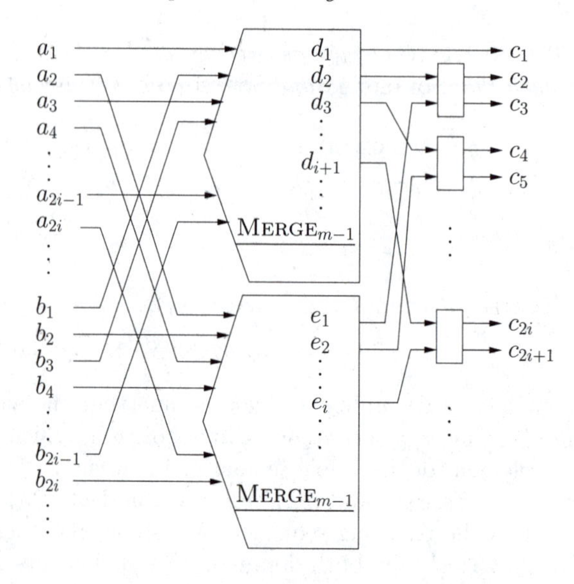

FIGURE 2.3: Recursive design of MERGE$_m$.

Figure 2.3 shows the MERGE$_m$ network, with its two copies of the MERGE$_{m-1}$ network whose outputs are connected with $2^m - 1$ comparators.

The processing time t_m of the MERGE$_m$ network is defined as the maximum number of comparators an input must traverse. Therefore, we have $t_1 = 2$ because some data have to go through two comparators (even though some only go through one) and $t_2 = 3$. Of course, many comparators can be active at the same time (this book is about parallel computing after all).

LEMMA 2.1. *The processing time t_m and the number of comparators p_m of* MERGE$_m$ *satisfy the following recursions:*

$$\begin{array}{llll} t_0 = 1 & t_1 = 2 & t_m = t_{m-1} + 1 & (i.e.,\ t_m = m + 1) \\ p_0 = 1 & p_1 = 3 & p_m = 2p_{m-1} + 2^m - 1 & (i.e.,\ p_m = 2^m m + 1). \end{array}$$

Proof. These recursions follow directly from Proposition 2.1. The expression $p_m = 2^m m + 1$ can be proved via a simple induction. □

When expressed as a function of the length $n = 2^m$ of the input sequence, these quantities are, respectively, $O(\log n)$ and $O(n \log n)$. The efficiency of the network is rather low: By multiplying the number of comparators with the processing time, one obtains the total work (see Section 1.2): $W_n = p_n \times t_n = O(n(\log n)^2)$. This is far beyond the total work of a sequential merge: Using a single comparator and $O(n)$ steps, the total work of the sequential merge is $O(n)$. Note, however, that the overall processing time, t_m, is very short.

The rather poor efficiency of this network can easily be explained: Each comparator is used only once during the merge. The efficiency of the network could be improved by merging many different pairs of sequences in sequence. Indeed, the same network can start processing a new pair of sequences at each time unit, with multiple sequences traversing the network concurrently in a *pipelined* usage. After d time units (where d is the depth of the network) all comparators are active and a merged sequence is output by the network at each time unit (the *period* of the network is said to be equal to 1).

2.1.2 Sorting Network

It is now easy to design a $\underline{\text{SORT}_m}$ network recursively to sort $n = 2^m$ elements. All we need to do is connect the output of two $\underline{\text{SORT}_{m-1}}$ networks with the input of a $\underline{\text{MERGE}_{m-1}}$ network, as shown in Figure 2.4.

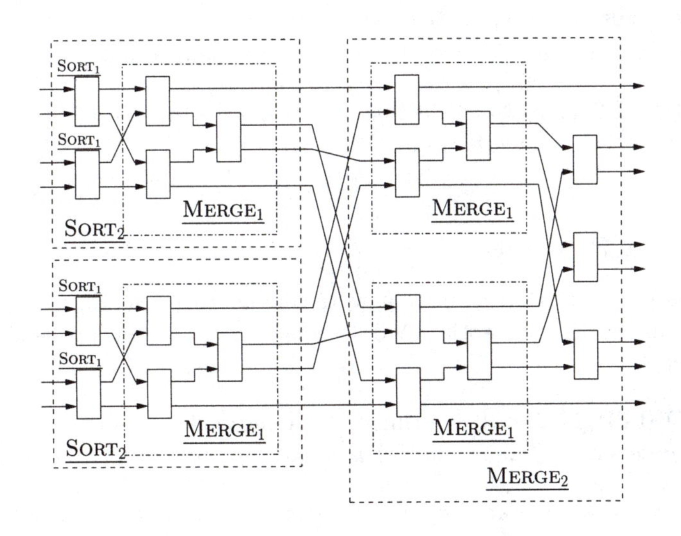

FIGURE 2.4: Recursive design of Batcher's $\underline{\text{SORT}_m}$ network ($m = 3$).

LEMMA 2.2. *The processing time t'_m and the number of comparators p'_m of $\underline{\text{SORT}_m}$ satisfy the following recursions:*

$$t'_1 = 1 \qquad t'_m = t'_{m-1} + t_{m-1} \qquad (i.e., t'_m = O(m^2))$$
$$p'_1 = 1 \qquad p'_m = 2p'_{m-1} + p_{m-1} \qquad (i.e., p'_m = O(2^m m^2)).$$

Proof. These recursions follow directly from the recursive design of the network. Since $t'_m = t'_{m-1} + t_{m-1} = t'_{m-1} + m$, we get $t'_m = O(m^2)$. For the second equation, $p'_m = 2p'_{m-1} + 2^{m-1}(m-1) + 1 = 2^{m-1}(1 + 2 + 3 + \cdots + (m-1)) + (1 + 2 + 4 + \cdots + 2^{m-1}) = 2^{m-1}(\frac{m(m-1)}{2}) + 2^m - 1$. Therefore, $p'_m = 2^{m-1}(\frac{m(m-1)}{2} + 2) - 1 = O(2^m m^2)$. □

When expressed as a function of the length $n = 2^m$ of the input sequence, the processing time is $O((\log n)^2)$ and the number of comparators is $O(n(\log n)^2)$. The total work is thus $O(n(\log n)^4)$. One may think that the number of comparators is prohibitively large given the processing time, at least for practical purposes. But here again, this very fast network can be used in a pipelined fashion.

One may wonder if there exists a sorting network with a $O(\log n)$ processing time for sequences of length n. Such a network could then be thought of as an equivalent of Cole's PRAM algorithm (Section 1.4), but for the less flexible sorting network model. Such a network was designed somewhat recently. In 1983, Ajtai, Komlos, and Szemeredi [2] proposed a network for sorting sequences of length n with $O(n \log n)$ comparators in $O(\log n)$ time. Unfortunately, the constants hidden in the $O(\dots)$ notations are so enormous that this result is of no practical use.

As a conclusion to this section, we recall the main result:

THEOREM 2.1. *Batcher's merge sort network sorts a sequence of length n with $O(n(\log^2 n))$ comparators in $O(\log^2 n)$ time.*

2.1.3 0–1 Principle

In this section, we present a new technique to prove the correctness of merging and sorting networks. A 0–1 sequence is a sequence whose elements are either 0 or 1.

PROPOSITION 2.2 (0–1 Principle). *A network is a sorting network for arbitrary sequences if and only if it is a sorting network for 0–1 sequences.*

Proof. If R sorts arbitrary sequences correctly, it obviously also sorts 0–1 sequences correctly. Let us now prove that if R does not sort arbitrary sequences correctly, then it does not sort 0–1 sequences correctly.

Let us first note that for any increasing function f a comparator has the same behavior when the input is $\langle x_1, x_2 \rangle$ as when the input is $\langle f(x_1), f(x_2) \rangle$.

Consider a given network R applied to a given sequence $\langle x_1, \dots, x_n \rangle$. The final position of x_i, i.e., the wire where x_i is output, does not depend on the value x_i but solely on its relative position in the sequence. Therefore, when applying R to $\langle f(x_1), \dots, f(x_n) \rangle$, where f is an increasing function, $f(x_i)$ is output on the same wire as x_i.

Let us assume that R does not sort correctly. Then, there is a sequence $x = \langle x_1, \dots, x_n \rangle$ and a position k such that $R(x)_k > R(x)_{k+1}$. Now let us define an increasing function $f : \{x_1, \dots, x_n\} \rightarrow \{0, 1\}$ as follows:

$$f(y) = \begin{cases} 0 & \text{if } y < R(x)_k \\ 1 & \text{if } y \geqslant R(x)_k \end{cases} .$$

It is easy to see that R does not sort the 0–1 sequence $\langle f(x_1), \dots, f(x_n) \rangle$ correctly. Indeed, $f(R(x_k)) = 1$ is output at position k and $f(R(x_{k+1})) = 0$ is output at position $k+1$. Therefore, R does not sort 0–1 sequences correctly, which completes the proof of the proposition. □

The 0–1 principle can be used to prove the correctness of $\underline{\text{MERGE}_m}$. Let us provide a new, less cumbersome proof of Proposition 2.1.

New proof of Proposition 2.1. Using the 0–1 principle we can now restrict the proof of Proposition 2.1 to 0–1 sequences.

Let us denote by $\text{ZEROS}(x)$ the number of zeros in sequence $x = \langle x_1, \dots, x_m \rangle$. A sorted 0–1 sequence is always structured as $x = 0^r 1^{m-r}$, where $r = \text{ZEROS}(x)$. Let $p = \text{ZEROS}(\langle a_1, \dots, a_{2n} \rangle)$ and $q = \text{ZEROS}(\langle b_1, \dots, b_{2n} \rangle)$. We distinguish four different cases:

- $p = 2p'$ and $q = 2q'$. We have

$$\text{ZEROS}(\langle d_1, \dots, d_{2n} \rangle) = \text{ZEROS}(\langle e_1, \dots, e_{2n} \rangle) = p' + q'.$$

  $\langle d_1, \dots, d_{2n} \rangle$ and $\langle e_1, \dots, e_{2n} \rangle$ have the same number of 0's. The last column of comparators receives as input $p' + q' - 1$ pairs of 0's, followed by a 10 (the first 1 of d and the last 0 of e) and $2n - p' - q' - 1$ pairs of 1's. The sorted sequence is thus obtained thanks to the $(p' + q')$-th comparator.

- $p = 2p'$ and $q = 2q' - 1$. We have

$$\text{ZEROS}(\langle d_1, \dots, d_{2n} \rangle) = p' + q'$$
$$\text{ZEROS}(\langle e_1, \dots, e_{2n} \rangle) = p' + q' - 1.$$

  The last column of comparators receives as input $p' + q' - 1$ pairs of 0's, followed by $2n - p' - q'$ pairs of 1's. None of the $2n - 1$ comparators is really useful as the resulting sequence $0^{p+q} 1^{4n-p-q}$ is obtained with a simple interleaving.

- $p = 2p' - 1$ and $q = 2q'$. This case is similar to the previous one.

- $p = 2p' - 1$ and $q = 2q' - 1$. We have

$$\text{ZEROS}(\langle d_1, \dots, d_{2n} \rangle) = p' + q'$$
$$\text{ZEROS}(\langle e_1, \dots, e_{2n} \rangle) = p' + q' - 2.$$

  The last column of comparators receives as input $p' + q' - 1$ pairs of 0's, followed by a 01 and $2n - p' - q'$ pairs of 1's. Once again, none of the $2n - 1$ comparators performs truly useful work. □

2.2 Sorting on a One-Dimensional Network

2.2.1 Odd-Even Transposition Sort

In this section, we present a very simple network known as an "odd-even transposition sorting network." To simplify the depiction of the "folding" of this network in the next section, we rotate the comparators as depicted below:

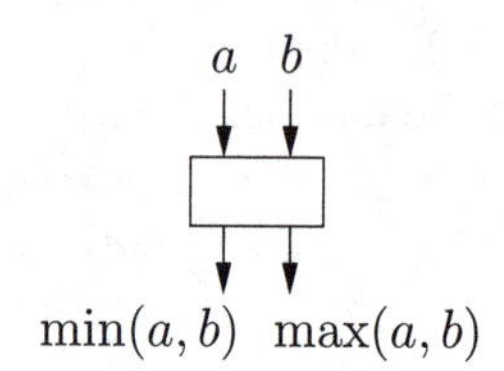

The odd-even transposition sorting network consists of a sequence of lines of comparators. More precisely, to sort a sequence of $n = 2p$ elements, we use p copies of a sub-network that consists of two lines. See Figure 2.5 for a depiction of this sub-network for $n = 8$. The first line contains p comparators whose inputs are the p pairs of wires $2i - 1$ and $2i$, $1 \leqslant i \leqslant p$ ("odd" step). The second line contains $p - 1$ comparators whose inputs are the $p - 1$ pairs of wires $2i$ and $2i + 1$, $1 \leqslant i \leqslant p - 1$ ("even" step). The whole network thus contains a total of $p(2p-1) = \frac{n(n-1)}{2}$ comparators. The layout is similar when n is odd (see Figure 2.5 for $n = 7$) and also contains $\frac{n(n-1)}{2}$ comparators.

PROPOSITION 2.3. *The odd-even transposition network is a sorting network.*

Proof. We use the 0–1 principle. Let $\langle a_1, \ldots, a_n \rangle$ denote a 0–1 sequence. Let us denote by k the number of 1's in this sequence and let j_0 denote the position of the last 1 (i.e., the rightmost 1). In Figure 2.6, we show an example for $n = 7$, $k = 3$, and $j_0 = 4$. Let us first note that a 1 never moves to the left: The only possible move is when it is compared with a 0 on its right, in which case it moves to the right.

Let us follow the last 1's moves. If j_0 is even, then it does not move in the first step but it moves to the right in the second step. If j_0 is odd, it moves to the right in the first step. In both cases, it moves to the right from step 2 and for all following steps until it is at position n. Before step 2, the last 1 is at least at position 2 and thus it always has enough time to arrive at position n in $n - 1$ steps.

Let us now follow the moves of the next-to-last 1. At step 0 it is at position j_1 ($j_1 = 2$ in Figure 2.6). As the last 1 moves to the right from step 2 on (at least), the next-to-last 1 is never blocked by the last 1 when moving to the right. Therefore, from step 3 on, the next-to-last 1 moves to the right until it reaches position $n - 1$. More generally, the i-th 1 (counting from the right)

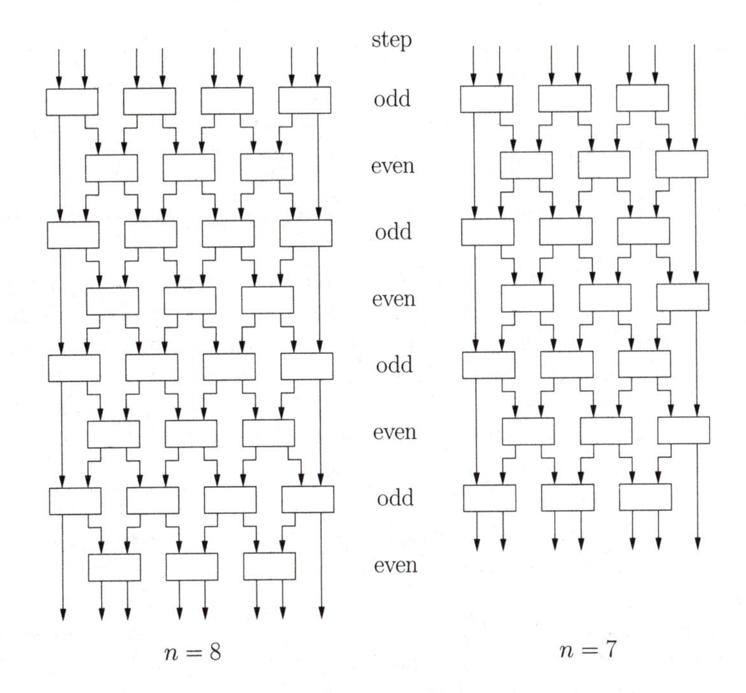

FIGURE 2.5: Layout of the odd-even transposition network.

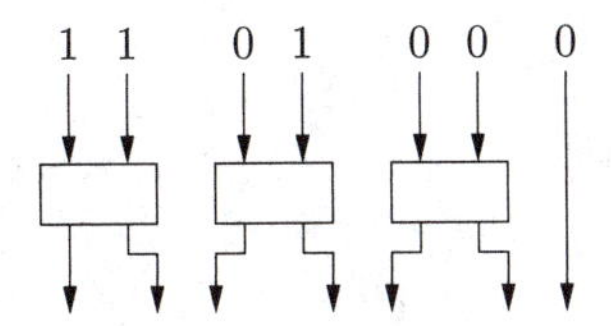

FIGURE 2.6: Illustration of the proof that the odd-even transposition network is a sorting network.

moves to the right starting (at least) at step $i+1$ until it reaches position $n-i+1$: It is never blocked by another 1 on its right.

At last, the leftmost 1 (the k-th, as there are k 1's in total) moves to the right at step $k+1$ until it reaches its final position $n-k+1$. The $k-1$ remaining 1's are on its right and the sequence is thus sorted. □

Another proof of Proposition 2.3. Another proof was proposed by Knuth [74] in Exercises 36 and 37, Section 5.3.4. It relies on "primitive sorting networks." A primitive sorting network is a sorting network such that comparisons are only made between neighbors (the odd-even transposition network is thus a primitive network).

Formally, a sorting network α can be modeled by a sequence of comparators, i.e., $\alpha = [x_k, y_k] \circ \ldots \circ [x_2, y_2] \circ [x_1, y_1]$, where k is the number of comparators.

A primitive network is such that for all i: $y_i = x_i + 1$. The proof relies on two lemmas:

LEMMA 2.3. *Let α be an n-entry primitive network. α is a sorting network if and only if α sorts $\langle n, n-1, \dots, 2, 1\rangle$.*

Proof. Proving the implication is trivial. Let us prove the reciprocal by contradiction.

Let x denote an input sequence such that $\alpha(x)_i > \alpha(x)_j$, with $i < j$. Let $y = \langle n, n-1, \dots, 2, 1\rangle$. We prove that $\alpha(y)_i > \alpha(y)_j$, which establishes the lemma. We prove this by induction on the number of comparators, k. More precisely, our induction hypothesis is

$$H_{i,j,x,y}(q) : \text{For any primitive sorting } \beta \text{ network of size } q,$$
$$\beta(x)_i > \beta(x)_j \Rightarrow \beta(y)_i > \beta(y)_j.$$

Let us first prove $H_{i,j,x,y}(0)$: A primitive network β of size 0 does not sort anything, so we necessarily have $\beta(y)_i = n + 1 - i > n + 1 - j = \beta(y)_j$, hence the result.

Let us now assume that $H_{i,j,x,y}(q-1)$ is true and consider an arbitrary primitive network γ of size q such that $\gamma(x)_i > \gamma(x)_j$. We have $\gamma = [p, p+1] \circ \beta$, where β is a primitive sorting network of size $q - 1$.

We need to distinguish between a few cases depending on p:

- If $p = i$: We have $\gamma(x)_i = \gamma(x)_p = \min(\beta(x)_p, \beta(x)_{p+1}) > \gamma(x)_j$ and $\gamma(x)_p \leqslant \gamma(x)_{p+1} = \max(\beta(x)_p, \beta(x)_{p+1})$, implying that $j \neq p + 1$. Therefore, we have $j > p + 1$ and $\gamma(x)_j = \beta(x)_j$. Thus, $\beta(x)_p > \beta(x)_j$ and $\beta(x)_{p+1} > \beta(x)_j$. Therefore, $\beta(x)_i > \beta(x)_j$ and using the induction hypothesis $H_{i,j,x,y}(q-1)$, we obtain $\beta(y)_p > \beta(y)_j$ and $\beta(y)_{p+1} > \beta(y)_j$, which gives us $\gamma(y)_i > \gamma(y)_j$.

- If $p = i - 1$, $p = j$ or $p = j - 1$: Similar arguments can be used.

- Other cases: The result is obvious. □

LEMMA 2.4. *A primitive network for sequences of length n requires at least $n(n-1)/2$ comparators to be a sorting network.*

Proof. Each comparator reduces the number of inversions of the input x. As the permutation $\langle n, n-1, \dots, 2, 1\rangle$ has exactly $n(n-1)/2$ inversions, a primitive sorting network has at least $n(n-1)/2$ comparators. □

Let us now end the proof of the odd-even transposition network. It is made of exactly $n(n-1)/2$ comparators. We can observe the effect of the transposition network on $\langle n, n-1, \dots, 2, 1\rangle$ and check that each comparator performs a real inversion. The resulting sequence is thus sorted, proving the result. □

The processing time t_n of the odd-even transposition network is $t_n = n$ and the number of comparators is $p_n = n(n-1)/2$. The total work is then $W_n = O(n^3)$, which is very high. The only nice feature of this network is its simplicity.

2.2.2 Odd-Even Sorting on a One-Dimensional Network

In this last section we focus on sorting using a 1-D network of processors. The algorithm we design is directly inspired by the previous odd-even transposition network. The idea is to "fold" the network to get a 1-D network of processors. Each processor communicates alternately with its left neighbor and its right neighbor.

	P_1		P_2		P_3		P_4		P_5		P_6
init	{8,3,12}		{10,16,5}		{2,18,9}		{17,15,4}		{1,6,13}		{11,7,14}
local sort	{3,8,12}		{5,10,16}		{2,9,18}		{4,15,17}		{1,6,13}		{7,11,14}
odd	{3,5,8}	↔	{10,12,16}		{2,4,9}	↔	{15,17,18}		{1,6,7}	↔	{11,13,14}
even	{3,5,8}		{2,4,9}	↔	{10,12,16}		{1,6,7}	↔	{15,17,18}		{11,13,14}
odd	{2,3,4}	↔	{5,8,9}		{1,6,7}	↔	{10,12,16}		{11,13,14}	↔	{15,17,18}
even	{2,3,4}		{1,5,6}	↔	{7,8,9}		{10,11,12}	↔	{13,14,16}		{15,17,18}
odd	{1,2,3}	↔	{4,5,6}		{7,8,9}	↔	{10,11,12}		{13,14,15}	↔	{16,17,18}
even	{1,2,3}		{4,5,6}	↔	{7,8,9}		{10,11,12}	↔	{13,14,15}		{16,17,18}

FIGURE 2.7: Odd-even merge sort on a 1-D network of processors ($p = 6$).

Since general purpose processors are more powerful than comparators, we only use $p \ll n$ processors to sort sequences of length n. For simplicity, we assume that p divides n. Each processor is given a sub-sequence of length n/p. These sub-sequences are sorted in parallel, and thus in time $O(\frac{n}{p}\log\frac{n}{p}) = O(\frac{n}{p}\log n)$. After this initial sorting, p steps of odd-even transpositions are performed. However, instead of exchanging a single element, processors exchange sub-sequences of length n/p. When two processors exchange two sub-sequences of length n/p they are merged and the leftmost processor keeps the first half of the resulting sequence, i.e., the n/p smaller elements, while the rightmost processor keeps the second half. This algorithm is illustrated in Figure 2.7 on an example.

The computation time needed for a transposition is that of the sequential merge, i.e., $O(n/p)$. The time for all the transpositions is thus $O(n)$ and the overall sorting time is $O(\frac{n}{p}\log n + n)$. The total work is $O(n(p + \log n))$ and the algorithm is therefore optimal for $p \leqslant \log n$. Note that our analysis does not account for the overhead due to communication between the processors (this is left for upcoming chapters).

The proof of correctness comes directly from Proposition 2.3. In the end, we have proved the following result:

PROPOSITION 2.4. *On a 1-D network of p general purpose processors, a sequence of length n can be sorted in time $O(\frac{n}{p}\log n + n)$. This algorithm is optimal for $p \leqslant \log n$.*

Bibliographical Notes

This chapter draws inspiration from the book by Gibbons and Rytter [61] for the Batcher network and from the book by Akl [3] for the section on the 1-D network. For additional information on sorting networks, we refer the curious reader to the seminal book by Knuth [74].

2.3 Exercises

The first exercise is a straightforward warm-up. The second exercise deals with bitonic sorting and is classical (see [61] or [44]). The third exercise presents a more sophisticated example of a sorting network. Many other (complex) examples can be found in [80].

◇ Exercise 2.1 : Particular Sequences

We have seen in Section 2.2.1 that a primitive network is a sorting network if and only if it sorts the sequence $\langle n, n-1, \ldots, 1\rangle$ correctly.

Prove that a network α sorts the sequence $\langle n, n-1, \ldots, 1\rangle$ correctly if and only if it sorts sequences $\langle 1^i 0^{n-i}\rangle$ for any $i \in \{1, \ldots, n\}$.

◇ Exercise 2.2 : Bitonic Sorting Network

DEFINITION 2.1. A **bitonic sequence** is composed of two subsequences, one monotonically non-decreasing and the other monotonically non-increasing. A $\vee$-frame (e.g., $\langle 12, 5, 10, 11, 19\rangle$) and a $\wedge$-frame (e.g., $\langle 2, 3, 7, 7, 4, 1\rangle$) are examples of bitonic sequences. 0–1 bitonic sequences are either of the form $0^i 1^j 0^k$ or of the form $1^i 0^j 1^k$.

DEFINITION 2.2. A **bitonic sorting network** is a comparator network that can sort any bitonic sequence

DEFINITION 2.3. A **split network** of length n, with n even, is a column of comparators such that input i is compared with input $i+\frac{n}{2}$ for $i \in \{1, 2, ..., \frac{n}{2}\}$.

1. Build a bitonic sorting network using split networks. Give the depth of your network and the number of comparators needed.

2. Design a network to merge two sorted sequences using bitonic sorting networks. Compute the size and the depth of a general sorting network built from these merging networks.

◇ Exercise 2.3 : Sorting on a Two-Dimensional Grid

This exercise presents an extension of the 1-D odd-even sorting network to a two-dimensional grid.

DEFINITION 2.4. A 2-D grid $A = (a_{i,j})$ of size $n \times n$, $n = 2^m$, is snake-ordered if its elements are ordered as follows:

$$
\begin{aligned}
a_{2i-1,j} &\leqslant a_{2i-1,j+1}, \text{ if } 1 \leqslant j \leqslant n-1, 1 \leqslant i \leqslant n/2,\\
a_{2i,j+1} &\leqslant a_{2i,j}, \text{ if } 1 \leqslant j \leqslant n-1, 1 \leqslant i \leqslant n/2,\\
a_{2i-1,n} &\leqslant a_{2i,n}, \text{ if } 1 \leqslant i \leqslant n/2,\\
a_{2i,1} &\leqslant a_{2i+1,1}, \text{ if } 1 \leqslant i \leqslant n/2-1.
\end{aligned}
$$

Note that this snake order induces an embedded 1-D network in the grid (see Figure 2.8 for an example).

$$\begin{array}{l} a_{1,1}\rightarrow a_{1,2}\rightarrow a_{1,3}\rightarrow a_{1,4} \\ \hphantom{a_{1,1}\rightarrow a_{1,2}\rightarrow a_{1,3}\rightarrow}\ \downarrow \\ a_{2,1}\leftarrow a_{2,2}\leftarrow a_{2,3}\leftarrow a_{2,4} \\ \downarrow \\ a_{3,1}\rightarrow a_{3,2}\rightarrow a_{3,3}\rightarrow a_{3,4} \\ \hphantom{a_{1,1}\rightarrow a_{1,2}\rightarrow a_{1,3}\rightarrow}\ \downarrow \\ a_{4,1}\leftarrow a_{4,2}\leftarrow a_{4,3}\leftarrow a_{4,4} \end{array}$$

FIGURE 2.8: Snake order on a 4×4 grid.

DEFINITION 2.5. The *shuffle* operation transforms the sequence $\langle z_1, \ldots, z_{2p}\rangle$ into $\langle z_1, z_{p+1}, z_2, z_{p+2}, \ldots, z_p, z_{2p}\rangle$. For example,

$$\textsc{Shuffle}(\langle 1,2,3,4,5,6,7,8\rangle) = \langle 1,5,2,6,3,7,4,8\rangle \ .$$

Let us study a merging algorithm for snake-ordered sequences. The algorithm merges four $2^{m-1} \times 2^{m-1}$ snake-ordered grids into a $2^m \times 2^m$ snake-ordered grid, and is shown below.

```
1  SNAKE_MERGE(A)
2  | SHUFFLE each row of the network (using odd-even
   | transpositions based on the elements' indices).
3  |                { Note that this amounts to SHUFFLE the columns. }
4  | Sort pairs of columns, i.e., snake-ordered n × 2 grids, using
   | 2n steps of odd-even transpositions on the induced linear
   | subnetwork.
5  | Apply 2n steps of odd-even transpositions on the induced
   L linear network of size n².
```

1. Apply this algorithm to the 4×4 array A defined by $a_{i,j} = 21 - 4i - j$ for $1 \leqslant i, j \leqslant 4$.

2. A time unit is defined as the time needed to perform an exchange between neighbors. Exchanges between different neighbors can be done in parallel though. By applying odd-even transpositions on a clever set of indices, prove that the first step of the merging algorithm can be done in $n/2 - 1$ time units. Conclude that the execution time of the merging algorithm is smaller than $\frac{9}{2}n$.

3. Assuming the merging algorithm is correct, design a network to sort a sequence of length 2^{2m} on a $2^m \times 2^m$ grid. Compute the execution time of this network.

4. Prove that the odd-even transposition sort on a grid is correct. This amounts to proving that the $2n$ steps of odd-even transpositions in the third step of the algorithm are sufficient to obtain a snake-ordered sequence.

2.4 Answers

Exercise 2.1 (Particular Sequences)

$\boxed{\Rightarrow}$ Let us first assume that α sorts the sequence $\langle n, n-1, \ldots, 1\rangle$. Then (see the proof of Proposition 2.2) α also sorts the sequences $\langle f_i(n), f_i(n-1), \ldots, f_i(1)\rangle$ for any $i \in \{1, \ldots, n\}$, where

$$f_i(x) = 0 \text{ if } x < i \text{ and } 1 \text{ otherwise.}$$

$\boxed{\Leftarrow}$ Let us now assume that α does not sort $\langle n, n-1, \ldots, 1\rangle$ correctly. We show that there exists an integer $i \in \{1, \ldots, n\}$ such that α does not sort $\langle 1^i 0^{n-i}\rangle$ correctly.

Let j denote the largest element of $\{1, \ldots, n\}$ whose position is incorrect in $\alpha(\langle n, n-1, \ldots, 1\rangle)$. Let us denote by σ a permutation of $\{1, \ldots, j\}$ such that

$$\alpha(\langle n, n-1, \ldots, 1\rangle) = \langle \sigma(1), \ldots, \underbrace{\sigma(k)}_{=j}, \ldots, \underbrace{\sigma(j)}_{<j}, j+1, \ldots, n\rangle .$$

Let us consider the previously defined function f_j. Since f_j is an increasing function, we have

$$\begin{aligned}\alpha(\langle 1^j, 0^{n-j}\rangle) &= \alpha(f_j(\langle n, n-1, \ldots, 1\rangle)) = f_j(\alpha(\langle n, n-1, \ldots, 1\rangle)) \\ &= \langle 0, \ldots, 0, 1, 0, \ldots, 0, 1, \ldots, 1\rangle.\end{aligned}$$

Therefore, the network does not sort sequences $\langle 1^i 0^{n-i}\rangle$ correctly for all $i \in \{1, \ldots, n\}$.

Exercise 2.2 (Bitonic Sorting Network)

▷ **Question 1.** Applying a split network to a bitonic sequence yields two equal-length bitonic sequences. At least one of these sequences is "clean," i.e., consisting only of 0's or only of 1's (see Figure 2.9).

Let us prove the above result more formally. Assume that there are more 1's than 0's in the initial bitonic sequence b.

- If b is of the form $1^i 0^j 1^k$.
  - If $k \geqslant \frac{n}{2}$, then the output sequence is $1^i 0^j 1^k$.
  - If $i \geqslant \frac{n}{2}$, then the output sequence is $1^{i-\frac{n}{2}} 0^j 1^k 1^{\frac{n}{2}}$.
  - If $i < \frac{n}{2}$ and $k < \frac{n}{2}$ (note that we then necessarily have $i + k \geqslant \frac{n}{2}$ since there are more 1's than 0's), then the output sequence is of the form $0^{\frac{n}{2}-k} 1^{\frac{n}{2}-j} 0^{\frac{n}{2}-i} 1^{\frac{n}{2}}$.

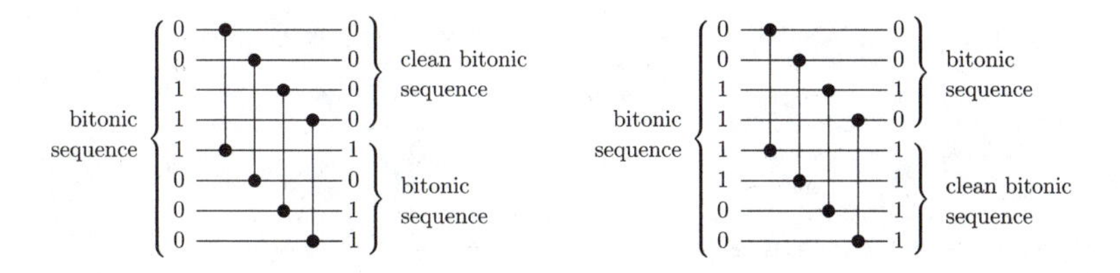

FIGURE 2.9: Applying a split network of size 8 to two different bitonic sequences.

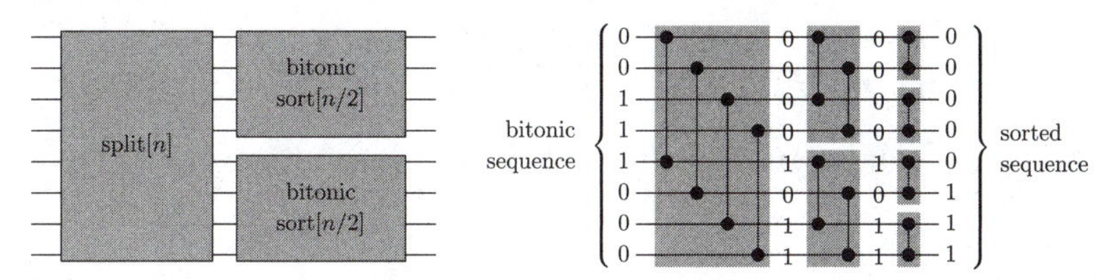

FIGURE 2.10: Using split networks to build a bitonic sorting network for sequences of length n.

- If b is of the form $0^i1^j0^k$, then since $j \geqslant \frac{n}{2}$ the output sequence is of the form $0^i1^{j-\frac{n}{2}}0^k1^{\frac{n}{2}}$.

We can thus easily check that in all cases the output sequence consists of two equal-length bitonic subsequences with at least one of them clean. The proof is similar when there are more 0's than 1's.

We can now use this property to build a bitonic sorting network from split networks (see Figure 2.10). Let t_m and p_m denote the depth and the size of bitonic sorting networks for inputs of length $n = 2^m$. We can easily check that

$$t_1 = 1,\ t_m = t_{m-1} + 1, \text{ which implies that } t_m = m,$$
$$p_1 = 1,\ p_m = t_m 2^{m-1}, \text{ which implies that } p_m = m2^{m-1}.$$

This bitonic sorting network thus contains $O(n \log n)$ comparators and its depth is $O(\log n)$.

▷ **Question 2.** Let us assume that we have two sorted sequences of length n: 0^i1^{n-i} and 0^k1^{n-k}. By reversing the second sequence (this amounts to inversing wires but does not require additional comparators) we obtain a bitonic sequence $0^i1^{2n-i-k}0^k$ that can be sorted with the previous bitonic sorting network. The merging network is thus easily built by inverting the first column of comparators in the bitonic sorting network (see Figure 2.11).

The general sorting network is then designed by recursively stacking merging networks. Let t'_m and p'_m denote the depth and the size of this sorting

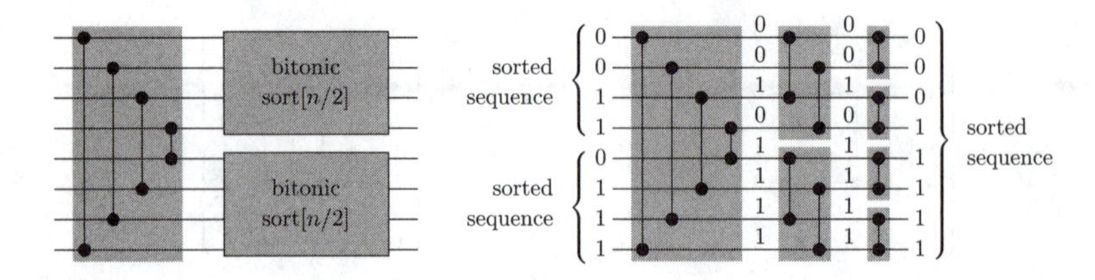

FIGURE 2.11: Building a merging network out of bitonic sorting networks.

network for inputs of length $n = 2^m$. We can easily check that

$$t'_1 = 1,\ t'_m = t'_{m-1} + t_m, \quad \text{which implies that } t'_m = O(m^2),$$
$$p'_1 = 1,\ p'_m = 2p'_{m-1} + p_m, \text{ which implies that } p'_m = O(m^2 2^m).$$

Hence, this sorting network contains $O(n(\log n)^2)$ comparators and its depth is $O((\log n)^2)$.

Exercise 2.3 (Sorting on a Two-Dimensional Grid)

▷ **Question 1.** Figure 2.12 depicts the different values of A after applying the steps of the snake merging algorithm.

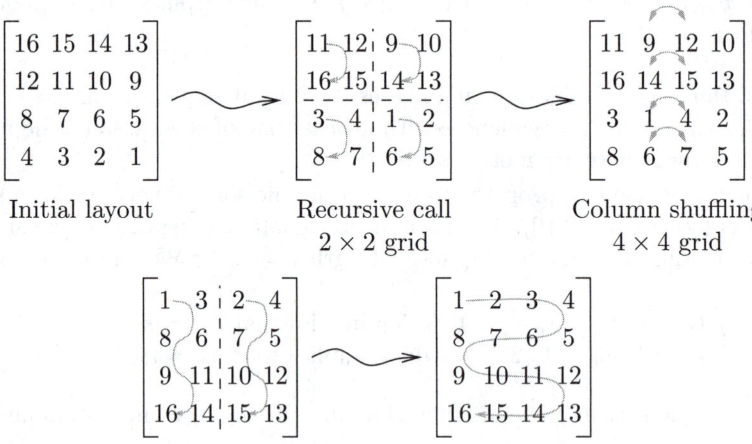

$$\begin{bmatrix} 16 & 15 & 14 & 13 \\ 12 & 11 & 10 & 9 \\ 8 & 7 & 6 & 5 \\ 4 & 3 & 2 & 1 \end{bmatrix} \rightsquigarrow \begin{bmatrix} 11 & 12 & 9 & 10 \\ 16 & 15 & 14 & 13 \\ 3 & 4 & 1 & 2 \\ 8 & 7 & 6 & 5 \end{bmatrix} \rightsquigarrow \begin{bmatrix} 11 & 9 & 12 & 10 \\ 16 & 14 & 15 & 13 \\ 3 & 1 & 4 & 2 \\ 8 & 6 & 7 & 5 \end{bmatrix}$$

Initial layout — Recursive call 2×2 grid — Column shuffling 4×4 grid

$$\begin{bmatrix} 1 & 3 & 2 & 4 \\ 8 & 6 & 7 & 5 \\ 9 & 11 & 10 & 12 \\ 16 & 14 & 15 & 13 \end{bmatrix} \rightsquigarrow \begin{bmatrix} 1 & 2 & 3 & 4 \\ 8 & 7 & 6 & 5 \\ 9 & 10 & 11 & 12 \\ 16 & 15 & 14 & 13 \end{bmatrix}$$

Column sorting 4×4 grid — $2n$ odd-even 4×4 grid

FIGURE 2.12: Snake merging.

▷ **Question 2.** For i between 1 and $n = 2p$, we define the index c_i of column i as the index this column should have after the SHUFFLE:

$$c_i = 2i - 1 \text{ if } 1 \leqslant i \leqslant p \text{ and } 2(i - 2^{m-1}) \text{ otherwise.}$$

This index set should then be sorted using local comparisons. Let us now consider the primitive network α made of $p-1$ stages. Stage i performs the following comparisons:

$$\langle p-i+1, p-i+2\rangle, \langle p-i+3, p-i+4\rangle, \ldots, \langle p+i-1, p+i\rangle \ .$$

For example, for $n = 8$ and $p = 4$, the algorithm would sort $\langle 1, 3, 5, 7, 2, 4, 6, 8\rangle$. The three stages of α consist of the following comparators: $(4, 5)$ in the first stage; $(3, 4), (5, 6)$ in the second stage; and $(2, 3), (4, 5), (6, 7)$ in the third stage.

Using a simple induction, one can prove that α makes it possible to sort the sequence c and thus can be used to SHUFFLE the columns.

The time needed for the SHUFFLE operation is $\frac{n}{2} - 1$. The time needed for the other steps is $2 \times 2n$, and the execution time of the merging algorithm is thus smaller than $\frac{9}{2}n$.

▷ **Question 3.** Let t_m denote the time needed to sort on a $2^m \times 2^m$ grid. Using the previous question, we have $t_m \leqslant t_{m-1} + \frac{9}{2}2^m$. We also have $t_0 = 0$ since a 1×1 grid is always snake-ordered. Summing up all these inequalities for t_m, t_{m-1}, t_{m-2}, ..., we obtain

$$t_m \leqslant \frac{9}{2}\sum_{k=1}^{m} 2^k = \frac{9}{2}(2^{m+1} - 2) \leqslant 9 \cdot 2^m \ .$$

The time needed to sort a sequence of length $n = 2^{2m}$ on a grid is thus $O(\sqrt{n})$.

▷ **Question 4.** We use the 0–1 principle to simplify the proof. The odd-even transposition sort on a grid is correct if and only if the third step of the algorithm can be done with only $2n$ odd-even transpositions on 0–1 sequences.

The two possible layouts of snake-ordered 0–1 sequences are depicted in Figure 2.13. The index i of the last row that starts with a 0 is always odd. There are two cases depending on whether the number of remaining 0's does not exceed a row or spills over to the following row.

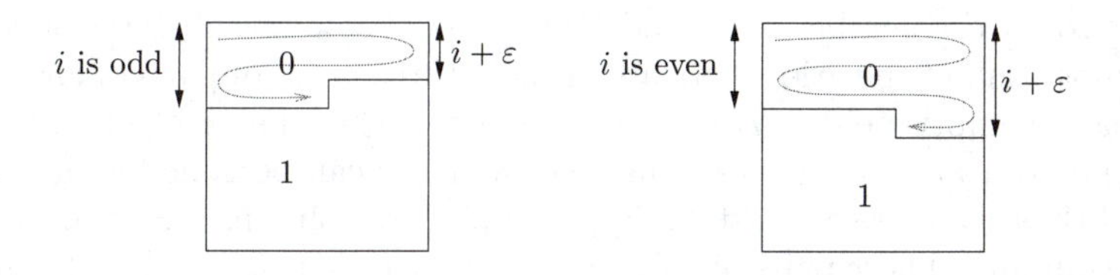

FIGURE 2.13: Two possible layouts for snake-ordered 0–1 sequences: i is always odd and $\varepsilon \in \{-1, 1\}$.

We only have to show that after the second step of the algorithm (sorting on linear networks of size $2n$) at most two rows of the grid are not clean, i.e., consisting only of 0's or only of 1's (see Figure 2.14).

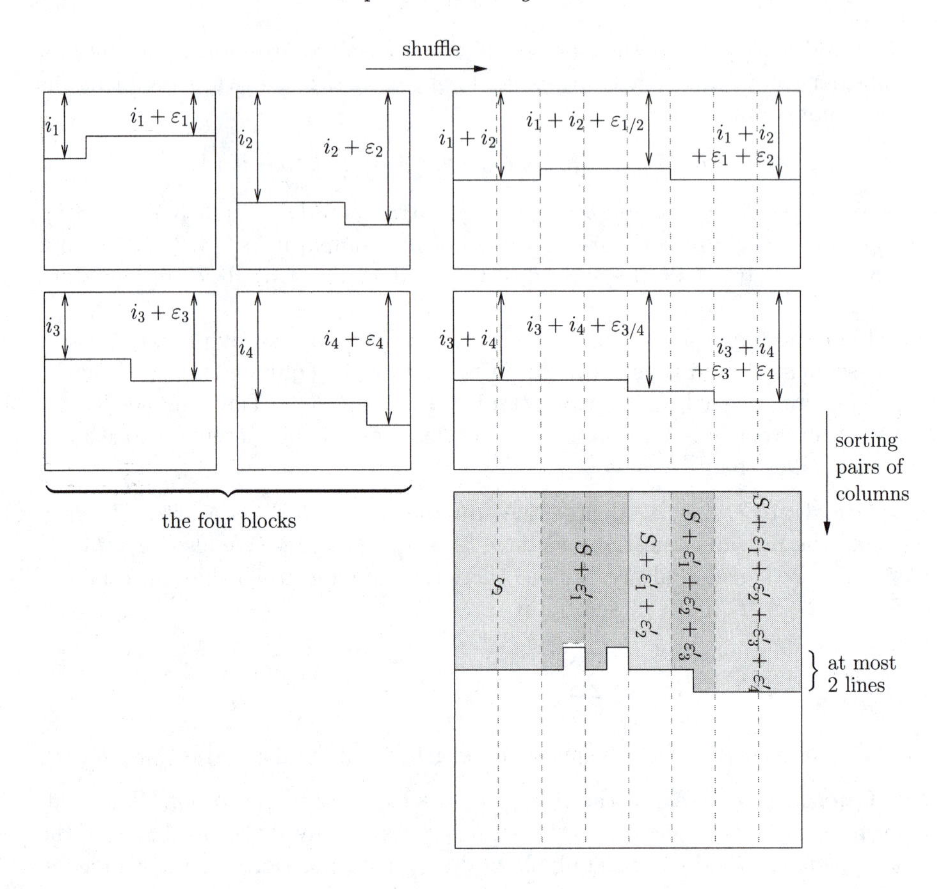

FIGURE 2.14: Merging two snake-ordered 0–1 sequences.

Using the notations of Figure 2.14, i_1, i_2, i_3, i_4 are odd integers and $\varepsilon_1, \varepsilon_2, \varepsilon_3, \varepsilon_4$ are in $\{-1, 1\}$.

Let us count the number of 0's and 1's in pairs of columns after the SHUFFLE. In the upper half of the grid, there are at most three types of pairs of columns. These pairs comprise either $i_1 + i_2$, $i_1 + i_2 + \varepsilon_{1/2}$ ($\varepsilon_{1/2}$ being equal either to ε_1 or ε_2) or $i_1 + i_2 + \varepsilon_1 + \varepsilon_2$ 0's. The same analysis can be done for the lower half, and thus there are at most five types of pairs of columns after the sort on pairs of columns. These pairs of columns comprise either S, $S+\varepsilon'_1$, $S+\varepsilon'_1+\varepsilon'_2$, $S + \varepsilon'_1 + \varepsilon'_2 + \varepsilon'_3$, or $S + \varepsilon'_1 + \varepsilon'_2 + \varepsilon'_3 + \varepsilon'_1$ 0's, where $S = i_1 + i_2 + i_3 + i_4$ and $\varepsilon'_1, \varepsilon'_2, \varepsilon'_3, \varepsilon'_4$ are in $\{-1, 1\}$.

S is even since it is the sum of four odd numbers. Therefore, one can check that in all cases at most two rows are not clean. Hence, at most $2n$ steps of odd-even transpositions are needed to sort the whole sequence correctly.

Chapter 3

Networking

In this chapter, we present network design and operation principles that are relevant to the study of parallel algorithms. We start with a brief description of classical network topologies (Section 3.1). Next, we present common message passing mechanisms along with a few performance models (Section 3.2). Then, we focus on the routing problem for two classical topologies: the ring (Section 3.3) and the hypercube (Section 3.4). More precisely, we discuss how to implement point-to-point communications as well as global communications on these topologies. Finally, we present some recent applications of some of these techniques to peer-to-oeer networks (Section 3.5).

3.1 Interconnection Networks

Let us we briefly review several classical interconnection networks used in *distributed memory* parallel platforms. In these platforms each processor has its own private memory. Processors need to exchange messages to communicate. By contrast, in *shared memory* parallel platform processors can communicate through simple reads and writes to a single shared memory. As a result, shared memory platforms are typically easier to program than distributed memory platforms. Unfortunately, the connection between the processors and the shared memory quickly becomes a bottleneck for most applications. As a result, shared memory platforms do not scale to as many processors as distributed memory platforms and become very expensive when the number of processors increases.

An alternative to a purely shared or purely distributed memory platform is a *distributed shared memory* platform. By extending the virtual memory implementation to use physically distributed local memories, one can implement a virtual global access to all of them, thereby emulating a large shared memory. This approach raises a number of difficult, but not unsolvable, technical issues. And, in fact, most of these issues arise in shared memory platforms anyway! Indeed, shared memory platforms are implemented using a hierarchy of multi-level caches. The memory is thus divided into parts whose access time is not uniform across the processors. In such non-uniform memory access

(NUMA) platforms, access time depends on the memory location relative to the processor: A processor can access its own local memory much faster than the memory of other processors. Furthermore, whenever a processor updates data in his own private cache, the change may need to be propagated to other processors. Otherwise, processors could be working with stale data, leading to an incoherent view of the shared memory. It is therefore necessary to ensure *cache coherence*, which requires careful implementation. We refer the reader to the books by Culler and Singh [48] and by El-Rewini and Abd-El-Barr [51] for more details on parallel architectures.

Beyond the architectural differences between shared and distributed memory platforms, there is a difference in programming style. Distributed memory platforms are generally programmed using the single program multiple data (SPMD) approach and using message-passing, while shared memory platforms are generally programmed with sequential imperative languages extended with parallelization primitives (parallel loops, critical sections, etc.). These two programming models can be mixed and used on almost any type of platform provided adequate software support is available (and understanding that performance may suffer if the programming model is poorly adapted to the underlying architecture).

3.1.1 Topologies

The processors in a distributed memory parallel platform are connected through an interconnection network. Nowadays all computers have specialized coprocessors dedicated to communication that route messages and put data in local memories (i.e., network cards). In what follows we say that a parallel platform comprises *nodes.* Each node consists of a (computing) processor, a memory, and a communication coprocessor. When there is no ambiguity we sometimes use the term "processor" to denote the entire node.

Nodes can be interconnected in arbitrary ways. Since the 70s, many industrial and academic projects have tried to determine the best way to interconnect nodes efficiently and at low cost. There are two main approaches:

- **Static Topologies**: The interconnection network is fixed by the constructor and cannot be changed. Nodes are connected directly to other nodes via point-to-point communication links. Most parallel machines (e.g., the Intel Paragon) of the previous decades were built in this way;

- **Dynamic Topologies**: The topology can change at runtime and one or more nodes can request that direct connections be established between them. This is accomplished by connecting nodes via *switches.* The SP series from IBM, Myrinet, and ATM networks are examples of such dynamic topologies.

3.1.2 A Few Static Topologies

The most common static topologies are depicted in Figure 3.1: (a) a fully-connected network also known as a clique; (b) a ring; (c) a two-dimensional grid; (d) a torus; and (e) a hypercube. A dynamic topology is also shown: (f) a *fat-tree*, which we will discuss later.

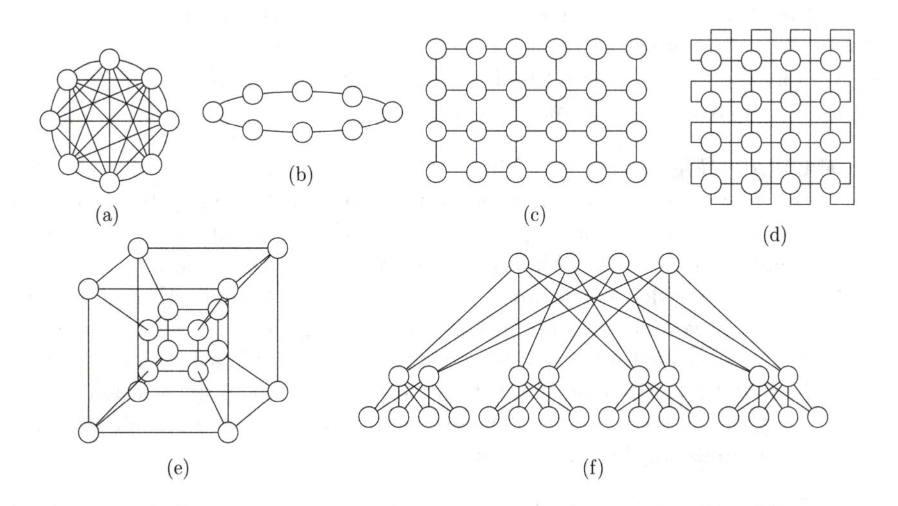

FIGURE 3.1: A few examples of interconnection network topologies: (a) clique; (b) ring; (c) grid; (d) torus; (e) hypercube; (f) fat-tree.

Static topologies are defined by graphs where vertices represent nodes of the platform and edges represent communication links. Such networks are generally characterized by the following quantities:

- **the number of nodes (p)** – This is a key quantity since it bounds the potential parallelism.
- **the degree (k)** – This is the number of edges incident to a node. For non-regular topologies, we can distinguish between the smallest degree δ and the largest degree Δ. In the following, we denote by $k = \Delta$ the number of communication links of the communication coprocessor.
- **the diameter (D)** – The distance between two nodes is the length of the shortest path between two nodes. The diameter is the greatest such distance across all node pairs.
- **the number of links (N_l)** – This is the total number of edges.
- **the bisection width (L_B)** – This is the minimum number of edges that have to be removed to partition the network into two equal-sized disconnected networks.

Let us first note that the number of processors p of a static topology with degree Δ is bounded as follows:

$$p \leqslant 1 + \Delta + \Delta(\Delta - 1) + \Delta(\Delta - 1)^2 + \ldots + \Delta(\Delta - 1)^{D-1} = \frac{\Delta(\Delta - 1)^D - 2}{\Delta - 2} .$$

This bound is known as Moore's bound. It is derived by counting the neighbors of a processor, the uncounted neighbors of neighbors, and so on, until the farthest processors are reached (at distance D).

A short discussion of the topologies from Figure 3.1 follows along with a summary of their main characteristics in Table 3.1.

- **Clique** (Figure 3.1(a)). This is the ideal network as the distance between any two processors is equal to 1. However, the number k of links per processor is equal to $p - 1$, and thus the total number of links N_l is equal to $p(p - 1)/2$. This type of network is only feasible with a limited number of nodes at reasonable cost. Thanks to routing techniques, some of which we discuss later in this chapter, any interconnection network can always be thought of as a clique by the application programmer. But, in this case, communication times are larger and there is a possibility of network contention when two messages travel through the same communication link at the same time.

- **Ring** (Figure 3.1(b)). This is one of the simplest topologies and is in fact used as a model for developing many useful parallel algorithms (see Chapter 4).

- **Two-Dimensional Grid** (Figure 3.1(c)). The maximum degree of processors is equal to 4. The main drawback of a grid is its lack of symmetry. Indeed, border processors and central processors have different characteristics. The two-dimensional (2-D) grid is well suited to, for instance, image processing problems where computations are local and communications are typically performed between neighbors. Another interesting aspect of 2-D grids is their scalability with regard to practical implementation: Processors only need to be added at the edges.

- **Two-Dimensional Torus** (Figure 3.1(d)). This one is easily derived from the 2-D grid by connecting edge processors with each others. The diameter is thus halved with the same degree. The connectivity can even be increased with a three-dimensional (3-D) torus like the one in the Cray T3D platform.

- **Hypercube** (Figure 3.1(e)). This topology has been used extensively in the previous decades. Due to its recursive definition, one can design simple yet efficient algorithms that operate dimension-by-dimension. Another attractive aspect of this topology is its low diameter (logarithmic in the number of nodes). However, the total number of links grows too

TABLE 3.1: Main characteristics of classical topologies.

Topology	**Num. of proc.** p	**Degree** k	**Diameter** D	**Num. of links** N_l	**Bisec. Width** L_B
Clique	p	$p-1$	1	$p(p-1)/2$	$(p/2)^2$
Ring	p	2	$\lfloor p/2 \rfloor$	p	2
2-D Grid	$\sqrt{p}\sqrt{p}$	$2 \rightarrow 4$	$2(\sqrt{p}-1)$	$2p - 2\sqrt{p}$	$\sqrt{p}$
2-D Torus	$\sqrt{p}\sqrt{p}$	4	$2\lfloor \sqrt{p}/2 \rfloor$	$2p$	$2\sqrt{p}$
Hypercube	$p = 2^d$	$d = \log(p)$	$d = \log(p)$	$p \log(p)/2$	$p/2$

quickly to be of practical interest for massively parallel machines. A hypercube is characterized by its dimension d, where $p = 2^d$.

3.1.3 Dynamic Topologies

The *fat-tree* dynamic topology depicted in Figure 3.1(f) was chosen by Thinking Machine Corporation to build the CM-5 [81]. This figure is different from the others because compute nodes are located only at the leaves and the nodes at higher levels in the tree do not perform computation but merely connect other nodes together. The topology is a binary tree with larger bandwidth near the root than near the leaves. This layout ensures that there is no bandwidth loss when increasing the number of processors: When processors in different subtrees communicate they have more available network links than when they are in the same subtree. Note that link redundancy also provides good fault tolerance.

The ideal way in which one can connect p processors dynamically is to use a crossbar switch, or *crossbar* for short. A schematic view of a crossbar is shown in Figure 3.2. The crossbar is built with p^2 switches. These switches are shown at the intersection of each vertical and horizontal "wire" in the picture, and can be controlled to either connect or not connect a vertical wire with a horizontal wire. The pictures shows an arbitrary example in which processor 4 is connected to processor 3, processor 3 to processor 2, processor 2 to processor 4, and processor 1 to itself. The problem with crossbars is that cost rises with the number of switches, and thus quadratically with the number of processors connected. To remedy this problem one can build a network to connect p processors by using multiple crossbars that each connect (far) fewer than p processors. Beneš or Omega networks use this approach (see Exercise 3.5). They are built from small crossbars that are arranged in stages. Only crossbars in adjacent stages are connected together. One then talks of multi-stage networks. Multi-stage networks are much cheaper to manufacture than a full crossbar (which can be thought of as a single-stage network). However, configuring a multi-stage network is generally much more difficult than configuring a crossbar. Figure 3.3 depicts a Beneš network: The p inputs are connected to the p outputs through $2(\log_2 p - 1)$ stages that each contain $p/2$ 2×2 crossbars.

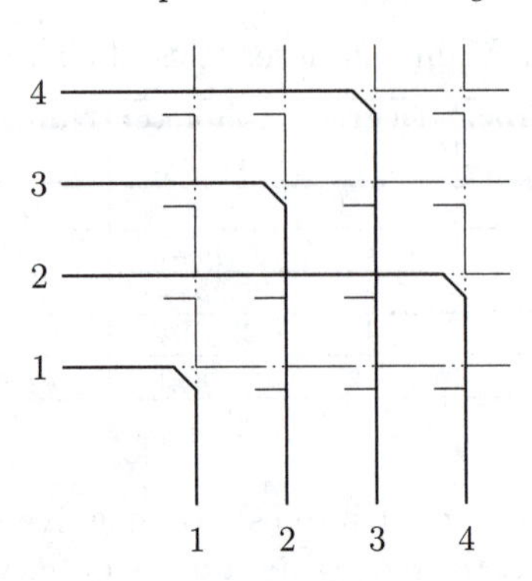

FIGURE 3.2: Crossbar network for $p = 4$.

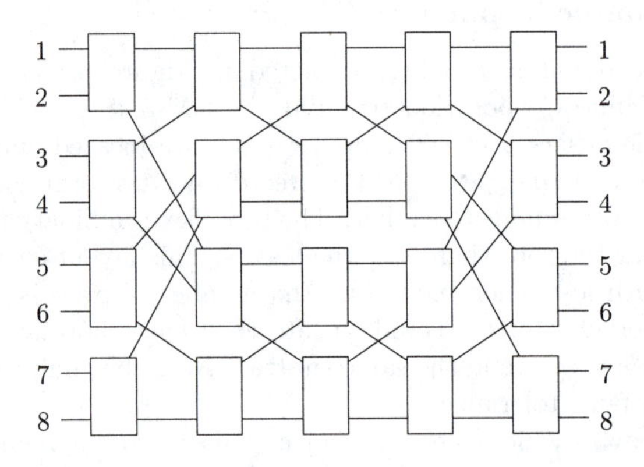

FIGURE 3.3: Beneš network for $p = 8$.

Unlike 15 years ago, dynamic topologies are now the most commonly used topologies. We refer the interested reader to the book by Culler and Singh [48] for an interesting discussion and perspectives on the topic of topologies.

3.2 Communication Model

In this section, we present different approaches to transfer a message along a given route in a topology. We analyze the performance of these approaches at first with a very simple model. We then discuss several possible extensions of this model.

3.2.1 A Simple Performance Model

Let us assume that a node, P_i, wishes to send a message M of size m to another node, P_j. The time needed to transfer a message along a network link is roughly linear in the message length. Therefore, the time needed to transfer a message along a given route is also linear in the message length. This is why the time $c_{i,j}(m)$ required to transfer a message of size m from P_i to P_j is often modeled as

$$c_{i,j}(m) = L_{i,j} + m/B_{i,j} = L_{i,j} + m.b_{i,j},$$

where $L_{i,j}$ is the start-up time (also called latency) expressed in seconds and $B_{i,j}$ is the bandwidth expressed in bytes per second. For convenience, we also define $b_{i,j}$ as the inverse of $B_{i,j}$. $L_{i,j}$ and $B_{i,j}$ depend on many factors: the length of the route, the communication protocol, the characteristics of each network link, the software overhead, the number of links that can be used in parallel, whether links are half-duplex or full-duplex, etc. This model was proposed by Hockney [67] to model the performance of the Intel Paragon and is probably the most commonly used.

3.2.2 Point-to-Point Communication Protocols

There are two main approaches for point-to-point communication: store-and-forward protocols and cut-through protocols.

Store-and-Forward (SF) – In this protocol each intermediate node receives and stores the whole message M before transmitting it toward the destination node. This model was implemented in the earliest parallel machines, in which nodes were not equipped with communication coprocessors. Consequently, intermediate nodes are interrupted to handle messages and route them to their destination.

If $d(i,j)$ is the number of links on the route between P_i and P_j, we have

$$\begin{aligned} c_{i,j}(m) = d(i,j) \cdot \left(L + \frac{m}{B}\right) &= d(i,j)L + \frac{m}{B/d(i,j)} \\ &= d(i,j)L + m.d(i,j)b\ . \end{aligned}$$

Such a protocol results in a rather poor latency and bandwidth. The communication cost can be reduced by using *pipeline*. The message is split into r packets of size $\frac{m}{r}$. Packets are sent one after the other from P_i to P_j. The first packet reaches node j after $c_{i,j}(m/r)$ time units and the remaining $r-1$ packets arrive immediately after, one after another, in $(r-1)(L + \frac{m}{r}b)$ time units. The total communication time is thus equal to

$$\left(d(i,j) - 1 + r\right)\left(L + \frac{m}{r}b\right).$$

One can then seek the value of r that minimizes the previous expression. This is a minimization problem of the form $(\gamma + \alpha.r)(\delta + \beta/r)$ with four constants α, β, γ, and δ. Removing all constant terms, this boils down to minimizing $\alpha\delta \cdot r + \gamma\beta/r$, and hence the optimal value of r, r_{opt}, is

$$r_{opt} = \sqrt{\frac{\gamma\beta}{\alpha\delta}} .$$

Indeed, the sum of two terms whose product is constant is minimized when the two terms are equal. This is the famous theorem of the goat in a pen whose perimeter, if its area is fixed, is minimal if the pen is a square. (The reader unimpressed by bucolic analogies can use the function's derivative to calculate r_{opt} analytically.) The minimum value obtained with r_{opt} is

$$\begin{aligned}((\gamma + \alpha.r)(\delta + \beta/r))[\sqrt{\gamma\beta/\alpha\delta}] &= ((\gamma/\delta)(\delta + \sqrt{\delta\beta\alpha/\gamma}))(\delta + \sqrt{\alpha\delta\beta/\gamma}) \\ &= (\sqrt{\gamma/\delta}(\delta + \sqrt{\delta\beta\alpha/\gamma}))^2 \\ &= (\sqrt{\gamma\delta} + \sqrt{\beta\alpha})^2 .\end{aligned}$$

Therefore, replacing α, β, γ, and δ by their values, we obtain $r_{opt} = \sqrt{\frac{(d(i,j)-1)mb}{L}}$ and the optimal time is

$$\left(\sqrt{(d(i,j)-1)L} + \sqrt{mb}\right)^2 = mb + O(\sqrt{mb}) = \frac{m}{B} + O\left(\sqrt{\frac{m}{B}}\right) .$$

Cut-Through (CT): Circuit Switching and Wormhole – In this model the message does not need to be stored on an intermediate node before being forwarded. The commonly used performance model for such networks is

$$c_{i,j}(m) = L + d(i,j) \cdot \delta + m/B,$$

where δ is the routing management overhead. Generally $\delta \ll L$ because most routing management is performed by the hardware whereas L comprises software overhead. Two different protocols rely on this model:

- **Circuit-switching (CC)** – A route is created before sending the first bytes of the message. Then, the message is directly sent to the destination. Note that all nodes belonging to the route are busy during the whole transmission and cannot be used for any other communication. This approach was first used in the Intel iPSC.

- **Wormhole (WH)** – In this protocol, the destination address is recorded in the header of the message and routing is performed dynamically at each node. The output of a node is computed when reading the destination address. The message is split into small packets called *flits*. In case of conflict at a node (e.g., two flits arriving at the same time), flits are stored in intermediate nodes' internal registers. There are many routing algorithms for this model (see Section 3.4). The Intel Paragon used this protocol.

Discussion on Point-to-Point Communications – Because of technology and limited knowledge of routing, the *store-and-forward* approach was the first to be used for message passing in parallel platforms. This protocol is still used nowadays but at the application level in so-called overlay networks (e.g., peer-to-peer networks). It is, however no longer used in physical networks. Cut-through protocols are more efficient because they hide the distance between nodes and avoid large buffer requirements on intermediate nodes. The *wormhole* approach is generally preferred to the *circuit switching* approach because latency is typically much lower (see Figure 3.4). These protocols

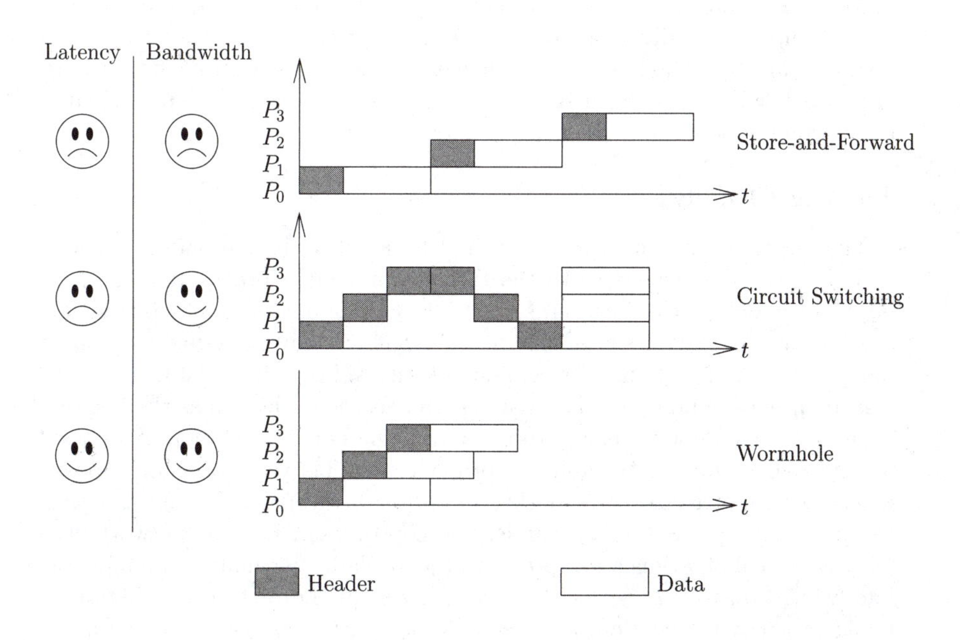

FIGURE 3.4: Illustration of various communication protocols.

have been used extensively for the dedicated networks in parallel platforms, with good buffer management, almost no message loss, and thus no need for flow-control mechanisms. By contrast, in large networks (e.g., on the Internet) there is a strong need for congestion avoidance mechanisms. Protocols like TCP split messages just like wormhole protocols but do not send all packets at once. Instead, they wait until they have received some acknowledgments from the destination node. To this end, the sender uses a sliding window of pending packet transmissions whose size changes over time depending on network congestion. The size of this window is typically bounded by some operating system parameter W_{max} and the achievable bandwidth is thus bounded by $W_{max}/$(round trip time). Such flow-control mechanisms involve a behavior called *slow start*: It takes time before the window gets full, and the maximum bandwidth is thus not reached instantly. One may thus wonder whether the Hockney model is still valid in such networks. It turns out that in most situations the model remains valid but $L_{i,j}$ and $B_{i,j}$ cannot be determined via simple measurements.

3.2.3 More Precise Models

Many aspects are not taken into account in the Hockney model. One of them is that at least three components are involved in a communication: the sender, the network, and the receiver. There is a priori no reason for all these components to be fully synchronized and busy at the same time. It may thus be important to evaluate precisely when each one of these components is active and whether it can perform some other task when not actively participating in the communication.

The LogP Family

The LogP [47] model, and other models based on it, have been proposed as more precise alternatives to the Hockney model. More specifically, these models account for the fact that some of the communications may be partially overlapped. Messages, say of size m, are split in small packets whose size is bounded by the Maximum Transmission Unit (MTU), w. In the LogP model, L is an upper bound on the latency. o is the *overhead*, defined as the length of time for which a node is engaged in the transmission or reception of a packet. The gap g is defined as the minimum time interval between consecutive packet transmissions or consecutive packet receptions at a node. During this time, the node cannot use the communication coprocessor, i.e., the network card. The reciprocal of g thus corresponds to the available per-node communication bandwidth. Lastly, P is the number of nodes in the platform. Figure 3.5 depicts the communication between nodes each equipped with a network card under the LogP model. In this model, sending m bytes with packets of size w takes time

$$c(m) = 2o + L + \left\lceil \frac{m-1}{w} \right\rceil \cdot g\ .$$

The processor occupation time on the sender and on the receiver is equal to

$$o + \left(\left\lceil \frac{m-1}{w} \right\rceil - 1 \right) \cdot g\ .$$

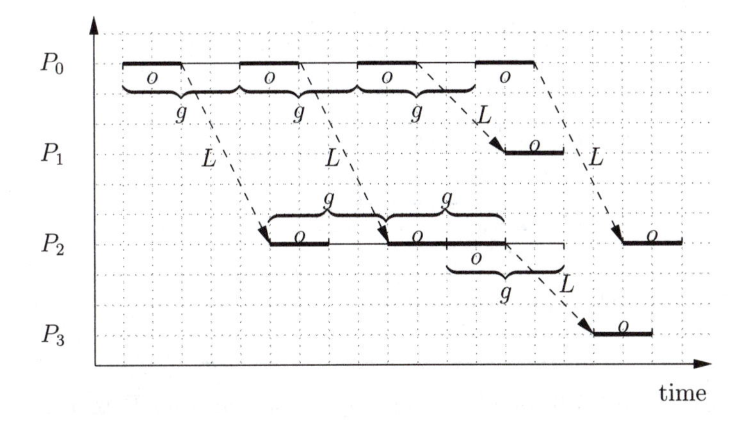

FIGURE 3.5: Communication under the LogP model.

One goal of this model was to summarize in a few numbers the characteristics of parallel platforms. It would thus have enabled the easy evaluation of complex algorithms based only on the above simple parameters. However, parallel platforms are very complex and their architectures are often hand-tuned with many optimizations. For instance, often different protocols are used for short and long messages. LogGP [4] is an extension of LogP, where G captures the bandwidth for long messages. One may, however, wonder whether such a model would still hold for average-size messages. pLogP [73] is an extension of LogP where L, o, and g depend on the message size m. This model also introduces a distinction between the sender and receiver overhead (o_s and o_r).

Affine Models

One drawback of the LogP models is the use of floor functions to account for explicit MTU. The use of these non-linear functions causes many difficulties for using the models in analytical/theoretical studies. As a result, many fully linear models have also been proposed. We summarize them through the general scenario depicted in Figure 3.6.

We first define a few notations to model the communication from P_i to P_j:

- The time during which P_i is busy sending the message is expressed as an affine function of the message size: $c^{(s)}_{i,j}(m) = L^{(s)}_{i,j} + m \cdot b^{(s)}_{i,j}$. The start-up time $L^{(s)}_{i,j}$ corresponds to the software and hardware overhead paid to initiate the communication. The quantity $b^{(s)}_{i,j}$ corresponds to the inverse of the transfer rate that can be achieved by the processor when sending data to the network (say, the data transfer rate from the main memory of P_i to a network card able to buffer the message).
- Similarly, the time during which P_j is busy receiving the message is ex-

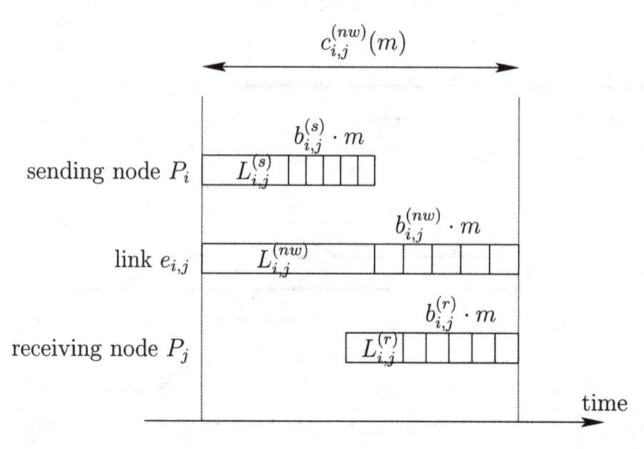

FIGURE 3.6: Sending a message of size m from P_i to P_j.

pressed as an affine function of the message length m, namely, $c_{i,j}^{(r)}(m) = L_{i,j}^{(r)} + m \cdot b_{i,j}^{(r)}$.

- The total time for the communication, which corresponds to the total occupation time of the link $e_{i,j} : P_i \to P_j$, is also expressed as an affine function, $c_{i,j}^{(nw)}(m) = L_{i,j}^{(nw)} + m \cdot b_{i,j}^{(nw)}$. The parameters $L_{i,j}^{(nw)}$ and $b_{i,j}^{(nw)}$ correspond respectively to the start-up cost and to the inverse of the link bandwidth.

Simpler models (e.g., Hockney) do not make the distinction between the three quantities $c_{i,j}^{(s)}(m)$ (emission by P_i), $c_{i,j}^{(nw)}(m)$ (occupation of $e_{i,j} : P_i \to P_j$), and $c_{i,j}^{(r)}(m)$ (reception by P_j). Such models use $L_{i,j}^{(s)} = L_{i,j}^{(r)} = L_{i,j}^{(nw)}$ and $b_{i,j}^{(s)} = b_{i,j}^{(r)} = b_{i,j}^{(nw)}$. This amounts to assuming that the sender P_i and the receiver P_j are blocked throughout the communication. In particular, P_i cannot send any message to another processor P_k during $c_{i,j}^{(nw)}(m)$ time units. However, some system/platform combinations may allow P_i to proceed to another send operation before the entire message has been received by P_j. To account for this situation, more complex models would use different functions for $c_{i,j}^{(s)}(m)$ and $c_{i,j}^{(nw)}(m)$, with the obvious condition that $c_{i,j}^{(s)}(m) \leqslant c_{i,j}^{(nw)}(m)$ for all message sizes m (which implies $L_{i,j}^{(s)} \leqslant L_{i,j}^{(nw)}$ and $b_{i,j}^{(s)} \leqslant b_{i,j}^{(nw)}$). Similarly, P_j may be involved only at the end of the communication, during a time period $c_{i,j}^{(r)}(m)$ smaller than $c_{i,j}^{(nw)}(m)$.

Here is a summary of the general framework, assuming that P_i initiates a communication of size m to P_j at time $t = 0$:

- Link $e_{i,j} : P_i \to P_j$ is busy from $t = 0$ to $t = c_{i,j}^{(nw)}(m) = L_{i,j}^{(nw)} + L \cdot b_{i,j}^{(nw)}$.

- Processor P_i is busy from $t = 0$ to $t = c_{i,j}^{(s)}(m) = L_{i,j}^{(s)} + m \cdot b_{i,j}^{(s)}$, where $L_{i,j}^{(s)} \leqslant L_{i,j}^{(nw)}$ and $b_{i,j}^{(s)} \leqslant b_{i,j}^{(nw)}$.

- Processor P_j is busy from $t = c_{i,j}^{(nw)}(m) - c_{i,j}^{(r)}(m)$ to $t = c_{i,j}^{(nw)}(m)$, where $c_{i,j}^{(r)}(m) = L_{i,j}^{(r)} + m \cdot b_{i,j}^{(r)}$, $L_{i,j}^{(r)} \leqslant L_{i,j}^{(nw)}$ and $b_{i,j}^{(r)} \leqslant b_{i,j}^{(nw)}$.

Banikazemi et al. [12] propose a model that is very close to the general model presented above. They use affine functions to model the occupation time of the processors and of the communication link. The only minor difference is that they assume that the time intervals during which P_i is busy sending (of duration $c_{i,j}^{(s)}(m)$) and during which P_j is busy receiving (of duration $c_{i,j}^{(r)}(m)$) do not overlap. Consequently, they write

$$c_{i,j}(m) = c_{i,j}^{(s)}(m) + c_{i,j}^{(nw)}(m) + c_{i,j}^{(r)}(m) .$$

In [12] a methodology is proposed to instantiate the six parameters of the affine functions $c_{i,j}^{(s)}(m)$, $c_{i,j}^{(nw)}(m)$, and $c_{i,j}^{(s)}(m)$ on a heterogeneous platform. The authors show that these parameters actually differ for each processor pair and depend upon the CPU speeds.

A simplified version of the general model was proposed by Bar-Noy et al. [14]. In this variant, the time during which an emitting processor P_i is blocked does not depend upon the receiver P_j (and similarly the blocking time in reception does not depend upon the sender). In addition, only fixed-sized messages are considered in [14], so that one obtains

$$c_{i,j} = c_i^{(s)} + c_{i,j}^{(nw)} + c_j^{(r)} . \tag{3.1}$$

Modeling Concurrent Communications

As we have seen in Section 3.1 there is a wide variety of interconnection networks. There is an even wider variety of network technologies and protocols. All these factors greatly impact the performance of point-to-point communications. They impact even more the performances of concurrent point-to-point communications. The LogP model can be seen as an attempt to model concurrent communications even though it has not clearly been designed to this end. In this section, we present a few models that account for the interference between concurrent communications.

Multi-port – Sometimes a good option is simply to ignore contention. This is what the multi-port model does. All communications are considered to be independent and do not interfere with each other. In other words, a given node can communicate with as many other nodes as needed without any degradation in performance. This model is often associated to a clique interconnection network, leading to the so-called macro-dataflow model (see Chapter 7). This model is widely used in scheduling theory since its simplicity makes it possible to prove that problems are hard and to know whether there is some hope of proving interesting results. (See Chapters 7 and 8.)

The major flaw of the macro-dataflow model is that communication resources are not limited. The number of messages that can simultaneously circulate between nodes is not bounded, and hence an unlimited number of communications can simultaneously occur on a given link. In other words, the communication network is assumed to be contention-free. This assumption is of course not realistic as soon as the number of nodes exceeds a few units. Some may argue that the continuous increase in network speed favors the relevance of this model. Note that the previously discussed models by Banikazemi et al. [12] and by Bar-Noy et al. [14] are called *multi-port* because they allow a sending node to initiate another communication while a previous one is still taking place. However, both models impose an overhead to pay before engaging in another communication. As a result, these models do not allow for fully simultaneous communications.

Bounded multi-port – Assuming an application that uses threads on, say, a node that uses multi-core technology, the network link could be shared by several incoming and outgoing communications. Therefore, the sum of the bandwidths allotted by the operating system to all communications cannot exceed the bandwidth of the network card. The bounded multi-port model proposed by Hong and Prasanna [68] is an extension of the multi-port model with a bound on the sum of the bandwidths of concurrent communications at each node. An unbounded number of communications can thus take place simultaneously, provided that they share the total available bandwidth.

Note that with this model there is no degradation of the aggregate throughput. Such a behavior is typical for protocols with efficient congestion control mechanisms (e.g., TCP). Note, however, that this model does not express how the bandwidth is shared among the concurrent communications. It is generally assumed in this model that the application is allowed to define the bandwidth allotted to each communication. In other words, bandwidth sharing is performed by the application and not by the operating system. While technology exists to achieve application-level bandwidth sharing, it is not the standard way in which networks and operating systems operate.

1-port (unidirectional or half-duplex) – To avoid unrealistically optimistic results obtained with the multi-port model, a radical option is simply to forbid concurrent communications at a node. In the 1-port model, a node can either send data or receive data, but not simultaneously. This model is thus very pessimistic as real-world platforms can achieve some concurrency of computation. On the other hand, it is straightforward to design algorithms that follow this model and thus to determine their performance a priori.

1-port (bidirectional or full-duplex) – Nowadays most network cards are full-duplex, which means that emissions and receptions are independent. It is thus natural to consider a model in which a node can be engaged in a single emission *and* in a single reception at the same time.

The bidirectional 1-port model is used by Bhat et al. [34, 35] for fixed-sized messages. They advocate its use because "current hardware and software do not easily enable multiple messages to be transmitted simultaneously." Even if non-blocking multi-threaded communication libraries allow for initiating multiple send and receive operations, they claim that all these operations "are eventually serialized by the single hardware port to the network." Experimental evidence of this fact has recently been reported by Saif and Parashar [105], who report that asynchronous sends become serialized as soon as message sizes exceed a few megabytes. Their results hold for two popular implementations of the MPI message-passing standard, MPICH on Linux clusters and IBM MPI on the SP2.

The 1-port model fully accounts for the heterogeneity of the platform, as each link has a different bandwidth. It generalizes a simpler model studied by Banikazemi et al. [11], Liu [84], and Khuller and Kim [72]. In this simpler model, the communication time $c_{i,j}(m)$ only depends on the sender, not on the receiver; in other words, the communication speed from a node to all its neighbors is the same.

k-ports – A node may have $k > 1$ network cards and a possible extension of the 1-port model is thus to consider that a node cannot be involved in more than an emission and a reception on each network card. This model will be used in Chapters 4 and 5.

Bandwidth sharing – The main drawback of all these models is that they only account for contention on nodes. However, other parts of the network can limit performance, especially when multiple communications occur concurrently. It may thus be interesting to write constraints similar to the bounded multi-port model on each network link.

Consider a network with a set $\mathcal{L}$ of network links and let B_l be the bandwidth of link $l \in \mathcal{L}$. Let a route r be a non-empty subset of $\mathcal{L}$ and let us consider $\mathcal{R}$ the set of active routes. Set $A_{l,r} = 1$ if $l \in r$ and $A_{l,r} = 0$ otherwise. Lastly, let us denote by ϱ_r the amount of bandwidth allotted to connection r. Then we have

$$\forall l \in \mathcal{L} : \sum_{r \in \mathcal{R}} A_{l,r} \varrho_r \leqslant B_l \ .$$

We also have

$$\forall r \in \mathcal{R} : \varrho_r \geqslant 0 \ .$$

The network protocol will eventually reach an equilibrium that results in an allocation ϱ such that $A\varrho \leqslant B$ and $\varrho \geqslant 0$. Most protocols actually optimize some metric under these constraints. Recent works [88, 85] have successfully characterized which equilibrium is reached for many protocols. For example, ATM networks recursively maximize a weighted minimum of the ϱ_r whereas some versions of TCP optimize a weighted sum of the $\log(\varrho_r)$ or even more complex functions.

Such models are very interesting for performance evaluation purposes, but they almost always prove too complicated for algorithm design purposes. For this reason, in the rest of this book we use the Hockney model or even simplified versions of it (e.g., with no latency) under the multi-port or the 1-port model. These models represent a good trade-off between realism and tractability.

3.3 Case Study: The Unidirectional Ring

We consider a computing platform that consists of p processors arranged in a logical unidirectional ring, as seen in Figure 3.7. The processors are denoted by P_k for $k = 0, \ldots, p-1$. Each processor can determine its logical index by calling function MY_NUM().

A processor can find out the total number of processors p by calling function NUM_PROCS(). Note that we use the term "processor" to denote both a physical compute device (or node) and a running program on that device, i.e., a process. There is no confusion because we always assume one process per device.

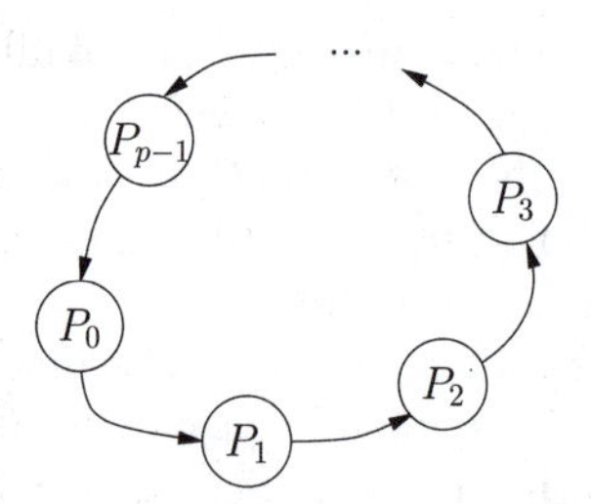

FIGURE 3.7: A unidirectional ring with p processors.

Each processor has its own local memory. All processors execute the same program, which operates on the data in their respective local memories. This mode of operation is common and termed *single program multiple data* (SPMD). To access data that are not in their local memories, processors must communicate via explicit sending and receiving of messages. This mode of operation is also common and is termed *message passing*. Each processor can send a message to its successor on the ring by calling the following function:

$$\text{SEND}(addr, m),$$

where *addr* is the address (in the memory of the sender processor) of the first data element to be sent and m is the message length expressed as a number of data elements. For simplicity we assume that the data elements of a message must be contiguous in memory, starting at base address *addr*. Note that message passing implementations, for instance, those of the MPI standard [109], typically provide sophisticated capabilities to communicate non-contiguous data with a single function call. To receive a message, a processor must call the following function:

$$\textsc{Receive}(addr, m).$$

Both functions above and others allowing processors to communicate are often termed "communication primitives." A few important points must be noted:

- Calls to communication primitives must match: If processor P_i executes a RECEIVE, then its predecessor (processor $P_{i-1 \bmod p}$) must execute a SEND. Otherwise, the program cannot terminate.

- Since each processor has a single outgoing communication link there is no need for specifying the destination processor; it is always the successor of the sending processor. Similarly, when executing a RECEIVE, the source of the incoming message is always the predecessor processor. For a more complex logical topology, one may have to specify the destination processor in the SEND and the source processor in the RECEIVE.

- The address of the first data element in a message, which is passed to SEND, is an address in the local memory of the sender processor. Similarly, the address passed to RECEIVE is in the local memory of the receiver processor. Therefore, even if the program uses the same variable for storing both addresses (remember that we are in an SPMD execution), the value of this variable will most likely be different in both processors. One can of course use two different variables, as shown below:

  $q \leftarrow$MY_NUM()
  if $q = 0$ **then** SEND($addr1, m$)
  if $q = 1$ **then** RECEIVE($addr2, m$)

So far we have not said anything regarding the semantics of the communication primitives. There are three standard assumptions:

- A rather restrictive assumption consists in assuming that each SEND and RECEIVE is blocking, i.e., the processor that calls one of these communication primitives can continue its execution only once the communication has completed. This completely synchronous message passing mode, which is also called "rendez-vous," is typical of first generation parallel computing platforms.

- A classical assumption is to keep the RECEIVE blocking but to assume that the SEND is non-blocking. This allows a processor to initiate a send but to continue execution while the data transfer takes place. Typically, this is implemented via two functions: one to initiate the communication, and the other to check whether the communication has completed. In this chapter, in order to not clutter the pseudo-code of our algorithms, we do not use the second function. Instead, in the description of our algorithms we mention which operations are blocking and which ones are non-blocking.

- A more recently proposed assumption is that both communication primitives are non-blocking; a single processor can send data, receive data, and compute simultaneously. Of course the three should occur concurrently only if there is no race condition. Again, this is implemented via two functions for each communication primitive: initialization and completion check. It is then convenient to think of each program running on a processor as three *logical threads* of control, one for computing, one for sending data, and one for receiving data (even though the implementation may not be multi-threaded or at least may not appear multi-threaded to the programmer).

The implications of these different assumptions will be clear in the course of writing our first parallel programs in the upcoming sections of this chapter and in the upcoming chapters. We almost always use the least restrictive third assumption above, both when writing the programs and when analyzing their performance. It is straightforward to adapt both the programs and the performance analyses to the first two more restrictive assumptions if they seem more appropriate for the underlying computing platform. Our goal here is to convey general principles of parallel algorithm design and of parallel algorithm performance analysis, and these principles hold regardless of the underlying assumptions.

As explained earlier, we use one of the simplest performance models: The time to send/receive a message of length m is $L + mb$, where L and b are two constants that depend on the platform. L is the *startup cost*, in seconds, due to the physical network latency and the software overhead involved in a network communication. b is the inverse of the data transfer rate, and measures the raw speed of the communication in steady-state.

3.3.1 Broadcast

For a given processor index k, we wish to write a program by which processor P_k sends the same message, of length m, to all other processors; this operation is called a *broadcast*. This is a fundamental collective communication primitive. For instance, the sender processor can be a "master" that broadcasts general information (e.g., problem size, input data) to all other "worker" processors.

At the beginning of the program the message is stored at address *addr* in the memory of the sender, processor P_k. At the end of the program the message will be stored at address *addr* in the memory of each processor. All processors must call the following function:

$$\text{BROADCAST}(k, addr, m).$$

The main idea is to have the message go around the ring, from processor P_k to processor P_{k+1}, then from processor P_{k+1} to processor P_{k+2}, etc. For the remainder of the chapter, we will often implicitly assume that processor indices are to be taken modulo p. There is no parallelism here since the SEND and the RECEIVE executed by each processor are not independent. Therefore, the whole program is written as shown in Algorithm 3.1. It is important to note that the predecessor of the sender processor, i.e., processor P_{k-1}, must not send the message. This is not only because processor P_k already has the message, since it is the sender processor, but also because it does not execute a RECEIVE in the program as we have written it.

```
1  BROADCAST(k, addr, m)
2      q ←MY_NUM()
3      p ←NUM_PROCS()
4      if q = k then
5          SEND(addr, m)
6      else
7          if q = k − 1 mod p then
8              RECEIVE(addr, m)
9          else
10             RECEIVE(addr, m)
11             SEND(addr, m)
```

ALGORITHM 3.1: Broadcast on a ring of processors.

For the program to be correct, we assume that the RECEIVE is blocking, since processors forward the message with a SEND immediately after the RECEIVE. For this program, the semantics of the SEND does not matter. Since we have a sequence of $p-1$ communications, the time needed to broadcast a message of length m is $(p-1)(L+mb)$.

Note that message passing implementations, for instance, those of the MPI standard [109], typically do not use a ring topology for implementing collective communication primitive such as a broadcast but rather use various tree topologies. Logical tree topologies are more efficient on modern parallel platforms for the purpose of collective communications. Nevertheless, we use a ring topology in this chapter for two reasons. First, a linear topology such as a ring makes for straightforward collective communication algorithms while highlighting key general concepts such as communication pipelining. Second, we will see in Chapter 4 that when designing algorithms for a logical ring topology it is sometimes beneficial to implement collective communications "by hand."

3.3.2 Scatter

We now turn to the *scatter* operation by which processor P_k sends a different message to each processor. To simplify, we assume that all sent messages have the same length, m. A scatter is useful, for instance, to distribute different data to worker processors, such as matrix blocks or parts of an image.

At the beginning of the execution, processor P_k holds the message to be sent to processor P_q at address $addr[q]$. To keep things uniform, we assume that there is a message at address $addr[k]$ to be sent by processor P_k to itself. (Such a convention is used in standard communication libraries as it turns out to be convenient in practice.) At the end of the execution, each processor holds its own message at address msg.

The simple but key idea to implement the scatter operation is to pipeline message sending, starting with the message to be sent to the furthest processor, that is, processor P_{k-1}. While this message is on its way, other messages to be sent to closer processors can be sent as well.

```
1  SCATTER(k, msg, addr, m)
2      q ←MY_NUM()
3      p ←NUM_PROCS()
4      if q = k then
5          for i = 1 to p − 1 do
6              SEND(addr[k + p − i mod p], m)
7          msg ←addr[k]
8      else
9          RECEIVE(tempR, m)
10         for i = 1 to k − 1 − q mod p do
11             tempS ↔tempR
12             SEND(tempS, m) || RECEIVE(tempR, m)
13         msg ←tempR
```

ALGORITHM 3.2: Scatter on a ring of processors.

The program is shown in Algorithm 3.2. In this program, each processor uses two buffers containing addresses *tempS* and *tempR*, so that the sending of a message and the receiving of the next message can occur in parallel. This parallelization is denoted by the notation "... || ..." and is possible if we assume that the SEND is non-blocking while the RECEIVE is blocking. We will make this common assumption for the vast majority of our algorithms. Finally, note that the instructions $tempS \leftrightarrow tempR$ and $msg \leftarrow tempR$ are mere pointer updates, and not memory copies.

The time for the scatter is the same as for the broadcast, i.e., $(p-1)(L+mb)$. Indeed, pipelining of the communications on the rings allows for several communication links to be used simultaneously. Therefore, it is possible to send $p-1$ messages in the same time as one, provided that their destinations are $p-1$ consecutive processors along the networks.

3.3.3 All-to-All

A third important collective communication primitive is the *all-to-all* exchange. For all k from 0 to $p-1$, processor P_k wishes to send a message to all others, which amounts to p simultaneous broadcasts. Again, we assume that all messages have the same length, m. At the beginning, each processor holds the message it wishes to send at address *my_message*. At the end, each processor will hold an array *addr* of p messages, where $addr[k]$ holds the message from processor P_k. The algorithm is surprisingly straightforward and consists of $p-1$ steps, with p messages on the network during each step, as shown in Algorithm 3.3.

```
1 ALL_TO_ALL(my_message, addr, m)
2     q ←MY_NUM()
3     p ←NUM_PROCS()
4     addr[q] ←my_message
5     for i = 1 to p − 1 do
6         SEND(addr[q − i + 1 mod p], m) ||
          RECEIVE(addr[q − i mod p], m)
```

ALGORITHM 3.3: All-to-all on a ring of processors.

Considerations regarding the semantics of the communication primitives are identical to those for the scatter operation in the previous section. And again, the execution time is $(p-1)(L+mb)$ as all communication links are used at each step.

There is a last classical collective communication operation called *gossip*, in which each processor sends a different message to each processor. We leave the gossip algorithm as an exercise for the reader.

3.3.4 Pipelined Broadcast

Can a broadcast be executed faster? The implementation we have proposed runs in time $(p-1)(L+mb)$, for broadcasting a message of length m. The algorithm is simple, but it turns out to be rather inefficient for long messages.

To improve execution time, one idea is to split the message into r pieces of the same length (assuming that r divides m). The sender processor sends the r message pieces in sequence, allowing the pieces to travel on the ring simultaneously. Such pipelining is key for reducing communication time. At the beginning of the execution, the r message pieces are stored at addresses $addr[0], \ldots, addr[r-1]$ on processor P_k. At the end of execution, all message pieces are stored at all processors. At each step, while a processor receives a message piece, it also sends the piece it has previously received, if any, to its successor. The program is shown in Algorithm 3.4.

```
1  BROADCAST(k, addr, m)
2     q ←MY_NUM()
3     p ←NUM_PROCS()
4     if q = k then
5        for i = 0 to r − 1 do
6           SEND(addr[i], m/r)
7     else if q = k − 1 mod p then
8        for i = 0 to r − 1 do
9           RECEIVE(addr[i], m/r)
10    else
11       RECEIVE(addr[0], m/r)
12       for i = 0 to r − 2 do
13          SEND(addr[i], m/r) || RECEIVE(addr[i + 1], m/r)
14       SEND(addr[r − 1], m/r)
```

ALGORITHM 3.4: Pipelined broadcast on a ring of processors.

To determine the execution time, we just need to determine when the last processor, P_{k-1}, receives the last message piece. There must be $p-1$ communication steps for the first message piece to arrive at P_{k-1}, which amounts to time $(p-1)(L+\frac{m}{r}b)$. Then, all remaining $r-1$ message pieces arrive subsequently, in time $(r-1)(L+\frac{m}{r}b)$. Therefore, the overall execution time is the sum of the two:

$$(p + r - 2)\left(L + \frac{m}{r}b\right) .$$

We seek the value of r that minimizes the previous expression. Using the "goat in a pen" theorem, we obtain $r_{opt} = \sqrt{\frac{m(p-2)b}{L}}$ and the optimal execution time is:

$$\left(\sqrt{(p-2)L} + \sqrt{mb}\right)^2 . \tag{3.2}$$

In this expression, p, L, and b are all fixed. Therefore, for long messages, the expression tends to mb, which does not depend on p! This is the same principle as the one used in IP networks, which split messages into many small packets to improve throughput over multi-hop paths thanks to pipelining of communications over multiple communication links.

3.4 Case Study: The Hypercube

We have briefly mentioned the hypercube earlier. In this section, we describe a few interesting properties of this topology.

3.4.1 Labeling Vertices

Hypercubes are built using duplication. A 0-cube is made of a single vertex and an n-cube (also called an n-dimensional hypercube) is recursively built from two identical $(n - 1)$-cubes whose vertices are connected one-by-one. Figure 3.8 depicts this recursive construction: Two 3-cubes are connected to build a 4-cube.

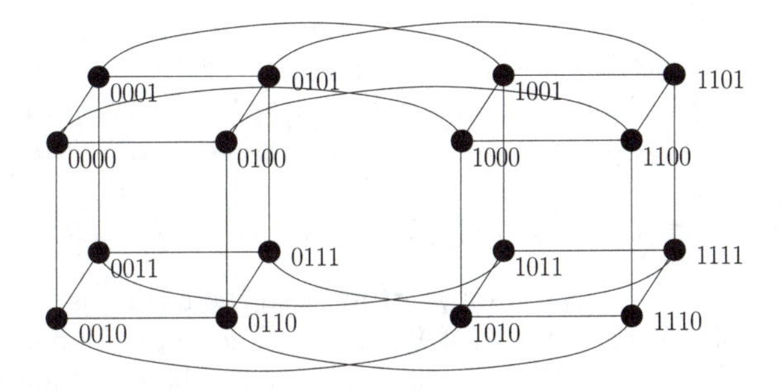

FIGURE 3.8: A four-dimensional hypercube.

Here is another equivalent definition: An n-cube is a graph made of 2^n vertices labeled from 0 to $2^n - 1$ and such that two vertices are connected if and only if their binary representations differ by a single bit. The recursive construction leads to a natural labeling of the vertices: Vertices from the first cube are labeled $0a_i$ and those from the second cube are labeled $1a_i$, where a_i is the binary representation of the "sibling" vertices. This is the labeling shown in Figure 3.8.

Using this recursive definition, it is easy to show that the diameter and the degree of an n-cube are equal to n: The diameter and the degree are increased by 1 at each step of the recursion.

3.4.2 Paths and Routing in a Hypercube

Paths With the hypercube vertex labeling one can build paths between any two arbitrary vertices. Let A and B denote two vertices from the n-cube. *Routing* a message from A to B is equivalent to finding a path from A to B in the n-cube whose length is minimum.

Let us define $H(A, B)$ as the Hamming distance between A and B, i.e., the number of bits that differ in the binary labels of A and B.

Let us consider a path from A to B whose length l is minimum. Exactly one bit is changed along this path from one vertex to another. Therefore, there are at most l different bits between A and B, which means that $l \geqslant H(A, B)$. Let us now build a path from A to B whose length is exactly $H(A, B) = i$. Let $A = a_{n-1} \dots a_1 a_0$ and $B = b_{n-1} \dots b_1 b_0$.

Without loss of generality, we can assume that the i different bits are the rightmost ones. Then we have

$$\begin{aligned} A &= a_{n-1}a_{n-2}\dots a_{i+1}a_i a_{i-1}a_{i-2}\dots a_2 a_1 a_0 \\ B &= a_{n-1}a_{n-2}\dots a_{i+1}a_i \overline{a_{i-1}a_{i-2}}\dots \overline{a_2 a_1 a_0}, \end{aligned}$$

where $\overline{x} = 1 - x$ is the bit complement of x. Here is a path from A to B that adjusts bits starting from the rightmost ones:

$$\begin{aligned} A = \text{vertex } 0 &= a_{n-1}a_{n-2}\dots a_{i+1}a_i a_{i-1}a_{i-2}\dots a_2 a_1 a_0 \\ \text{vertex } 1 &= a_{n-1}a_{n-2}\dots a_{i+1}a_i a_{i-1}a_{i-2}\dots a_2 a_1 \overline{a_0} \\ \text{vertex } 2 &= a_{n-1}a_{n-2}\dots a_{i+1}a_i a_{i-1}a_{i-2}\dots a_2 \overline{a_1 a_0} \\ \text{vertex } 3 &= a_{n-1}a_{n-2}\dots a_{i+1}a_i a_{i-1}a_{i-2}\dots \overline{a_2 a_1 a_0} \\ &\dots \\ B = \text{vertex } i &= a_{n-1}a_{n-2}\dots a_{i+1}a_i \overline{a_{i-1}a_{i-2}}\dots \overline{a_2 a_1 a_0}. \end{aligned}$$

There is no particular reason for starting with the rightmost bits. Actually, there are $i!$ paths from A to B of length i. They are obtained by simply adjusting the i different bits in an arbitrary order. The right question is, how many independent minimal length paths from A to B are there? (Two paths are said to be independent if they do not have any vertex in common except for A and B.) Such paths are of great interest as they allow one to route simultaneously several messages from A to B (or different parts of the same message!).

With the previous construction, we can build i independent paths. The j-th path γ_j, $0 \leqslant j < i$, is obtained by first adjusting the j-th distinct bit, then the $(j+1)$-th distinct bit, and so on until the $(i-1)$-th bit is adjusted. Then, the 0-th, 1-st, ..., $j-1$-th in this order. Let us call γ_0 the path we have just built. Let us assume that γ_{j_0} and γ_{j_1} share a common vertex X. The lengths of the subpaths from A to X in γ_{j_0} and γ_{j_1} are equal, otherwise we could build a shorter path from A to B. However, for any j and k, the first adjusted bits in γ_j are $S_{j,k} = \{j, \ldots, j+k-1\}$ if $j+k-1 < n$ and $S_{j,k} = \{j, \ldots, n-1\} \cup \{0, \ldots, k-(n-j)\}$ otherwise. Therefore, $S_{j_0,k} = S_{j_1,k}$ if and only if $j_0 = j_1$. Lastly, it is impossible to find more independent paths. Any path starts by adjusting a bit, say the j-th, and thus is in conflict with path γ_j. Note that i independent paths only reach $i(i-1)+2$ vertices and the bandwidth available from A is fully used only when $i = n$.

Routing Many routing algorithms can be designed to exploit the many paths in a hypercube. Let us start with a simple one: To route a message from processor A to processor B, we use the path that always adjusts the rightmost bit. At a hardware level, we simply need to route the message through the link numbered by the index of the first non-zero bit in A XOR B. The message is initially labeled with A XOR B and this label is modified along the route. Each vertex inspects this label. If this label is equal to 0, then the message has reached its destination; otherwise, the vertex updates the rightmost bit and routes the message along the corresponding link.

Here is an example with the route from $A = 1011$ to $B = 1101$ in the 4-cube. The message is originally labeled by A XOR $B = 0110$. Vertex A routes the message along link 1 (the link indices range from 0 to 3) with the new label 0100. Vertex 1001 receives this message and knows it should be forwarded along link 2. Vertex B receives this new message labeled by 0000 and thus knows it should store this message in its own local memory.

The simplicity of this routing algorithm is well suited to a hardware implementation of a *wormhole* or *cut-through* protocol. There are, however, two limitations: (i) The algorithm does not exploit redundant paths; and (ii) the routing is purely static. The first limitation is not critical. There are actually very few machines that are able to route efficiently along several links in parallel. After all, pipelining already makes it possible to obtain good performances using a single link. The second limitation is more of a concern: If another pair of processors has already reserved one of the links on the desired path, the message will stall until the end of the other communication. This could be avoided since many other paths are possible.

The routing algorithm can be made dynamic by simply adjusting the first bit that corresponds to an *available* link. The routers dynamically select links to use based on a link reservation table and on the message labels. Let us go back to our example on the 4-cube with $A = 1011$ and $B = 1101$. As A XOR $B = 0110$, the static routing suggests using link 1. If this link is already used by another ongoing communication, then the message is sent on link 2 to router 1111 and the new label is 0010. The message will thus be forwarded on link 1. Whenever a vertex determines that there is no more available (useful) links, it has no other choice but to wait for one of them to be available again. This algorithm is harder to implement than one may think at first glance. One needs to ensure the fairness of the routing protocol or at least to avoid starvation; a communication should not be delayed indefinitely on a loaded network. One also needs to rebuild a message upon reception since packets may follow different routes and thus may arrive out of order. Lastly, one needs to ensure that all messages are delivered to their destination, which is easy with the static algorithm but could become rather nightmarish if messages are allowed to travel backwards: If links 1 and 2 are busy at processor A in our previous example, one could try to use link 0 even if it extends the route.

The static algorithm was implemented by Intel on the iPSC2. The dynamic algorithm was also implemented, but on the Paragon's 2-D grid but not on a hypercube. It uses the same idea: A message sent from processor (x, y) to processor (x', y') is labeled with $(x' - x, y' - y)$, assuming that $x' > x$ and $y' > y$. A simple static routing procedure is obtained by forwarding the message horizontally $x' - x$ times and then vertically $y' - y$ times. It is however possible to use any Manhattan path. The message can use the vertical link whenever the horizontal one is busy, and conversely. The only restriction is to never go out of the rectangle defined by the source and the destination.

3.4.3 Embedding Rings and Grids into Hypercubes

Hypercubes are popular for their strong connectivity but also because classical topologies including rings, 2-D grids, and 3-D grids can be embedded into hypercubes while preserving locality. For instance, any two neighbors in a ring can be mapped to neighbor nodes in the hypercube. Topology embedding is very useful for designing algorithms on rings or grids and then using the powerful connectivity of hypercubes for global communications like broadcasts. It is not always possible to find an embedding that preserves locality. Sometimes, one has to resort to embeddings in which neighbor processors in the original topology are no longer neighbors in the hypercube. In this case one strives to minimize their distance in the hypercube.

In this section, we limit our study to rings and grids whose dimensions are powers of 2 and that use all the vertices of the hypercube. A well-adapted mathematical tool for such a study is the Gray code. The Gray code is an ordered sequence of binary codes whose successive values differ in a single bit. The Gray code for dimension n is denoted G_n and is defined recursively for $n \geqslant 2$ as

$$G_n = \{0G_{n-1}, 1G_{n-1}^{rev}\} ,$$

where

- xG is the sequence obtained by prefixing every element of G by x;
- G^{rev} is the sequence obtained by enumerating the elements of G in the reverse order (i.e., starting from the end).

The recursive construction is initiated with $G_1 = \{0, 1\}$. Thus, we get

$$\begin{aligned} G_2 &= \{00, 01, 11, 10\} \\ G_3 &= \{000, 001, 011, 010, 110, 111, 101, 100\} \\ G_4 &= \{0000, 0001, 0011, 0010, 0110, 0111, 0101, 0100, \\ &\quad 1100, 1101, 1111, 1110, 1010, 1011, 1001, 1000\}. \end{aligned}$$

We denote by $g_i^{(r)}$ the i-th element of the Gray code of dimension r.

The Gray code G_n makes it possible to define a ring of 2^n nodes in the n-cube. Let us assume that this is true for $n - 1$, with $n \geqslant 2$ (this is trivial for $n = 1$). Consecutive elements in positions 0 to $2^{n-1} - 1$ differ by a single bit (due to the recursion). Similarly, consecutive elements in positions 2^{n-1} to $2^n - 1$ also differ by a single bit. Finally, since the last element of G_{n-1} is equal to the first element of G_{n-1}^{rev}, elements $2^{n-1} - 1$ and 2^{n-1} also differ by a single bit. All elements are thus connected.

To embed a $2^r \times 2^s$ 2-D torus in a n-cube, with $r + s = n$, we can simply use the Cartesian product $G_r \times G_s$. A processor with coordinates (i, j) on the grid is mapped to processor $f(i, j) = (g_i^{(r)}, g_j^{(s)})$ in the n-cube. Horizontal neighbors $(i \pm 1, j)$ are mapped to $f(i \pm 1, j) = (g_{i\pm1}^{(r)}, g_j^{(s)})$, where the index $i \pm 1$ is taken modulo 2^r; these nodes are indeed neighbors of $f(i, j)$ in the n-cube. We can prove likewise that the vertical neighborhood is preserved. This construction can easily be generalized to a $2^r \times 2^s \times 2^t$ 3-D torus, where $r + s + t = n$.

3.4.4 Collective Communications in a Hypercube

The goal of this section is to highlight the difficulty of designing collective communications in an interconnection network slightly more complicated than a ring, in our case a hypercube. We only study the simplest operation: the broadcast. We assume processor 0 wants to broadcast a message to all other processors. The naïve algorithm works as follows: Processor 0 sends the message to all its neighbors; then every neighbor of 0 sends the message to all its neighbors; and so on. There is redundancy in this algorithm: The same processor receives the same message several times. For example, processor 0 receives the message from all its neighbors; mismatched SENDS and RECEIVES are thus very likely to happen!

We seek a strategy in which each processor receives the message only once and in which the number of steps is minimal. Note that the optimal number of steps strongly depends on the communication model (e.g., multi-port, k-ports, 1-port). The main idea is to build one or more spanning trees of the hypercube. The functions SEND and RECEIVE need an additional argument: the dimension in which the communication takes place (there was no need for such an argument in the unidirectional ring). Therefore, we use the following new functions:

$$\text{SEND}(cube\_link, send\_addr, m) \text{ and } \text{RECEIVE}(cube\_link, recv\_addr, m).$$

The broadcast algorithm works as follow: There are n steps numbered from $n-1$ to 0. All processors will receive their message on the link corresponding to their rightmost 1 and forward it on the links whose index is smaller than this rightmost 1. At step i all processors such that their rightmost 1 is strictly larger than i forward the message on link i. We assume that processor 0 has a fictitious 1 at position n. Let us sketch the beginning of the algorithm for $n = 4$:

- At step 3, processor 0000 is the only processor whose rightmost 1 is in position four. It sends the message on link 3 to 1000.
- At step 2, 0000 and 1000 have their rightmost 1 in position at least 3. They both send the message on link 2 (to processors 0100 and 1100 respectively).
- And so on until step 0, at which each even-numbered processor sends the message on link 0.

The corresponding spanning tree is depicted in Figure 3.9.

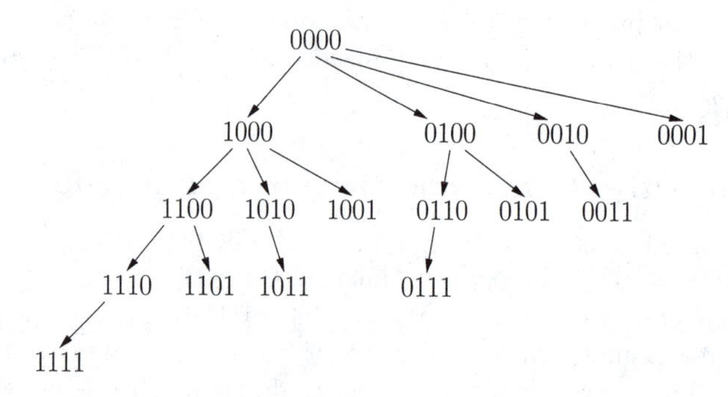

FIGURE 3.9: Broadcasting using a spanning tree on a hypercube for $n = 4$.

Let us denote by BIT(A, b) the value of the b-th bit of processor A. The broadcast of a message of length m by processor k can be done with Algorithm 3.5.

As there are n steps, the execution time with the *store-and-forward* model is $n(L + mb)$. At each step, a processor communicates with only one other processor, making the algorithm valid for the 1-port model. We could consider splitting the message in packets and pipelining them to improve the execution

```
1  BROADCAST(k, addr, m)
2     q ← MY_NUM()
3     n ← log(TOT_PROC_NUM())
         { Update pos to work as if P0 was the root of the broadcast }
4     pos ←q XOR k
         { Find the rightmost 1 }
5     first1 ←0
6     while ((BIT(pos, first1) = 0) And (first1 < n)) do
7        first1←first1 + 1
         { Core of the algorithm }
8     for phase= n − 1 to 0 do
9        if (phase=first1) then RECEIVE(phase, addr, m)
10       else if (phase<first1) then SEND(phase, addr, m)
```

ALGORITHM 3.5: Broadcast algorithm in a hypercube.

time as seen in Sections 3.2.2 and 3.3.4. It would, however, only work in the n-port model. Indeed, as soon as the root of the broadcast reaches steady-state, it needs to send its packets simultaneously to all its neighbors.

The n-port pipelined algorithm could still be improved using n edge-disjoint spanning trees. The n parts of the message would then all be pipelined in parallel on the n trees. Such a "forest" exists, but the resulting spanning tree is slightly taller. Its height is equal to $n+1$ instead of n for the previous tree:

1. The previous spanning tree is rooted at 0 and is denoted by T_0 (see Figure 3.10(a)).

2. Let $s = s_{n-1}s_{n-2}\dots s_1 s_0$ be the binary label of vertex s. The rotation (to the left) $R(s)$ is defined by $R(s) = s_{n-2}s_{n-3}\dots s_0 s_{n-1}$. We build new trees $T_1, \dots, T_{n-1}$ by rotating (to the left) the binary representations of T_0. If we denote $R^{(1)} = R$ and $R^{(i)} = R \circ R^{(i-1)}$ for $2 \leqslant i \leqslant n-1$, then T_i is obtained by applying $R^{(i)}$ to the nodes of T_0.

3. The n trees T_i are unfortunately not edge-disjoint. For example, the neighbors of 0 are the same on all trees. To this end, we inverse the i-th bit (modulo n) of T_i's vertices (see Figure 3.10(a)).

4. After this XOR operation, node 0 is a leaf in each tree T_i. We can then merge these trees in a single tree rooted at 0 (see Figure 3.10(b)).

This broadcast works only when assuming all links are bidirectional: 101 communicates with 100 in the leftmost subtree while 100 communicates with 101 in the rightmost subtree (see Figure 3.10(b)).

Using optimal pipelining on a single tree, this new algorithm improves the execution time to

$$\left(\sqrt{mb} + \sqrt{(n-1)L}\right)^2 .$$

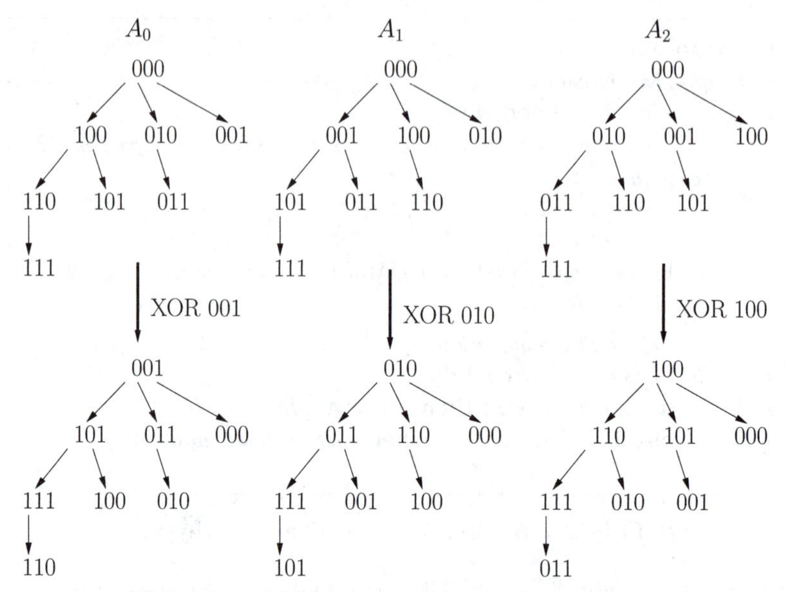

(a) Rotate and XOR operation on spanning trees, $n = 3$.

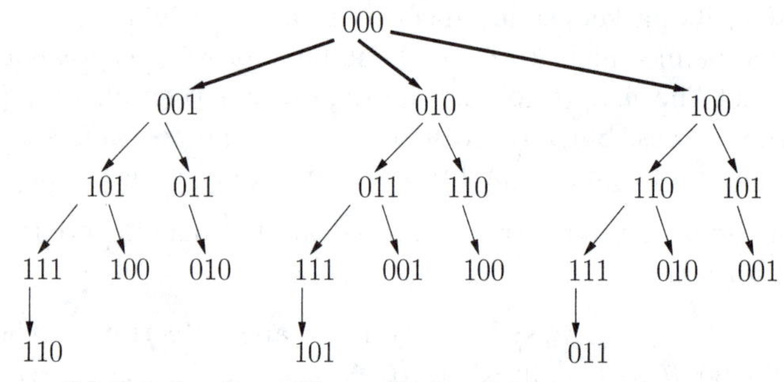

(b) Broadcasting with $n = 3$ edge-disjoint spanning trees.

FIGURE 3.10: Building edge-disjoint spanning trees.

When optimally pipelining n messages of length m/n on the n edge-disjoint spanning trees, the execution time is further improved to

$$\left(\sqrt{\frac{m}{n}b} + \sqrt{nL}\right)^2 .$$

This execution time is only twice as large as the absolute lower bound on the time needed to broadcast a message of size m on an n-cube under the n-port moadel. This bound is $nL + (\frac{m}{n} + n - 1)b$. Indeed, the first bit of the message needs at least $n(L + b)$ time units to reach the furthest processor. At this time, this processor has received at most n bits of the message. Therefore, there are at least $m - n$ bits left to receive on each link, which requires at least $(\frac{m}{n} - 1)b$ time units.

This bound cannot be reached: Only one bit has been initially sent so how could the end of the message arrive right after the first bit at no cost? The study of the complexity of collective communications has led to many research articles published in the 80s: The difficulty comes from the non-linear term L that precludes the use of flow theory.

3.5 Peer-to-Peer Computing

The last decade has witnessed the rapid development and growing popularity of peer-to-peer networks. This rapid growth was originally motivated by the will of Internet users to share music and videos freely. More generally, such networks enable users to share data, storage, computation, bandwidth, etc. Peer-to-peer networks make it possible to avoid relying on centralized entities or servers. This property is clearly useful for conducting illegal activities like sharing copyrighted materials. But it is also a key advantage as centralized servers are never 100% reliable and often become performance bottlenecks, Peer-to-peer networks are inherently fault tolerant and scalable, two fundamental properties for enabling novel distributed applications with many participating computers.

A peer-to-peer network is an application-level network, or *overlay*, that consists of peers that simultaneously act both as "clients," "servers," and "forwarders" to the other peers in the network; they may consume resources, provide resources, or forward requests for resources. Typically a peer is aware of only a very small number of other peers in the network. These peers are called the peer's neighbors and one talks of the *neighbor set* of a peer. The (average) number of peers in the neighbor set of a peer is called the (average) *degree*. If a participating peer knows the location (e.g., the IP address) of another peer in the P2P network, then there is a directed edge from the former to the latter in the overlay network. This edge is only in the application-level overlay network, in the sense that it has nothing to do with the underlying physical network links.

Peer-to-peer networks differ from the topologies discussed in the previous sections in many ways. First, they are generally enormous: from a few thousands to millions of peers. Second, their structure generally changes over time very quickly: Peer arrivals and departures can happen at any time. This phenomenon is referred to as *churn*. Third, the proximity of two peers in the network is in principle completely unrelated to their end-to-end bandwidth and latency. Obviously, techniques exist to make the overlay network and the physical network somewhat congruent in a view to improving performance. In this context finding a peer corresponding to a particular criterion, e.g., a peer holding a particular information, is a very challenging problem. Note that this problem is not particularly difficult in, say, a dedicated parallel platform with 128 peers interconnected via a hypercube topology.

Peer-to-peer systems and applications have caught the interest of many computer science researchers in the last decade. Depending on how the peers in the overlay network are linked to each other, one can classify a P2P network as *unstructured* or *structured*. Unstructured networks are generally built using randomized protocols: New peers randomly connect to some peers and update their neighbor sets over time based on random walks. The graph of the corresponding overlay network is thus generally modeled well by random graphs. Data or peer localization is achieved via "flooding," i.e., by broadcasting the request to the whole network with a bound on the maximum number of hops through the network. Flooding is known to be very inefficient. Structured networks are built using a globally consistent and deterministic protocol. This protocol ensures that the graph of the overlay network has a particular structure that facilitates efficient routing. We present briefly a few important ideas underlying structured networks, and refer the reader to [111] for a comprehensive review of developments in peer-to-peer networking over the last decade.

3.5.1 Distributed Hash Tables and Structured Overlay Networks

A hash table is a data structure that associates *keys* (e.g., names or song titles) with *values* (e.g., phone numbers or files). It is often used as a dictionary or as yellow pages. A distributed hash table (DHT) is a distributed system that offers the same kind of functionality as a hash table. Each peer in the distributed system is assigned a single key called its identifier (ID). A peer with ID i is responsible for all the keys lying between i and the smallest peer ID larger than i. A routing mechanism in such a system is a means for any peer to find the peer associated to an arbitrary key (and thus the value associated to this key). DHTs have received a lot of attention because they provide a fundamental abstraction useful for popular peer-to-peer applications in which users wish to locate and retrieve particular items based on their names or on a specification of their properties. The desired features of a DHT system are the following:

- **Decentralization:** No central coordination is needed to build the network. Thus, no peer has global knowledge of the system, which implies that peers only have vastly reduced local (and different!) views of the system. Hence, the arrival or the departure of a peer only requires a small update of the network. Having a low maintenance cost of the network requires that the underlying overlay network have a **relatively small degree** compared to the size of the system. Note that too small a degree may not be desirable either as a peer could then easily be disconnected from the system.

- **Scalability:** A peer-to-peer network typically comprises thousands or millions of peers and its performance should not degrade with its size. In particular, the time needed to find the value associated to a key should be kept as small as possible. Such a property requires that underlying overlay network have a relatively **small diameter** compared to the size of the system.

- **Fault tolerance:** The system should be reliable even in the presence of the sudden failure of a large fraction of the peers. Such a property requires that the underlying overlay network have a relatively **large bisection** compared to the size of the system.

Let N, Δ, and D, respectively, denote the size, the maximum degree, and the diameter of the network. Using Moore's bound (see Section 3.1.2), we easily obtain

$$D = \Omega\left(\frac{\log N}{\log \Delta}\right).$$

Several structures can thus be considered:

- With degree $O(\log N)$, we obtain a route length of at least $\Omega\left(\frac{\log N}{\log \log N}\right)$.

  In practice, many systems have $D = \Theta(\log N)$, which is not optimal but convenient. Hypercubes typically have such a degree and such a diameter. They also have a large bisection and have thus been a great source of inspiration when designing peer-to-peer protocols (e.g., Chord [112], Pastry [103], and Tapestry [120]). Overlay networks with an optimal diameter have later been proposed [70], but the routing protocol is very complex and difficult to implement.

- With degree $O(1)$, the diameter is at least $\Omega(\log N)$.

  Some networks like CAN [99] are based on d-dimensional grids. Therefore, they have degree $O(d)$, a suboptimal route length $\Theta(d.N^{1/d})$, and a good bisection $\Theta(N^{(d-1)/d})$.

  The optimal diameter is attained by a tree, for example. However, trees suffer from a very small bisection that makes them poorly adapted to this context; removing a single peer in a tree causes it to become partitioned. Many graphs like cube-connected cycles, butterfly graphs, or De Bruijn graphs (see Exercise 3.3, 3.5, and 3.6) achieve a bounded degree, a logarithmic diameter, and a large bisection. These structures have inspired some overlay networks like Viceroy [86] (butterfly) and D2B [56] (de Bruijn). Viceroy is the first known constant-degree overlay network with logarithmic diameter. Its construction and management are, however, relatively complex and require sophisticated procedures that might be difficult to implement in a practical setting. Although these structures offer advantages over hypercube-based ones, they have been used and advertised slightly later and have thus not received as much attention.

- With degree $O(N^\alpha)$ we obtain a diameter of $\Omega(1)$, but such a high degree is far too large to be of practical interest.

In the following, we focus on hypercube-based overlay networks, which are the most commonly used.

3.5.2 Chord

In the chord system [112], the set of keys is $\{0, \ldots, 2^m - 1\}$ and the peers are organized in a virtual ring. If peers n_1 and n_2 are two consecutive peers on this ring, then peer n_2 is responsible for all keys that fall between n_1 and n_2. Finding a key in this network would obviously be very time consuming since its diameter is $O(N)$. This is why there are "shortcuts." Node n is connected to all peers responsible for key $n + 2^i[2^m]$ for all $i \in \{0, \ldots, m-1\}$ (see Figure 3.11). The lookup is thus easily done in $O(\log N)$ hops by always forwarding the request to the peer with the closest smallest ID (the distance to the destination is halved at each hop). Also, the routing table, meaning the data structure that holds the neighbor set, remains relatively small ($O(\log N)$).

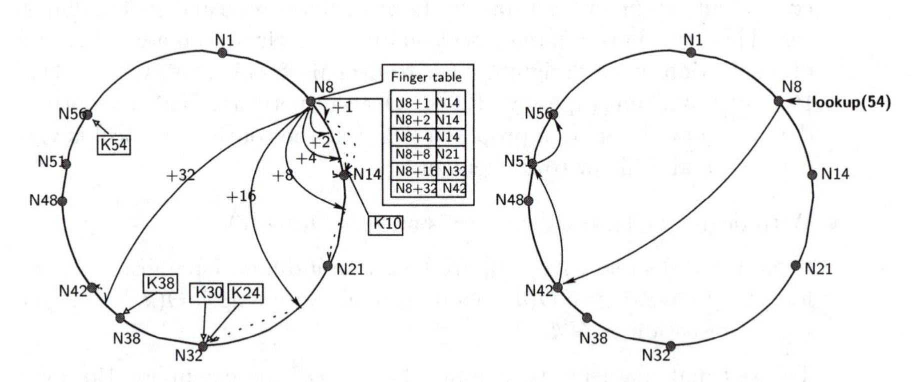

FIGURE 3.11: Routing in the Chord network.

Using this hypercube-inspired organization, it is possible to design join/leave protocols that require only $O((\log N)^2)$ messages, that balance the load between peers, and that are robust to the sudden failure of a large fraction of peers. The main drawback of this approach is that there is absolutely no freedom on the choice of neighbor sets. Peer n is connected to peer $n + 2^i$ even if this peer is physically very far away. It could, however, be the case that another peer whose ID is close to $n + 2^i$ is also physically close to n and would thus be a much better shortcut in practice. This is one of the ideas described in the next section.

3.5.3 Plaxton Routing Algorithm

Plaxton et al. [96] proposed a suffix routing algorithm similar to the one used in a hypercube. Like in Chord, the peers and the keys are ordered in a circular fashion and shortcuts are created to allow efficient routing. The system uses an alphabet B of size b. Each peer is identified by a key n of size $m = \log_b N$ and for each letter $x \in B$ and for each $i \in \{0, \dots, m-1\}$, peer n is connected to a peer whose ID ends with $xn_{m-i} \dots n_{m-1}x$ (the size of the neighbor set is thus $O(b \log_b(N))$). Other redundant links (to $n \pm 1, n \pm 2, \dots, n \pm d$) are often added to replicate the data managed by n in case of sudden failure of n. The neighbor set is then of size $O(d + b \log_b(N))$, which is still reasonable. Table 3.2 shows an example of a possible routing table for peer 2310 with $B = \{0, 1, 2, 3\}$.

Routing is then performed by increasing the length of the common suffix at each hop. For example, the route from peer 3745 to peer 3BA8 is ???8 $\rightarrow$??$A8$ $\rightarrow$?$BA8$ $\rightarrow$ $3BA8$. As expected, the routing length is $O(\log_b N)$ but now there is much more freedom (at least when the suffix is short) on the choice of the shortcuts, making it possible to chose peers that are geographically close. As a consequence, adjacent peers in the key-space are generally geographically diverse and thus have a small probability of simultaneous failure, which strengthens the fault tolerance. Besides, the first hops of a route are generally fast, whereas the last ones are more likely to be slow. Therefore, the redundant links (with $n \pm 1, n \pm 2, \dots, n \pm d$) not only improve fault tolerance but also lookup efficiency because the last steps are not always needed.

The two most famous overlay networks using such an approach are Pastry [103] and Tapestry [120].

3.5.4 Multi-Casting in a Distributed Hash Table

A publish/subscribe system is a system in which any participant can subscribe to a set of topics and receive an event notification as soon as these topics are updated. Implementing such a system using a DHT is an appealing idea that has been used for example in SCRIBE [104]. A topic t is identified by the hash $h(t)$ of the corresponding string. As previously, the peer n_t whose ID is closer to $h(t)$ is responsible for topic t. Any participant willing to subscribe to this topic can thus route a message to $h(t)$ to inform n_t about its

TABLE 3.2: A possible routing table for peer 2310.

suffix \ letter	**0**	**1**	**2**	**3**
∅	-	120**1**	320**2**	212**3**
0	320**0**0	-	322**2**0	213**3**0
10	3**0**10	3**1**10	2**2**10	-
310	**0**310	**1**310	-	**3**310

subscription. Let us denote by $S(t)$ the set of subscribers to topic t. Merging the set of routes from $s \in S(t)$ to n_t we obtain a tree rooted at n_t that can be used backward to multi-cast notification messages from n_t to every subscriber $s \in S(t)$. In practice, when a peer n subscribes to topic t the routing ends as soon as it encounters a peer belonging to the tree.

This approach has many interesting features. First, n_t does not need to know the whole set of subscribers. Indeed, routes from the subscribers to the owner of the topic often merge very soon and thus the tree is well balanced in practice. Peers responsible for hot topics are thus overloaded neither by requests nor by notifications. Second, the upper part of the tree is made of geographically diverse peers, whereas the leaves are generally close to each other within a subtree. Therefore, the more communication-intensive part of the execution consists of a large group of efficient and mostly independent communications.

Bibliographical Notes

The famous PVM message passing library [58] provided primitives for both synchronous and asynchronous communications. The current standard for message passing is MPI [109], which offers a rich set of collective communication primitives. A large part of the material in this chapter is inspired by Desprez's thesis [50] and by the book by Gengler, Ubéda, and Desprez [59]. The book by Culler and Singh [48] is a great reference to find other examples of collective communication algorithms on grids or hypercube and more information on commutation networks and distributed memory architectures. Most work regarding peer-to-peer architectures is very recent. Our section on this topic takes a very particular point of view that is strongly related to the preceding sections on interconnection networks and routing. [111] provides a comprehensive overview of developments in peer-to-peer networking over the last decade.

3.6 Exercises

◇ Exercise 3.1 : Torus, Hypercubes, and Binary Trees

1. What is the difference between a 6-cube and a $4 \times 4 \times 4$ 3-D torus?

2. Explain how to embed a complete binary tree with $2^n - 1$ vertices in a $r \times r$ 2-D grid?

Exercise 3.2 : Torus, Hypercubes, and Binary Trees (again!)

1. What is the diameter of a binary tree of depth h?. What is the average length of a path in a binary tree?

2. What is the number of paths (with minimal length) from (x_1, y_1) to (x_2, y_2) in an $n \times n$ torus?

3. Prove that the bisection width of a d-cube is 2^{d-1}.

◇ Exercise 3.3 : Cube-Connected Cycles

A cube-connected cycle $CCC(m)$ is obtained by replacing each processor of an m-cube by a ring of size m, where each processor of the ring is associated to a dimension of the hypercube (see Figure 3.12).

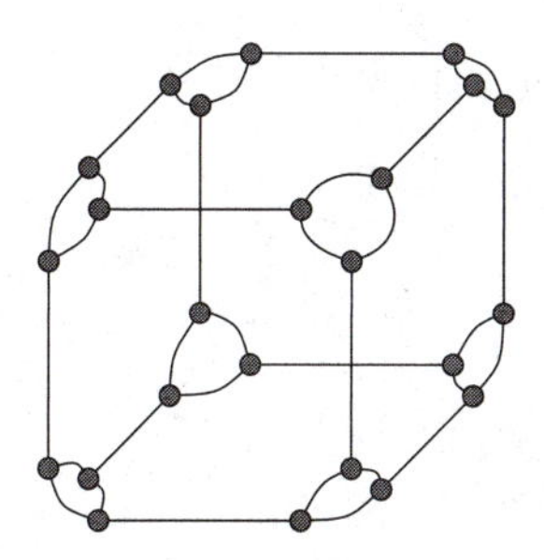

FIGURE 3.12: Cube-connected cycles of order 3.

1. How many vertices are there in a $CCC(m)$? Give a formal definition of a $CCC(m)$ and a simple upper bound on its diameter.

2. Prove that the diameter of $CCC(m)$ is exactly $D = 2m - 2 + \lfloor \frac{m}{2} \rfloor$ if $m > 3$, and is equal to 6 if $m = 3$.

◇ Exercise 3.4 : Matrix Transposition

In this exercise, we design a parallel algorithm to transpose an $n \times n$ matrix. We assume the matrix is already distributed on the processors. We also assume bidirectional communication links and a multi-port model.

1. Propose an algorithm on a ring of p processors and give its complexity. We assume a 1-D distribution: Each processor is responsible for a set of consecutive rows.

2. Propose an algorithm to transpose a matrix on a $p = q \times q$ torus and give its complexity. We assume a 2-D distribution: Each processor is responsible for a $\frac{n}{q} \times \frac{n}{q}$ sub-matrix.

3. Propose an algorithm to transpose a matrix on an m-cube and give its complexity. We assume a 2-D distribution: Each processor is responsible for a $\frac{n}{q} \times \frac{n}{q}$ sub-matrix.

◇ Exercise 3.5 : Dynamically Switched Networks

A butterfly network $BUT(r)$ of order r is a multi-stage network made of $(r+1)2^r$ vertices organized in 2^r rows of $r+1$ stages. A vertex is thus represented by a pair $\langle w, i \rangle$, where w is a sequence of r bits numbering the corresponding row, and i corresponds to the stage number ($0 \leqslant i \leqslant r$). Two vertices $\langle w, i \rangle$ and $\langle w', i' \rangle$ are connected if and only if $i' = i + 1$ and one of the following two condition is fulfilled:

- $w = w'$.
- w and w' only differ on the i-th bit.

$BUT(3)$ is depicted in Figure 3.13.

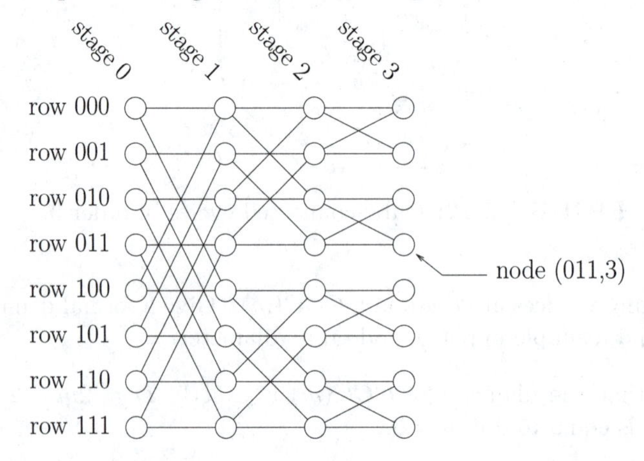

FIGURE 3.13: $BUT(3)$: the butterfly network of order 3.

1. What kind of network is obtained when all the vertices of a given row are grouped?

2. The butterfly network is built recursively. Give two ways to split a butterfly network of order r into two butterfly networks of order $r-1$.

3. Prove that there is a unique path of length r between any vertex $\langle w, 0\rangle$ from stage 0 and any vertex $\langle w', r\rangle$ from stage r. What is the diameter of $BUT(r)$?

A Beneš network of order r consists of two back-to-back butterfly networks. Vertices from stage r of each butterfly network are merged. There are thus $2r+1$ vertices on each row. Figure 3.14 depicts a Beneš network of order 3. As this graph has degree 4, we can replace all vertices by 2×2 switches that can be configured either in cross or in parallel to transmit data.

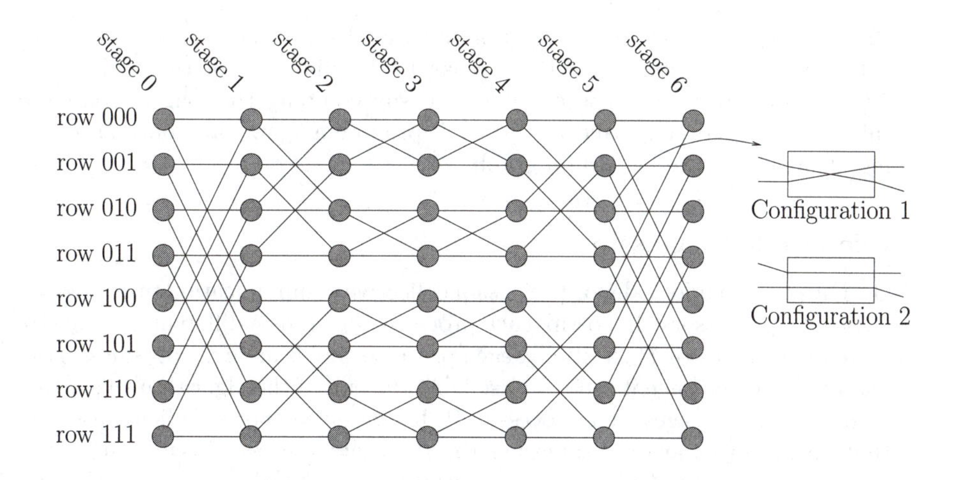

FIGURE 3.14: Beneš network of order 3.

4. Prove (by induction on the order r of the network) that Beneš networks can be configured to implement any permutation. Given an arbitrary permutation π of $\{0, \ldots, 2^r - 1\}$, there is a configuration of switches for establishing simultaneous routes from any input i to the corresponding output $\pi(i)$.

Figure 3.15 depicts a configuration for the Beneš network of order 2 implementing the permutation $\pi = (0, 1, 2, 3) \to (3, 1, 2, 0)$.

Exercise 3.6 : De Bruijn Network

The De Bruijn graph $B(m)$ of order m consists of the set of vertex $\{0,1\}^m$. For α and $\beta \in \{0,1\}$, and $x \in \{0,1\}^{m-2}$, the vertex $\alpha x \beta$ is connected as follows:

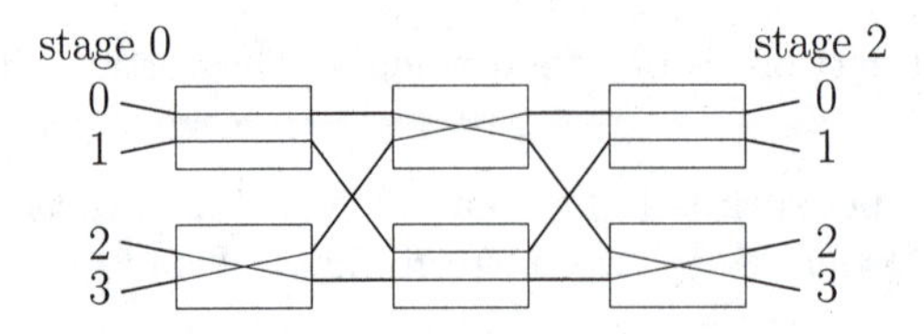

FIGURE 3.15: Configuration of Beneš network of order 2 for the permutation $\pi = (0,1,2,3) \rightarrow (3,1,2,0)$.

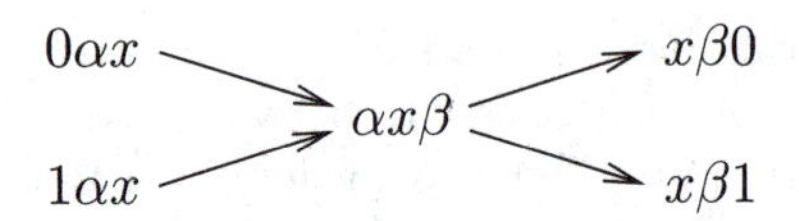

1. What are the degree and the diameter of $B(m)$? Propose a simple and optimal routing algorithm in the De Bruijn graph.

2. Assume now that the network is undirected (i.e., that $\alpha x\beta$ is connected with $0\alpha x$, $1\alpha x$, $x\beta 0$, and $x\beta 1$). What is the diameter of the undirected $B(m)$? Assuming that it is possible to easily compute the longest common sub-sequence between u and $v \in \{0,1\}^m$, propose a simple and optimal routing algorithm in the De Bruijn graph.

Exercise 3.7 : Gossip

1. Give the pseudo-code for the *gossip* collective communication primitive on a ring of p processors. In gossip, each processor P_k sends a different message to each other processor P_q. Initially, each processor holds an array of p messages, including one message it sends to itself. At the end of the algorithm this array holds all the messages that processor P_k has received. For simplicity, assume that all messages are the same length, L. Write the pseudo-code in a way similar to, e.g., the way we have written it for the all-to-all algorithm in Section 3.3.3.

2. Give a performance model for your algorithm.

3.7 Answers

Exercise 3.1 (Torus, Hypercubes, and Binary Trees)

▷ **Question 1.** There is no difference: Both architectures are the same! One can easily check that their diameter and degree are equal to 6 in both cases. The embedding of the torus in the hypercube with Gray codes, as seen in Section 3.4.3, is an isomorphism: Each processor has exactly 6 neighbors in both networks. One can also easily check that a 4-cube is strictly equivalent to a 4×4 2-D torus.

▷ **Question 2.** It is impossible as soon as n is large enough, even for large values of r. Indeed, in a complete binary tree with $2^n - 1$ vertices, all vertices are at most $n-1$ hops from the root. However, in a $r \times r$ torus (even a large one), there are exactly $2n(n+1)$ vertices at less than $n-1$ hops from a given vertex S (all vertices have to be in a square centered on S and rotated 45°). Considering for S the image of the root of the tree, we get a contradiction as soon as $2^n - 1 \geqslant 2n(n+1)$, i.e., for $n \geqslant 7$.

Exercise 3.3 (Cube-Connected Cycles)

▷ **Question 1.** There are clearly $p = m \cdot 2^m$ vertices. A vertex can be represented by the pair $s = \langle w, i \rangle$, where w is the hypercube part of s, i.e., a sequence of m bits, and $i \in \{0, \ldots, m-1\}$ is the ring part of s, i.e., the position of the vertex in the ring. Vertices $\langle w, i \rangle$ and $\langle w', i' \rangle$ are connected if and only if:

- $w = w'$ and $i - i' \equiv \pm 1[m]$ (ring link) or
- $i = i'$ and w and w' only differ on the i-th bit (hypercube link).

To bound the diameter, we can try to bound the distance between two arbitrary vertices. As we can XOR the hypercube components, this amounts to finding a path from vertex $\langle 0, 0 \rangle$ to a vertex $\langle w, i \rangle$.

Like for the hypercube, we adjust the bits of w starting, for example, from the rightmost ones. For each 1 in w, we go through the corresponding hypercube link. To this end, we will, however, have to move forward on one or more vertices of the ring to reach the desired hypercube link. We can thus reach the ring labeled by w in at most $2m - 1$ steps: at most m steps on the hypercube links and at most $m - 1$ steps on the ring's links. Then, we need to reach position i on the ring. The diameter of a ring of size m is $\lfloor \frac{m}{2} \rfloor$. The upper bound on the diameter of our $CCC(m)$ is thus $D \leqslant 2m - 1 + \lfloor \frac{m}{2} \rfloor$.

A nice feature is that the degree of such a network is bounded (it is equal to 3) while its diameter is logarithmic in the total number of processors (we have $m \leqslant \log p$).

▷ **Question 2.** The previous upper bound was almost tight. We can easily check in Figure 3.12 that the diameter of $CCC(3)$ is equal to 6.

Let us now focus on the case $m > 3$. To go from $\langle 0,0 \rangle$ to $\langle w,i \rangle$, we necessarily go through at least $|w|$ hypercube links, where $|w|$ is the number of 1's in w. Let us assume for now that a shortest path goes through exactly w hypercube links. Let us consider the "boxed" cycle associated to 0, where vertices corresponding to 1's in w have been boxed. For example, if $m = 8$ and $w = 11000110$, we have

$$\begin{array}{ccccccc} 0 & \rightarrow & \boxed{1} & \rightarrow & \boxed{2} & \rightarrow & 3 \\ \uparrow & & & & & & \downarrow \\ \boxed{7} & \leftarrow & \boxed{6} & \leftarrow & 5 & \leftarrow & 4 \end{array}$$

We will go through a hypercube link as soon as we arrive on a boxed vertex. Thus, we need to find a shortest path from 0 to i going through all boxed vertices. The total length of the path from $\langle 0,0 \rangle$ to $\langle w,i \rangle$ is then equal to $|w|$ plus the length of the path on the boxed ring. If $i = 5$, a shortest path on the boxed cycle is

$$0 \rightarrow 1 \rightarrow 2 \rightarrow 1 \rightarrow 0 \rightarrow 7 \rightarrow 6 \rightarrow 5,$$

and its length is 7. By adding $|w| = 4$ we get the total routing length.

With such a representation, it is clear that going through additional hypercube links is useless. One gets the worst case with $w = 11\ldots1$ (all bits are set to 1) and $i = \lfloor \frac{m}{2} \rfloor$; the shortest path from 0 to i that goes through all links is of length $m + \lfloor \frac{m}{2} \rfloor - 2$. Hence, the result. With our previous example ($m = 8$), a shortest path from 0 to 4 going through all vertices is

$$0 \rightarrow 7 \rightarrow 6 \rightarrow 5 \rightarrow 6 \rightarrow 7 \rightarrow 0 \rightarrow 1 \rightarrow 2 \rightarrow 3 \rightarrow 4,$$

whose length is $10 = 8 + 4 - 2$.

Exercise 3.4 (Matrix Transposition)

▷ **Question 1.** Let us first look at the kind of communications that are needed to transpose the matrix with four processors (see Figure 3.16).

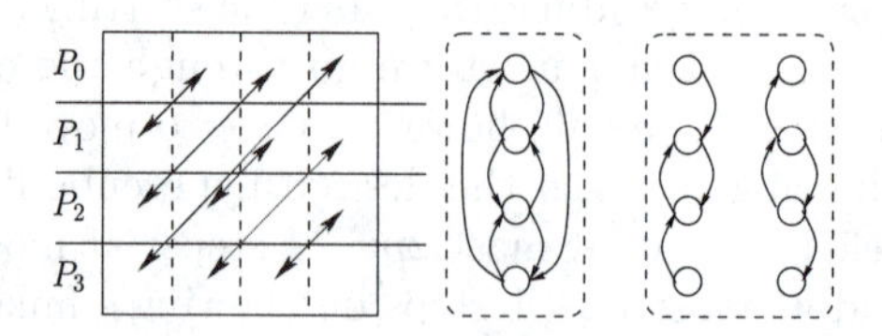

FIGURE 3.16: Data movement on a ring of size 4.

We assume that p divides n. Each processor needs to send a specific data (a $\frac{n}{p} \times \frac{n}{p}$ sub-matrix) to all other processors. It is an *all-to-all* operation.

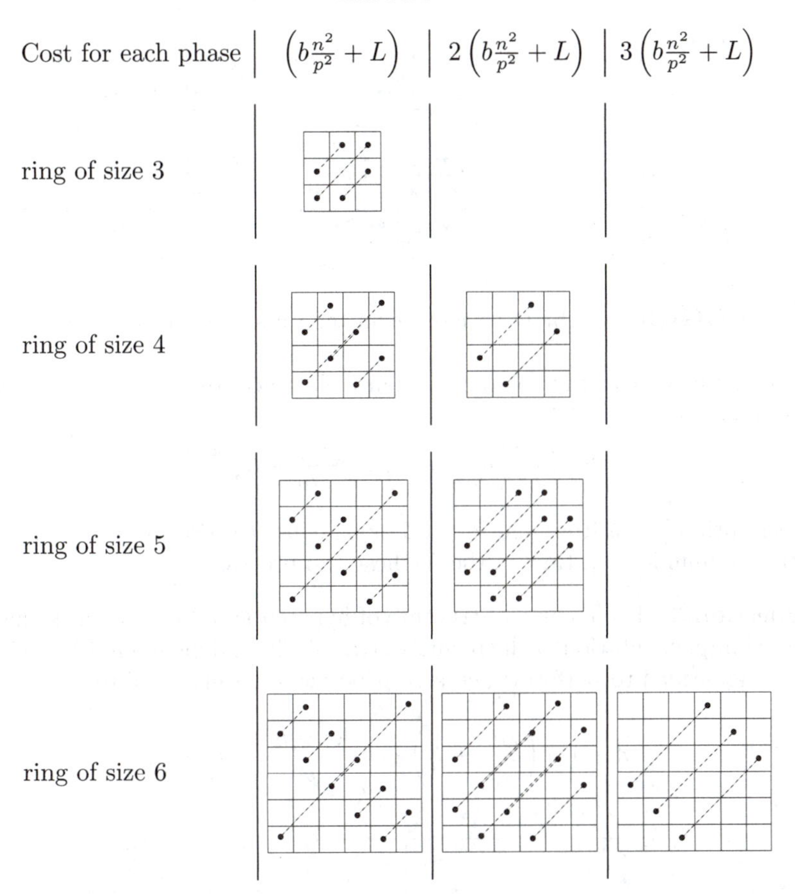

FIGURE 3.17: Organizing communication phases on rings.

By grouping communications by their numbers of hops, we can organize the execution in phases (see Figure 3.17). In each phase (except maybe for the last one) each processor exchanges data with exactly two processors. Thus, by pipelining the communications, we can use every link at any given time and all messages travel along their shortest path.

The total time needed to transpose a matrix on a ring is thus

$$\sum_{k=1}^{\lfloor \frac{p}{2} \rfloor} k\left(L + b\frac{n^2}{p^2}\right) = \frac{1}{2}\left(\left\lfloor\frac{p}{2}\right\rfloor\left(\left\lfloor\frac{p}{2}\right\rfloor + 1\right)\right)\left(L + b\frac{n^2}{p^2}\right) \sim \frac{p^2}{8}L + \frac{n^2}{8}b\,.$$

▷ **Question 2.** Let us group communications by their number of hops (see Figure 3.18). By analyzing communications along the network links, one can see that in a given phase, a processor is involved in only one communication.

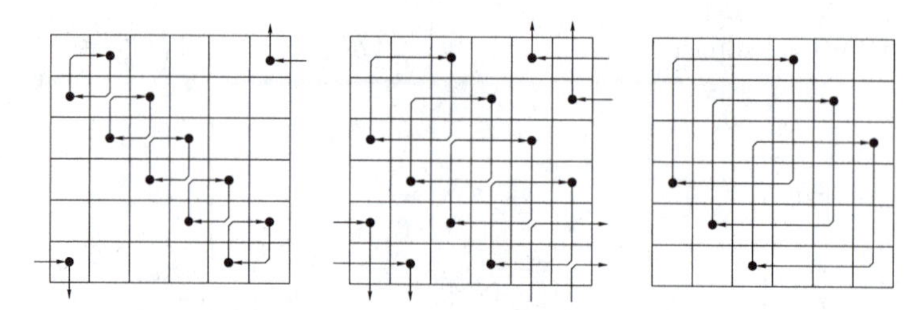

FIGURE 3.18: Data movement on a 6×6 processor grid.

All these phases can thus be done in parallel, hence we have a transposition time equal to

$$2 \left\lfloor \frac{q}{2} \right\rfloor \cdot \left(b\frac{n^2}{q^2} + L \right) \sim b\frac{n^2}{\sqrt{p}} + L\sqrt{p}\,.$$

This is optimal (in a store-and-forward mode) since the execution time is equal to the communication time of the farthest two processors.

▷ **Question 3.** The recursive structure of hypercubes allows for the straightforward implementation of the recursive transposition depicted in Figure 3.19. The time needed to perform such a transposition is thus equal to

$$(m-1) \cdot \left(b\frac{n^2}{p} + L \right) \sim b\frac{(\log p)n^2}{p} + L\log p.$$

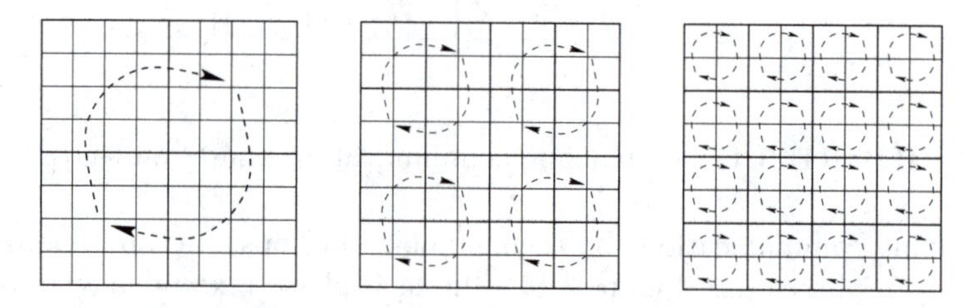

FIGURE 3.19: Data movement on a 4-cube.

Exercise 3.5 (Dynamically Switched Networks)

▷ **Question 1.** We obtain an n-cube: In this new architecture, w and w' are connected by a link if and only if they differ by a single bit. Therefore, any routing algorithm designed for a hypercube can be simulated on a butterfly network with the same execution time. This is another example of a bounded-degree network with hypercube-like capabilities.

▷ **Question 2.** By removing all vertices at stage 0, we obtain two butterfly networks: The first one consists of the 2^{r-1} first rows and the second one consists of the 2^{r-1} last rows. We can also remove all the vertices of stage r, but it is not at straightforward. The first network consists of all even rows and the second one consists of all odd rows. Indeed, removing the last level amounts to ignoring the last bit of the row indices.

▷ **Question 3.** At stage i, we choose either the horizontal link (thus, staying on the same row) if w and w' do not differ on the i-th bit, and the other link otherwise.

The longest path is thus of length $2r$ (e.g., consider the path from $\langle 0, 0\rangle$ to $\langle 2^r - 1, 0\rangle$).

▷ **Question 4.** If $r = 1$, then there is single switch and the result is trivial. Let us assume that the result holds for a Beneš network of order $r - 1$.

The proof relies on the following observation: Stages 1 to $2r - 1$ of a Beneš network of order r comprise two Beneš networks of order $r-1$ (see Figure 3.20).

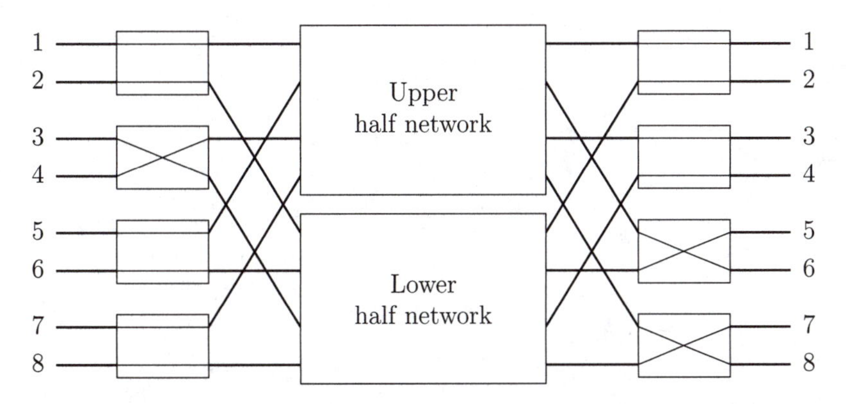

FIGURE 3.20: Recursive layout of Beneš network.

To implement permutation π, we simply route half of the messages through the upper network and the other half of the messages through the lower network. We need to ensure that all 2^r routes are edge-disjoint, but it is not difficult. We start by routing the first message from input $x_1 = 0$ to output $y_A = \pi(0)$ through the upper network (using the induction hypothesis). Output y_1 shares a switch with another output, say y_2. Let x_2 be the output corresponding to y_2 (i.e., $\pi(x_2) = y_2$). We route the message from x_2 to y_2 through the lower network (using the induction hypothesis). Output x_2 shares a switch with another input, say x_3. We route the message from x_3 to $y_3 = \pi(x_3)$ through the upper network, and so on. We keep routing messages in that way until we loop back to x_1. We can then repeat this procedure with a message that has not been routed yet. In the end, half of the messages are routed through the upper network and the other half are routed through the lower network.

Part II

Parallel Algorithms

Chapter 4

Algorithms on a Ring of Processors

When writing a parallel algorithm on a distributed-memory platform that consists of multiple processors (e.g., a cluster of workstations with some interconnect technology), it is convenient to abstract away the *physical* network topology of the platform and to design algorithms using a *logical* topology instead. This chapter presents several parallel algorithms that are designed for use with a logical ring topology. This topology is both simple, which makes it an ideal candidate for a first look at distributed-memory parallel algorithms, and popular. Indeed, a ring is a linear interconnection network: Each processor has a single predecessor and a single successor. We will see that linear networks make for a natural decomposition of regular data structures like arrays.

Our first algorithm is a simple matrix-vector multiplication (Section 4.1), followed by its extension to a matrix-matrix multiplication (Section 4.2). Next, we present a "stencil" algorithm such as those used, for instance, in many numerical methods (Section 4.3). We then provide an in-depth study of the famous LU factorization of a dense matrix (Section 4.4). Our last algorithm in this chapter is a stencil algorithm that motivates the use of a bidirectional ring (Section 4.5). We end the chapter with two sections about practical implementation considerations (Sections 4.6 and 4.7).

4.1 Matrix-Vector Multiplication

Let us write our first parallel algorithm on a unidirectional ring of processors: the multiplication $y = Ax$ of a matrix A of dimension $n \times n$ by a vector x of dimension n (we number indices from 0 to $n-1$). A matrix-vector multiplication is simply a sequence of n scalar products, as seen in the algorithm below:

```
for i = 0 to n − 1 do
    y_i ← 0
    for j = 0 to n − 1 do
        y_i ← y_i + A_{i,j} × x_j
```

Each iteration of the outer loop (the i loop) computes the scalar product of one row of A by vector x. Furthermore, all these scalar product computations are independent, in the sense that they can be performed in any order. Consequently, a natural way to parallelize a matrix-vector multiplication is to distribute the computations of these scalar products among the processors. Let us assume that n is divisible by p, the number of processors, and let us define $r = n/p$. Each processor will compute r scalar products to obtain r components of the result vector y. To do this, each processor must store r rows of matrix A. It is natural for a processor to store r contiguous such rows, which is often termed a block of rows, or simply a *block row*. For instance, one can allocate the first block row of matrix A to the first processor, the second block row to the second processor, and so on. The components of vector y (which are to be computed) are then distributed among the processors in a similar manner as the rows of matrix A (e.g., the first r components to the first processor, etc.). These kinds of data distributions are typically called "one-dimensional distributions," because arrays are partitioned along a single dimension. A one-dimensional (1-D) data distribution is a natural choice when developing an algorithm on a 1-D logical topology like a ring or processors.

If one assumes that vector x is fully duplicated across all processors, then the computations of the scalar products are completely independent since no input data are shared. But it is usual, in practice, to assume that vector x is distributed among the processors in the same manner as matrix A and vector y. This is for reasons of modularity and consistency. Parallel programs often consist of sequences of parallel operations. Assuming that the input vector is distributed in the same manner as the matrix and the output vector makes it possible to perform a second matrix-vector multiplication $z = By$, where the input vector is the output vector of the previous computation (assuming that matrix B is distributed similarly to matrix A). Furthermore, parallel algorithms are often easier to understand and modify when all data objects are distributed in a consistent manner.

Each processor holds r rows of matrix A, stored in an array A of dimension $r \times n$, and r elements of vectors x and y, stored in two arrays x and y of dimension r. More precisely, processor P_q holds rows qr to $(q+1)r - 1$ of matrix A, and components qr to $(q+1)r - 1$ of vectors x and y. Thus, one needs the following declarations in our parallel program:

```
var A:   array[0..r − 1,0..n − 1] of real;
var x, y:   array[0..r − 1] of real;
```

Using these declarations, element $A[0,0]$ on processor P_0 corresponds to element $A_{0,0}$ of the matrix, but element $A[0,0]$ on processor P_1 corresponds to element $A_{r,0}$ of the matrix. Typically the indices of matrix elements, which we denote via subscripts, are called the *global indices*, while the indices of array elements, which we denote with square brackets, are called the *local indices*. The mapping between global and local indices is one of the technical difficul-

$$
\begin{array}{c|c|c}
P_0 & \begin{pmatrix} A_{00} & A_{01} & A_{02} & A_{03} & A_{04} & A_{05} & A_{06} & A_{07} \\ A_{10} & A_{11} & A_{12} & A_{13} & A_{14} & A_{15} & A_{16} & A_{17} \end{pmatrix} & \begin{pmatrix} x_0 \\ x_1 \end{pmatrix} \\
\hline
P_1 & \begin{pmatrix} A_{20} & A_{21} & A_{22} & A_{23} & A_{24} & A_{25} & A_{26} & A_{27} \\ A_{30} & A_{31} & A_{32} & A_{33} & A_{34} & A_{35} & A_{36} & A_{37} \end{pmatrix} & \begin{pmatrix} x_2 \\ x_3 \end{pmatrix} \\
\hline
P_2 & \begin{pmatrix} A_{40} & A_{41} & A_{42} & A_{43} & A_{44} & A_{45} & A_{46} & A_{47} \\ A_{50} & A_{51} & A_{52} & A_{53} & A_{54} & A_{55} & A_{56} & A_{57} \end{pmatrix} & \begin{pmatrix} x_4 \\ x_5 \end{pmatrix} \\
\hline
P_3 & \begin{pmatrix} A_{60} & A_{61} & A_{62} & A_{63} & A_{64} & A_{65} & A_{66} & A_{67} \\ A_{70} & A_{71} & A_{72} & A_{73} & A_{74} & A_{75} & A_{76} & A_{77} \end{pmatrix} & \begin{pmatrix} x_6 \\ x_7 \end{pmatrix}
\end{array}
$$

FIGURE 4.1: Matrix-vector multiplication example ($n = 8$, $p = 4$, $r = 2$): initial data distribution.

ties encountered when writing parallel programs that operate over distributed arrays. However, for our particular program in this section, this mapping is straightforward: Global indices (i, j) correspond to local indices $(i - \lfloor \frac{i}{r} \rfloor, j)$ on processor $P_{\lfloor i/r \rfloor}$.

Note that faster execution is not the only motivation for the parallelization of a sequential algorithm on a distributed memory platform. Parallelization also makes it possible to solve larger problems. For instance, in the case of matrix-vector multiplication, the matrix is distributed over the local memories of p processors rather than being stored in a single local memory as in the sequential case. Therefore, the parallel algorithm can solve a problem (roughly) p times larger than its sequential counterpart.

The principle of the algorithm is depicted in Figure 4.1, which shows the initial distribution of matrix A and of vector x, and in Figure 4.2, which shows the p steps of the algorithm. At each step, the processors compute a partial result, i.e., the product of an $r \times r$ matrix by a vector of size r. In the beginning (step 0), each processor P_q holds the q-th block of vector x, and can calculate the partial scalar product that corresponds to a diagonal block of matrix A. We use a block notation by which $\widehat{x_q}$ (resp. $\widehat{y_q}$) is the q-th block of size r of vector x (resp. y), and by which $\widehat{A_{q,s}}$ is the $r \times r$ block at the intersection of the q-th block row and the s-th block column of matrix A. Using this notation, each processor P_q can compute $\widehat{y_q} = \widehat{A_{q,q}} \times \widehat{x_q}$ during the first algorithm step. While this computation is taking place, one can do a circular block shift of vector x among the processors: processor P_q sends $\widehat{x_q}$ to processor P_{q+1}. Note that we assume that processor indices are taken modulo p.

This is shown in Figure 4.2 with communicated blocks stored into buffer *tempR*. As mentioned earlier in Section 3.3, we assume that sending, receiving, and computing can all occur concurrently at a processor as long as they are independent of each other. By the beginning of step 1, processor P_q has received $\widehat{x_{q-1}}$. (Note that, like for processor indices, we implicitly assume that block indices are taken modulo p.) Processor P_q can therefore compute $\widehat{y_q} = \widehat{y_q} + \widehat{A_{q,q-1}} \times \widehat{x_{q-1}}$, and at the same time it can participate in

$$
\begin{array}{c|c}
P_0 & \begin{pmatrix} A_{00} & A_{01} & \bullet & \bullet & \bullet & \bullet & \bullet & \bullet \\ A_{10} & A_{11} & \bullet & \bullet & \bullet & \bullet & \bullet & \bullet \end{pmatrix} \begin{pmatrix} x_0 \\ x_1 \end{pmatrix} \quad tempR \leftarrow \begin{pmatrix} x_6 \\ x_7 \end{pmatrix} \\ \hline
P_1 & \begin{pmatrix} \bullet & \bullet & A_{22} & A_{23} & \bullet & \bullet & \bullet & \bullet \\ \bullet & \bullet & A_{32} & A_{33} & \bullet & \bullet & \bullet & \bullet \end{pmatrix} \begin{pmatrix} x_2 \\ x_3 \end{pmatrix} \quad tempR \leftarrow \begin{pmatrix} x_0 \\ x_1 \end{pmatrix} \\ \hline
P_2 & \begin{pmatrix} \bullet & \bullet & \bullet & \bullet & A_{44} & A_{45} & \bullet & \bullet \\ \bullet & \bullet & \bullet & \bullet & A_{54} & A_{55} & \bullet & \bullet \end{pmatrix} \begin{pmatrix} x_4 \\ x_5 \end{pmatrix} \quad tempR \leftarrow \begin{pmatrix} x_2 \\ x_3 \end{pmatrix} \\ \hline
P_3 & \begin{pmatrix} \bullet & \bullet & \bullet & \bullet & \bullet & \bullet & A_{66} & A_{67} \\ \bullet & \bullet & \bullet & \bullet & \bullet & \bullet & A_{76} & A_{77} \end{pmatrix} \begin{pmatrix} x_6 \\ x_7 \end{pmatrix} \quad tempR \leftarrow \begin{pmatrix} x_4 \\ x_5 \end{pmatrix}
\end{array}
$$

(a) First step

$$
\begin{array}{c|c}
P_0 & \begin{pmatrix} \bullet & \bullet & \bullet & \bullet & \bullet & \bullet & A_{06} & A_{07} \\ \bullet & \bullet & \bullet & \bullet & \bullet & \bullet & A_{16} & A_{17} \end{pmatrix} \begin{pmatrix} x_6 \\ x_7 \end{pmatrix} \quad tempR \leftarrow \begin{pmatrix} x_4 \\ x_5 \end{pmatrix} \\ \hline
P_1 & \begin{pmatrix} A_{20} & A_{21} & \bullet & \bullet & \bullet & \bullet & \bullet & \bullet \\ A_{30} & A_{31} & \bullet & \bullet & \bullet & \bullet & \bullet & \bullet \end{pmatrix} \begin{pmatrix} x_0 \\ x_1 \end{pmatrix} \quad tempR \leftarrow \begin{pmatrix} x_6 \\ x_7 \end{pmatrix} \\ \hline
P_2 & \begin{pmatrix} \bullet & \bullet & A_{42} & A_{43} & \bullet & \bullet & \bullet & \bullet \\ \bullet & \bullet & A_{52} & A_{53} & \bullet & \bullet & \bullet & \bullet \end{pmatrix} \begin{pmatrix} x_2 \\ x_3 \end{pmatrix} \quad tempR \leftarrow \begin{pmatrix} x_0 \\ x_1 \end{pmatrix} \\ \hline
P_3 & \begin{pmatrix} \bullet & \bullet & \bullet & \bullet & A_{64} & A_{65} & \bullet & \bullet \\ \bullet & \bullet & \bullet & \bullet & A_{74} & A_{75} & \bullet & \bullet \end{pmatrix} \begin{pmatrix} x_4 \\ x_5 \end{pmatrix} \quad tempR \leftarrow \begin{pmatrix} x_2 \\ x_3 \end{pmatrix}
\end{array}
$$

(b) Second step

$$
\begin{array}{c|c}
P_0 & \begin{pmatrix} \bullet & \bullet & \bullet & \bullet & A_{04} & A_{05} & \bullet & \bullet \\ \bullet & \bullet & \bullet & \bullet & A_{14} & A_{15} & \bullet & \bullet \end{pmatrix} \begin{pmatrix} x_4 \\ x_5 \end{pmatrix} \quad tempR \leftarrow \begin{pmatrix} x_2 \\ x_3 \end{pmatrix} \\ \hline
P_1 & \begin{pmatrix} \bullet & \bullet & \bullet & \bullet & \bullet & \bullet & A_{26} & A_{27} \\ \bullet & \bullet & \bullet & \bullet & \bullet & \bullet & A_{36} & A_{37} \end{pmatrix} \begin{pmatrix} x_6 \\ x_7 \end{pmatrix} \quad tempR \leftarrow \begin{pmatrix} x_4 \\ x_5 \end{pmatrix} \\ \hline
P_2 & \begin{pmatrix} A_{40} & A_{41} & \bullet & \bullet & \bullet & \bullet & \bullet & \bullet \\ A_{50} & A_{51} & \bullet & \bullet & \bullet & \bullet & \bullet & \bullet \end{pmatrix} \begin{pmatrix} x_0 \\ x_1 \end{pmatrix} \quad tempR \leftarrow \begin{pmatrix} x_6 \\ x_7 \end{pmatrix} \\ \hline
P_3 & \begin{pmatrix} \bullet & \bullet & A_{62} & A_{63} & \bullet & \bullet & \bullet & \bullet \\ \bullet & \bullet & A_{72} & A_{73} & \bullet & \bullet & \bullet & \bullet \end{pmatrix} \begin{pmatrix} x_2 \\ x_3 \end{pmatrix} \quad tempR \leftarrow \begin{pmatrix} x_0 \\ x_1 \end{pmatrix}
\end{array}
$$

(c) Third step

$$
\begin{array}{c|c}
P_0 & \begin{pmatrix} \bullet & \bullet & A_{02} & A_{03} & \bullet & \bullet & \bullet & \bullet \\ \bullet & \bullet & A_{12} & A_{13} & \bullet & \bullet & \bullet & \bullet \end{pmatrix} \begin{pmatrix} x_2 \\ x_3 \end{pmatrix} \quad tempR \leftarrow \begin{pmatrix} x_0 \\ x_1 \end{pmatrix} \\ \hline
P_1 & \begin{pmatrix} \bullet & \bullet & \bullet & \bullet & A_{24} & A_{25} & \bullet & \bullet \\ \bullet & \bullet & \bullet & \bullet & A_{34} & A_{35} & \bullet & \bullet \end{pmatrix} \begin{pmatrix} x_4 \\ x_5 \end{pmatrix} \quad tempR \leftarrow \begin{pmatrix} x_2 \\ x_3 \end{pmatrix} \\ \hline
P_2 & \begin{pmatrix} \bullet & \bullet & \bullet & \bullet & \bullet & \bullet & A_{46} & A_{47} \\ \bullet & \bullet & \bullet & \bullet & \bullet & \bullet & A_{56} & A_{57} \end{pmatrix} \begin{pmatrix} x_6 \\ x_7 \end{pmatrix} \quad tempR \leftarrow \begin{pmatrix} x_4 \\ x_5 \end{pmatrix} \\ \hline
P_3 & \begin{pmatrix} A_{60} & A_{61} & \bullet & \bullet & \bullet & \bullet & \bullet & \bullet \\ A_{70} & A_{71} & \bullet & \bullet & \bullet & \bullet & \bullet & \bullet \end{pmatrix} \begin{pmatrix} x_0 \\ x_1 \end{pmatrix} \quad tempR \leftarrow \begin{pmatrix} x_6 \\ x_7 \end{pmatrix}
\end{array}
$$

(d) Fourth (and last) step

FIGURE 4.2: Matrix-vector multiplication example ($n = 8$, $p = 4$, $r = 2$): steps of the parallel execution.

```
1  var A:  array[0..r − 1,0..n − 1] of real
2  var x, y:  array[0..r − 1] of real
3  q ←My_Num()
4  p ←Num_Procs()
5  tempS ← x
6  for step=0 to p − 1 do
7  |   Send(tempS, r)
8  |   ||
9  |   Receive(tempR, r)
10 |   ||
11 |   for i = 0 to r − 1 do
12 |   |   for j = 0 to r − 1 do
13 |   |   |   y[i] ← y[i] + A[i, (q − step mod p)r + j] × tempS[j]
14 |   tempS ↔tempR
```

ALGORITHM 4.1: Matrix-vector multiplication algorithm on a ring of processors.

the circular block shift of vector x. At the end of the p-th step, array y in processor P_q contains block $\widehat{y_q}$, i.e., the desired r components of the result: $y_{qr}, \ldots, y_{qr+r-1}$. This parallel algorithm is shown in Algorithm 4.1. Note our use of the || symbol to indicate actions that occur concurrently at a processor.

The reader will notice that the circular block shift of vector x in the last step is not absolutely necessary. But it allows all processors to hold at the end of the execution the block of x that they held initially, which may be desirable for subsequent operations.

The computation of the execution time of this algorithm on p processors is straightforward. There are p identical steps. Each step lasts as long as the longest of the three activities performed during the step: compute, send, and receive. Since the time to send and the time to receive are identical, $T(p)$ can be written as

$$T(p) = p \max\{r^2 w, L + rb\} ,$$

where w is the computation time for one basic operation (in this case multiplying two vector components and adding the result to another vector component), b is the inverse of the data transfer rate measured in number of basic data units (in this case one vector component) per second, and L is the communication startup cost. Since $r = n/p$, for a given number of processors p the computation component in the equation above is asymptotically larger than the communication component as n becomes large ($\frac{n^2}{p^2} w \gg L + \frac{n}{p} b$). This means that our algorithm achieves an asymptotic parallel efficiency of 1 as n becomes large. Note that if we had opted for a full duplication of vector x across all processors, the disadvantages of which were discussed earlier, then there would be no need for any communication at all. Therefore, the parallel efficiency would always be 1, only at the cost of a slight increase in memory usage.

4.2 Matrix-Matrix Multiplication

Now that we know how to multiply a matrix by a vector on a ring of processors, multiplying a matrix by a matrix is straightforward. Consider the multiplication $C = A \times B$, where A, B, and C are matrices of dimension $n \times n$. The multiplication consists in computing n^2 scalar products:

```
for i = 0 to n − 1 do
  for j = 0 to n − 1 do
    C_{i,j} ← 0
    for k = 0 to n − 1 do
      C_{i,j} ← C_{i,j} + A_{i,k} × B_{k,j}
```

Inspired by the matrix-vector multiplication, one distributes all three matrices among the p processors so that the first processor holds the first $r = n/p$ rows of each of the matrices, the second processor holds the next r rows of each of the matrices, etc. Unlike in the matrix-vector case, full duplication of matrix B across all processors would severely limit the size of the problems that can be solved, due to the increased memory usage. Thus, one needs the following declarations in our parallel program:

```
var A, B, C:  array[0..r − 1,0..n − 1] of real;
```

The principle of the algorithm is identical to that of our parallel matrix-vector multiplication and thus also proceeds in p steps. At each step, each processor computes p partial matrix multiplications, that is, p multiplications of an $r \times r$ matrix by an $r \times r$ matrix. Using the block notation again, at step 0 of the algorithm, each processor P_q holds blocks $\widehat{A_{q,l}}$, $\widehat{B_{q,l}}$, and $\widehat{C_{q,l}}$ of matrices A, B, and C for $l = 0, \ldots, p-1$. (Note that the blocks of matrix C are all initialized to zero.) Processor P_q computes the $\widehat{A_{q,q}} \times \widehat{B_{q,l}}$ products for all l. Each such product is added to block $\widehat{C_{q,l}}$. While this computation is being performed, a circular shift of the block rows of matrix B among the processors also takes place: Each processor P_q sends blocks $\widehat{B_{q,l}}$ to processor P_{q+1} for all l. By the beginning of step 1, processor P_q has thus received blocks $\widehat{B_{q-1,l}}$ for all l and can therefore compute all $\widehat{A_{q,q-1}} \times \widehat{B_{q-1,l}}$ products. As in the previous section, we assume that block indices are taken modulo p. These products are added to the blocks of matrix C held by processor P_q. After p such steps (and a final shift of the block rows of matrix B so that its distribution among the

```
1  var A,B,C:  array[0..r-1,0..n-1] of real
2  q ←MY_NUM()
3  p ←NUM_PROCS()
4  tempS ←B
5  for step=0 to p-1 do
6  |  SEND(tempS, r × n)
7  |  ||
8  |  RECEIVE(tempR, r × n)
9  |  ||
10 |  for l = 0 to p-1 do
11 |  |  for i = 0 to r-1 do
12 |  |  |  for j = 0 to r-1 do
13 |  |  |  |  for k = 0 to r-1 do
14 |  |  |  |  |  C[i, lr+j] ←C[i, lr+j] + A[i, r × ((q -
   |  |  |  |  |  step) mod p) + k] × tempS[k, lr+j]
15 |  tempS ↔tempR
```

ALGORITHM 4.2: Matrix-matrix multiplication algorithm on a ring of processors.

processors after the algorithm terminates is identical to its initial distribution), processor P_q holds the desired blocks of the result matrix C. The algorithm is shown in Algorithm 4.2. This algorithm is very similar to the one for matrix-vector multiplication, essentially replacing the scalar products by sub-matrix multiplications and replacing the circular shifting of vector components by circular shifting of matrix rows. Consequently, the performance analysis is also similar. The algorithm proceeds in p steps. Each step lasts as long as the longest of the three activities performed during the step: compute, send, and receive. Therefore, using the same notations as in the previous section, $T(p)$ can be written as

$$T(p) = p.\max\{nr^2w, L + nrb\} .$$

Here again, the asymptotic parallel efficiency when n is large is 1. It is interesting to note that the matrix-matrix multiplication could be achieved by executing the matrix-vector multiplication algorithm n times. With this more naïve algorithm the execution time would be simply the one obtained in the previous section multiplied by n:

$$T'(p) = p.\max\{nr^2w, nL + nrb\} .$$

The only difference between $T(p)$ and $T'(p)$ is the term L, which has become nL. While in the naïve approach processors exchange vectors of size r, in the algorithm developed in this section they exchange matrices of size $r \times n$, thereby reducing the overhead due to network startup cost L. This change does not reduce the asymptotic efficiency, but it can be significant in practice. This notion of sending data in bulk is a well-known and general principle used to reduce communication overhead in parallel algorithms, and we will see it again and again in this chapter. Furthermore, with modern processors, operating on blocks of data is often better for performance as it exploits a processor's memory hierarchy (registers, multiple levels of cache, main memory) to the best of its capabilities. As a result, operating on blocks of data improves both computation and communication performance. Note that in this algorithm the total number of data elements sent over the network is $p \times p \times n \times r = p \times n^2$ (recall that $r = n/p$). We will see how this total amount of communication can be reduced.

4.3 A First Look at Stencil Applications

In this section, we discuss the parallelization of a popular class of applications commonly called "stencil applications." These applications operate on a discrete domain, A, that consists of cells. Each cell holds some value(s) and has neighbor cells. The application uses an algorithm that applies pre-defined rules to update the value(s) of a cell using the values of its neighbor cells. The location of a cell's neighbors and the function used to update cell values form a "stencil" that is applied to all cells in the domain. Stencil applications arise in many areas of science and engineering. Examples include image processing algorithms in which cells are pixels; numerical methods used to compute approximate solutions of partial differential equations over a physical domain, in which case cells are contiguous small regions of the domain; and simulations of complex cellular automata, of which Conway's Game of Life is a well-known if simplistic example. For some of these applications the domain is large enough that one may have to resort to a distributed-memory implementation, rather than using a simple shared-memory implementation. In this section, we investigate how a distributed-memory parallel stencil application can be implemented using a logical ring topology.

4.3.1 A Simple Sequential Stencil Algorithm

We consider a stencil application that operates on a two-dimensional (2-D) domain of size $n \times n$ for $n \geqslant 0$. Each cell has eight neighbor cells, as shown in Figure 4.3.

The algorithm we consider updates the value of cell c as a function of its current value and of the *already updated* value of its West and North neighbors.

$$\begin{array}{ccc} NW & N & NE \\ W & \boxed{c} & E \\ SW & S & SE \end{array}$$

FIGURE 4.3: The eight neighbor cells of cell c.

The stencil is depicted in Figure 4.4 and can be formalized as follows:

$$c_{new} \leftarrow \textsf{UPDATE}(c_{old}, W_{new}, N_{new}) \; .$$

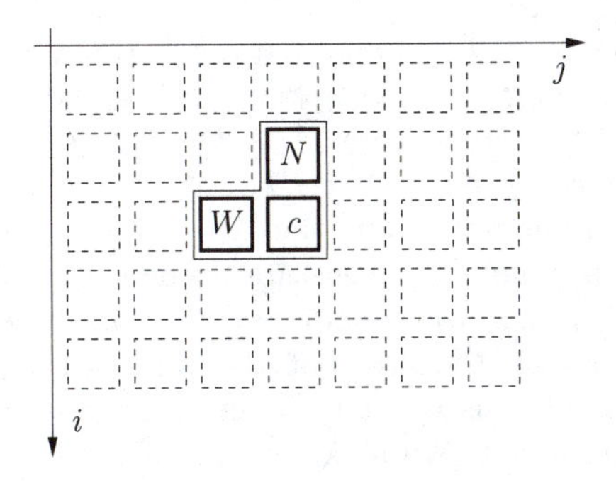

FIGURE 4.4: A simple stencil in which cell c is updated using the values of its West and North neighbors.

This stencil, albeit simple, is at the heart of important applications such as the Gauss-Seidel numerical method [97] and the Smith-Waterman biological string comparison algorithms [118]. Note that the stencil, as it is described above, cannot be applied to the cells in the topmost row or the leftmost column of the domain. Indeed, the former have no North neighbors and the latter have no West neighbors. This is handled by the UPDATE function, for instance, by using constant values for neighbors outside the domain or by using modified calculations when a cell has only one or even no neighbors as is the case for the top left cell. To indicate that no neighbor exists for a cell update, we pass a **Nil** argument to UPDATE. For instance, the call UPDATE(c, W_{new}, **Nil**) is used to update a cell on the upper edge of the domain, and the call UPDATE(c, **Nil**, **Nil**) is used to update the top left cell.

4.3.2 Parallelizations of the Stencil Algorithm

Consider a ring of p processors $P_0, \ldots, P_{p-1}$. We must distribute the cells in the domain among the processors. A natural solution is to allocate rows of cells to processors. The main challenge here is to come up with a distribution that balances the computational load among the processors without hindering performance due to overly expensive communications.

Greedy Algorithm

A first idea to parallelize our stencil algorithm is to use a greedy approach by which processors send the cells they compute to their neighbors as early as possible. Such a parallelization reduces the startup latencies (i.e., processors start computing as early as possible) and leads to a good load balance.

Let us assume for now that the number of processors p is equal to the domain's dimension n. In this case, we allocate row i of domain A to processor P_i and thus each processor has the following declaration:

```
var A:  array[0..n − 1] of real;
```

As soon as processor P_i has computed a cell value, it sends that value to processor P_{i+1} ($0 \leqslant i < p-1$). As in Sections 4.1 and 4.2, we use subscripts for global indices and we denote by $A_{i,j}$ the value of the cell on row i and column j of the domain ($0 \leqslant i, j \leqslant n-1$).

Given the shape of our stencil and the fact that the already updated values of the West and North neighbor cells are needed, initially only $A_{0,0}$ can be computed. Once $A_{0,0}$ has been computed, then both $A_{1,0}$ and $A_{0,1}$ can be computed. It can be seen easily that the computation proceeds in steps: At step k all values of cells on the k-th anti-diagonal are computed. The corresponding program is shown in Algorithm 4.3. Figure 4.5 shows each cell labeled by the step at which its value can be computed.

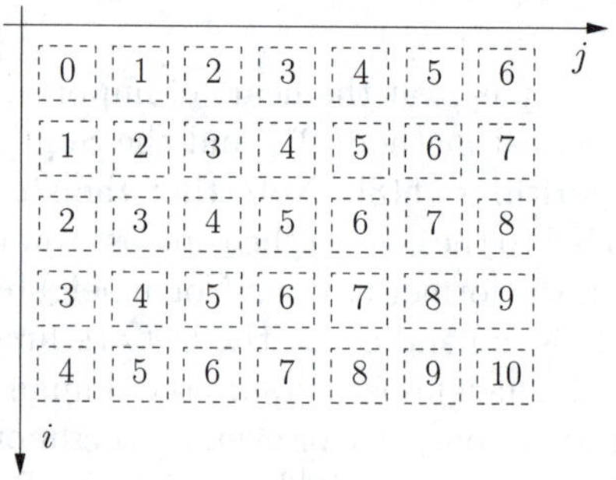

FIGURE 4.5: Steps of the greedy parallel algorithm.

At time $i + j$, processor P_i performs the following operations:

- it receives $A_{i-1,j}$ from P_{i-1};
- it computes $A_{i,j}$;
- then it sends $A_{i,j}$ to P_{i+1}.

Note that the above is not technically true because P_0 does not need to receive cell values to update its cells and P_{p-1} does not need to send cell values after

```
1  var A: array[0..n − 1] of real
2  q ←MY_NUM()
3  p ←NUM_PROCS()
4  if q = 0 then
5  |  A[0] ←UPDATE(A[0], Nil, Nil)
6  |  SEND(A[0], 1)
7  else
8  |  RECV(v, 1)
9  |  A[0] ←UPDATE(A[0], Nil, v)
10 for j = 1 to n − 1 do
11 |  if q = 0 then
12 |  |  A[j] ←UPDATE(A[j], A[j − 1], Nil)
13 |  |  SEND(A[j], 1)
14 |  else if q = p − 1 then
15 |  |  RECV(v, 1)
16 |  |  A[j] ←UPDATE(A[j], A[j − 1], v)
17 |  else
18 |  |  SEND(A[j − 1], 1) || RECV(v, 1)
19 |  |  A[j] ←UPDATE(A[j], A[j − 1], v)
```

ALGORITHM 4.3: Greedy algorithm for the stencil application on a ring of processors.

updating its cells. However, these two special cases have no influence on the performance analysis of the algorithm.

When p is lower than n, which is typical, then the question of which row to assign to which processor arises. Without loss of generality, let us assume that p divides n. For our matrix multiplication algorithm we assigned a block of contiguous rows to each processor. However, this approach goes against the main goal of our greedy algorithm. Consider processor P_0. If the first n/p rows are assigned to it, then it takes at least n/p steps before the value of the first cell of the last row assigned to P_0 is computed. This cell value is necessary for processor P_1 to start computing the first cell of its block of rows. Therefore, P_1 (and all other processors) stay idle for at least the first n/p steps of the algorithm. In a view to allowing processors to start computing as early as possible, one can instead assign rows to processors in an interleaved, or *cyclic*, manner: Row j is assigned to processor $P_{j \bmod p}$. Such a cyclic distribution is a classic technique to achieve a good load balance across processors. Each processor has the following declaration:

```
var A: array[0..n/p − 1,0..n − 1] of real;
```

This is a contiguous array of rows that are non-contiguous in the domain. We can now modify our previous algorithm to implement a cyclic distribution of rows among processors, as shown in Algorithm 4.4.

```
1  var A:  array[0..n/p-1,0..n-1] of real
2  q ←My_Num()
3  p ←Num_Procs()
4  for i = 0 to n/p - 1 do
5      if q = 0 And i = 0 then
6          A[0,0] ←Update(A[0,0], Nil, Nil)
7          Send(A[0,0], 1)
8      else
9          Recv(v, 1)
10         A[i,0] ←Update(A[i,0], Nil, v)
11     for j = 1 to n - 1 do
12         if q = 0 And i = 0 then
13             A[i,j]
               ←Update(A[i,j], A[i-1,j], Nil)
14             Send(A[i,j], 1)
15         else if q = p - 1 And i = n/p - 1 then
16             Recv(v, 1)
17             A[i,j] ←Update(A[i,j], A[i-1,j], v)
18         else
19             Send(A[i,j], 1) || Recv(v, 1)
20             A[i,j] ←Update(A[i,j], A[i-1,j], v)
```

ALGORITHM 4.4: Algorithm for the stencil application on a ring of processors using a cyclic data distribution.

Let us compute the execution time of this algorithm, $T(n,p)$. We assume that receiving a message is a blocking operation, while sending a message is not. Therefore, the sending of a message at step k of the algorithm occurs in parallel with the reception of a message at step $k+1$. Consequently, the time needed to perform one algorithm step is $w+b+L$, where w is the time needed to update a cell value, b is the rate at which cell values are communicated on a network link, and L is the communication startup cost. We now need to compute the number of such steps. The computation terminates when processor P_{p-1} finishes computing the rightmost cell value of its last row of cells. It takes $p-1$ algorithm steps before processor P_{p-1} can start doing its first computation. At this point, processor P_{p-1} computes one cell value at each step. There are n^2 cells in the domain, and each processor holds n^2/p cells. Therefore, processor P_{p-1} is engage in cell value computations for n^2/p steps, and the total number of steps is $p - 1 + n^2/p$, which gives

$$T(n,p) = \left(p - 1 + \frac{n^2}{p}\right)(w + b + L) \, .$$

This algorithm was designed so that the time between a cell value computation and its reception by the next processor is as short as possible. But a glaring problem is that the algorithm performs many communications of data items that may be small. In practice, the communication startup cost L can be orders of magnitude larger than b if the size of a cell value is small. In the case of stencil applications, the cell value is often as small as a single integer or a single floating-point number. Furthermore, w may not be large. It may be comparable to or smaller than L. Indeed, cell value computations can be very simple and involve only a few operations. This is the case for many numerical methods that update cell values based on simple arithmetic formulas. Therefore, for many applications, a large fraction of the execution time $T(n,p)$ could be due to the L terms in the previous equation. Spending a large fraction of the execution time in communication overhead reduces parallel efficiency significantly. This can be seen more plainly by simply computing the parallel efficiency. The parallel efficiency of the algorithm is equal to the sequential execution time, $n^2 w$, divided by $p \times T(n,p)$. When n gets large, the parallel efficiency tends to $w/(w + b + L)$, which may be well below 1. In the next section we present two techniques for addressing this problem.

Augmenting the Granularity of the Algorithm

A simple technique to decrease communication costs, or, more precisely, communication overhead due to startup latencies, is to send fewer but larger messages. Using the same allocation of rows to processors as before, a processor now computes k contiguous cell values in each row at each step, where $1 \leqslant k \leqslant n$, as opposed to just one. As before, the algorithm proceeds in several steps. For simplicity we assume that k divides n so that each row contains n/k segments of k contiguous cells. If k does not divide n, then the $(\lfloor n/k \rfloor + 1)$-th segment of the first row stored by each processor spills over to the second row stored by that processor, and so on. The last segment of the last row stored by that processor may contain fewer than k cells, which does not modify the spirit of the algorithm. We make this assumption to simplify the performance analysis of the algorithm. The data distribution is depicted in Figure 4.6 for four processors, where each segment of k consecutive cells is labeled by the step at which it is computed.

With this algorithm the number of cell values communicated among processors is the same as with the greedy algorithm, but cells values are communicated in bulk, k at a time. Larger values of k lead to lower communication overhead due to network latencies. However, larger values of k also imply that the time between a cell value computation and its reception by the next processor is larger. This goes against the principle guiding our design of the algorithm in the previous section. In this algorithm processors will start computing cell values later, leading to more idle time.

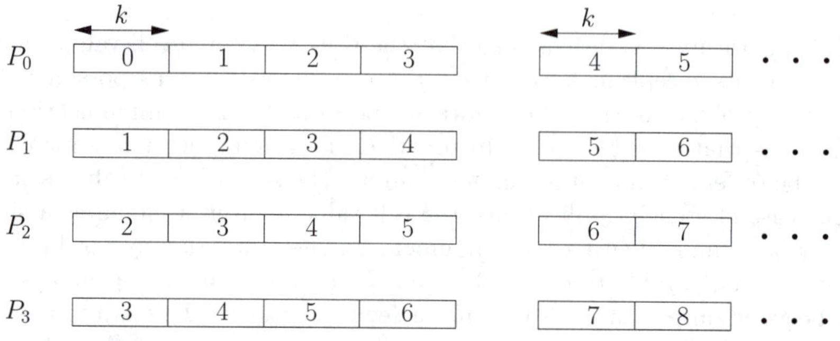

	First row of each processor (k per cell)				Second row (k per cell)		
P_0	0	1	2	3	4	5	...
P_1	1	2	3	4	5	6	...
P_2	2	3	4	5	6	7	...
P_3	3	4	5	6	7	8	...

FIGURE 4.6: Steps of the modified algorithm with $k > 1$ and $p = 4$ processors.

Another technique to reduce communication costs is to reduce the number of cell values that need to be communicated between the processors. This can be done by allocating blocks of r consecutive rows to processors ($rp \leqslant n$), still in a cyclic fashion. At each step each processor now computes $r \times k$ cell values. We assume that $r \times p$ divide n for simpler performance analysis. For instance, with $r = 3$, $n = 36$, and $p = 4$, we have the following allocation of rows to processors:

P_0	P_1	P_2	P_3
0, 1, 2	3, 4, 5	6, 7, 8	9, 10, 11
12, 13, 14	15, 16, 17	18, 19, 20	21, 22, 23
24, 25, 26	27, 28, 29	30, 31, 32	33, 34, 35

More formally, processor P_i holds rows j such that $i = \lfloor \frac{j}{r} \rfloor \bmod p$, $0 \leqslant j \leqslant n-1$. This notion of a *block-cyclic* allocation is classic and we will use it often.

The steps of the algorithm are similar to those shown in Figure 4.6, simply replacing individual rows by blocks of rows. We can easily obtain a sufficient condition for processors not to be idle. A processor, say P_0, computes all cell values in its first block of rows in n/k steps of the algorithm. Processor P_{p-1} sends k cell values to processor P_0 after p algorithm steps. These values are necessary for P_0 to start computing its second block of rows. Therefore, if $n \geqslant kp$, no processor stays idle, which is desirable, and which we will assume hereafter. Note that if $n > kp$, then processors have to temporarily store received cell values while finishing computing their block of rows for the previous step.

With this new distribution of rows to processors, the amount of data communicated between processors is r times smaller than with the previous algorithm. Indeed, the processors need to exchange data only at the boundaries between blocks of rows. So the larger r, the lower the communication cost, but the larger r, the larger the time between the computation of a cell value and its communication to the next processor. This is once again the same trade-off: lower communication costs or lower latency between a cell computation and its reception by the next processor.

One interesting question is, are there optimal values of k and r? We can answer this question via performance analysis of the algorithm. We assume that $n \geqslant kp$ so that no processor stays idle and leave the (less relevant in practice) $n < kp$ case as an exercise for the motivated reader. The algorithm proceeds in a sequence of steps. At each step at least one processor is involved in the following activities:

- it receives k cell values from its predecessor;
- it computes kr cell values;
- it sends k cell values to its successor.

The analysis is very similar to that for our simple greedy algorithm. Here again we assume that receiving a message is a blocking operation while sending a message is not. Consequently, the time needed to perform one step of the algorithm is $krw + kb + L$. The computation terminates when processor P_{p-1} finishes computing the rightmost segment of its last block of rows of cells. It takes $p - 1$ algorithm steps before processor P_{p-1} can start doing any computation. At this point, processor P_{p-1} will compute one segment of a block row at each step. There are $n^2/(kr)$ such segments in the domain, and each processor holds the same number of segments. Therefore, processor P_{p-1} computes cell values for $n^2/(pkr)$ steps. Overall, the algorithm runs for $p - 1 + n^2/(pkr)$ steps and the total execution time $T(n, p, r, k)$ is

$$T(n, p, r, k) = \left(p - 1 + \frac{n^2}{pkr}\right)(krw + kb + L)\ .$$

Let us compare the asymptotic parallel efficiency of this algorithm with that of the algorithm in Section 4.3.2, whose parallel efficiency was $w/(w+b+L)$. Dividing the sequential execution time $n^2 w$ by $p \times T(n, p, r, k)$ we obtain an asymptotic efficiency of $w/(w + b/r + L/rk)$. In essence, increasing r and k makes it possible to achieve significantly higher asymptotic efficiency by reducing communication costs.

While low asymptotic parallel efficiency is important, one may wonder what values of k and r should be used in practice. It turns out that, for n and p

fixed, one can easily compute the optimal value of k, $k_{opt}(r)$. Equating the derivative of $T(n,p,r,k)$ to zero and solving for k one obtains the value $k'(r)$:

$$k'(r) = n\sqrt{\frac{L}{p(p-1)r(rw+b)}} .$$

Since $k \leqslant n/p$, we obtain $k_{opt}(r) = \min(k'(r), n/p)$. Finally, for given values of L, w, and b, one can compute $k_{opt}(r)$ and inject it in the expression for $T(n,p,r,k)$. One can then determine the best value for r numerically.

4.4 LU Factorization

In this section, we develop a parallel algorithm that performs the classic LU decomposition of a square matrix A using the Gaussian elimination method, which in turn allows for the straightforward solution of linear systems of the form $Ax = b$. Readers familiar with the algorithm will notice that we make two simplifying assumptions so as not to make matters overly complicated. First, we use the Gaussian elimination method without any pivoting. This assumption is rather inconsequential at least in the case of partial pivoting: Since the columns of A will be distributed among the processors, partial pivoting would not add any extra communication. Second, our algorithm eliminates columns of A one after another. This is not realistic for modern computers as their memory hierarchies would not be exploited to the best of their capability. Instead, the algorithm should really process several columns at a time (i.e., column blocks), exactly as in our matrix-matrix multiplication algorithm (Section 4.2). But from a purely algorithmic standpoint there is no conceptual difference between computing a single column or a column block, and the algorithm's spirit is unchanged. Note, however, that when writing the corresponding programs in practice the utilization of column blocks requires many more lines of code. This is true even for sequential algorithms. For instance, see the sequential version of the Gaussian elimination algorithm in the publicly available LAPACK library [8].

4.4.1 First Version

The sequential algorithm for a matrix A of dimension $n \times n$ (with global indices from 0 to $n-1$) proceeds in $n-1$ steps, as shown in Algorithm 4.5. Note that we show two inlined functions, PREP and UPDATE. This is to identify and to refer to the two main computations performed by the algorithm conveniently, namely, the *preparation* of the current column and the *update* of the bottom right sub-matrix of matrix A. We depict the execution of the algorithm in Figure 4.7.

```
1 for k = 0 to n − 2 do
2     PREP(k) :
3         for i = k + 1 to n − 1 do
4             A_ik ← −A_ik / A_kk
5     for j = k + 1 to n − 1 do
6         UPDATE(k, j) :
7             for i = k + 1 to n − 1 do
8                 A_ij ← A_ij + A_ik × A_kj
```

ALGORITHM 4.5: Sequential LU factorization algorithm.

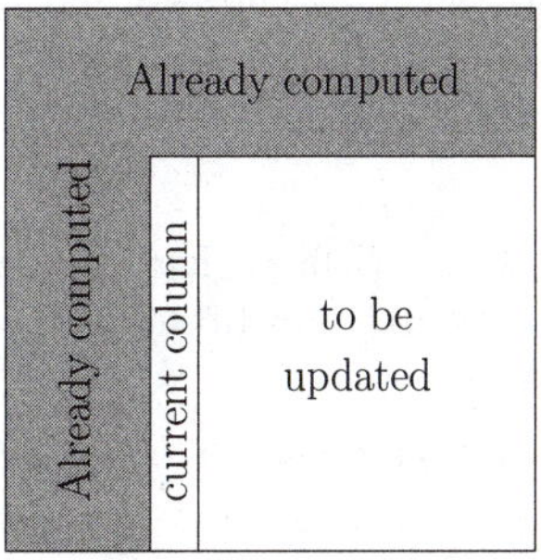

FIGURE 4.7: A step of the LU factorization.

A point of detail for the interested reader: The $A = LU$ decomposition is obtained after the above algorithm completes by defining $L_{ik} = -A_{ik}$ for $i > k$ (L is lower triangular with a unit diagonal) and $U_{kj} = A_{kj}$ for $k \leqslant j$ (U is upper triangular). This sign swapping for the L_{ik} values is due to the fact that these values are in fact the coefficients used to eliminate columns when making A triangular [63]. Also, as stated above, adding pivoting in functions PREP and UPDATE would be straightforward.

The question we wish to answer is, how can we parallelize this algorithm on a ring of p processors? Since the algorithm processes the matrix column-by-column, it is natural to distribute these columns among the processors. Rather than fixing the data distribution scheme at this point, let us just assume that we have defined a function ALLOC, which, given a column index k between 0 and $n - 1$, returns the index of the processor that stores column k in its memory. At step k of the algorithm, processor ALLOC(k) broadcasts column k to all processors, allowing them to update their own columns of matrix A. Using the BROADCAST function (see Section 3.3), we obtain the parallel algorithm shown in Algorithm 4.6.

Note that in the algorithm above we use a helper array called *buffer* in which we store updated column elements. The use of such an array is often necessary

```
1  var buffer:  array[0..n−1] of real
2  q ←MY_NUM()
3  p ←NUM_PROCS()
4  for k = 0 to n − 2 do
5  |  if ALLOC(k) = q then
6  |  |  PREP(k) :
7  |  |  |  for i = k + 1 to n − 1 do
8  |  |  |  |_ buffer[i − k − 1] ← A_ik ← −A_ik/A_kk
9  |  BROADCAST(ALLOC(k), buffer, n − k − 1)
10 |  for j = k + 1 to n − 1 do
11 |  |  UPDATE(k, j) :
12 |  |  |  for i = k + 1 to n − 1 do
13 |  |  |  |_ A_ij ← A_ij + buffer[i − k − 1] × A_kj
```

ALGORITHM 4.6: LU factorization algorithm on a ring of processors with the use of a broadcast.

due to the memory layout of 2-D arrays, in our case array A. Since computer memory is inherently 1-D, 2-D arrays (as well as arrays of higher dimension) must be "chopped up" into 1-D pieces. The two common approaches are *row-major* and *column-major*, depending on whether 2-D arrays are chopped along rows or along columns. So far in this chapter we have implicitly assumed row-major storage of 2-D arrays (as in C) rather than column-major storage (as in FORTRAN). This assumption implies that elements of a row are contiguous in memory, while elements of a column are not. The reader will recall that in all the algorithms we have seen so far processors were communicating rows of 2-D arrays, rather than columns. Consequently, the processors were always sending array elements that were contiguous in memory, which was possible with single calls to SEND. The LU decomposition algorithm, however, needs to broadcast columns of the array, that is, elements that are not contiguous in memory! Therefore, we place these elements into a helper array so that a single call to BROADCAST can send matrix elements in bulk. The alternative would have been to place n individual calls to BROADCAST, one for each element in the current column of matrix A. This alternative leads to high overhead due to network latencies (typically much higher than the overhead of an extra memory copy). Understanding memory layouts is always a good idea (for instance, to improve cache reuse) but is paramount when implementing distributed memory algorithms to ensure that data are communicated in bulk as much as possible.

One remaining issue is to map global indices to local indices for accessing elements of matrix A. Defining $r = n/p$, each processor stores r columns of

A in its local memory. As the algorithm makes progress, some columns stop being accessed: After step k, only columns with an index higher than k are read and/or updated. The classical idea here is for each processor to use a local index, l, which indicates the next column of the local array that should be accessed. At the beginning of the execution all processors set l to 0. At step $k = 0$, processor ALLOC(0) prepares column 0 by calling function PREP(0). Its local index will then be incremented to $l = 1$. When this processor updates its columns of matrix A, it only updates the $r - l$ last ones. The value of l is unchanged at other processors. At step $k = 1$, processor ALLOC(1) increments its own value of l after calling PREP(1). Using array declarations and array indices to replace the matrix element specifications in the previous algorithm, we now obtain the full program shown in Algorithm 4.7.

```
1  var A: array[0..n-1,0..r-1] of real
2  var buffer: array[0..n-1] of real
3  q ←MY_NUM()
4  p ←NUM_PROCS()
5  l ← 0
6  for k = 0 to n - 2 do
7      if ALLOC(k) = q then
8          PREP(k) :
9              for i = k + 1 to n - 1 do
10                 buffer[i-k-1] ← A[i,l] ← -A[i,l]/A[k,l]
11         l ← l + 1
12     BROADCAST(ALLOC(k), buffer, n - k - 1)
13     for j = l to r - 1 do
14         UPDATE(k, j) :
15             for i = k + 1 to n - 1 do
16                 A[i,j] ← A[i,j] + buffer[i-k-1] × A[k,j]
```

ALGORITHM 4.7: LU factorization algorithm on a ring of processors written with local indices.

The only remaining thing that needs to be specified is the function ALLOC, or, in other terms, which matrix columns are assigned to which processors? This was not a difficult issue for the matrix-vector or the matrix-matrix multiplication: Just allocate even blocks of consecutive matrix rows to processors. Although one could think of doing the same thing with matrix columns here, this simple solution is not adequate for two reasons:

- The amount of data to process varies throughout the algorithm's execution. At the core of the algorithm is the update of columns $k+1$ to $n-1$ at each step k. Therefore, there are fewer and fewer columns to process as the algorithm makes progress, that, is when k increases.

- The amount of computation is not proportional to the amount of data! Indeed, column k is updated k times. Therefore, from the perspective of a processor, holding column $n-1$ implies significantly more computation than holding, say, column $\lfloor n/2 \rfloor$.

Given the above, we need to find an allocation that balances both the memory consumption (all processors must hold the same number of columns so that matrices as large as possible can be processed) and the computation (processors must perform similar numbers of operations to ensure that no processing power is wasted due to processors being idle). Furthermore, computation must be balanced at every step of the algorithm since the amount and distribution of the computation vary throughout algorithm execution.

Perhaps expectedly, a cyclic allocation, which assigns column j to processor $P_{j \bmod p}$, gets the job done. Unlike for the stencil application, in which a cyclic allocation leads to a perfect balance of computation across processors, here processor P_{p-1} will have slightly more computations to perform than processor P_0, but it can be easily shown that the difference is asymptotically negligible. Indeed, consider a column j assigned to processor $P_{j \bmod p}$. This column is updated at algorithm steps $k = 0, \ldots, j-1$. More precisely, at step k, the elements of column j that are updated are the ones from position $k+1$ to position $n-1$. Therefore, at step k, the number of basic update operations performed by the processor holding column j is equal to $n-k-1$. Consequently, the total number of elemental updates performed by the processor holding column j on this column is

$$ops(j) = \sum_{k=0}^{j-1} (n-k-1) = -\frac{1}{2}j^2 + \left(n - \frac{1}{2}\right) j \,,$$

With a cyclic distribution, processor P_i holds columns $lp+i$ for $l = 0, \ldots, n/p-1$. Therefore, processor P_i performs a total of $Ops(i)$ update operations, where

$$Ops(i) = \sum_{l=0}^{\frac{n}{p}-1} ops(lp+i) \,.$$

Replacing $ops(lp+j)$ by its expression, separating terms, and using well-known formulas for the sums of consecutive integers and sums of the squares of consecutive integers, one obtains that $Ops(i) = \frac{n^3}{3p} + O(n^2)$. Therefore, asymptotically, $Ops(i)$ does not depend on i and all processors perform the same amount of update operations. It is easy to show that asymptotically all processors also perform the same number of column preparations. However, note that there are $O(n)$ times more update operations than column preparation operations, making the time spent doing column preparations asymptotically negligible. To modify our program to use a cyclic allocation of matrix columns to processors, one just has to replace the test $\textsf{Alloc}(k) = q$ by $k = q \bmod p$.

Let $p(k, P_q)$ and $u(k, j, P_q)$ denote the executions of functions $\textsf{Prep}(k)$ and $\textsf{Update}(k, j)$ on processor P_q, respectively. The execution time of the algorithm is that of the longest path of operations (also called the *critical path*), that is

$$p(0, P_0) \rightarrow \left(\sum_{j=1}^{r-1} u(0, 1, P_1), p(1, P_1) \right) \rightarrow \left(\sum_{j=2}^{r-1} u(1, j, P_2), p(2, P_2) \right) \rightarrow \ldots,$$

where each $\rightarrow$ symbol denotes a communication between neighbor processors. After expressing the individual function execution times as functions of the platform parameters, as we have done above for the number of column update operations, it is straightforward to obtain that the overall execution time is the sum of:

- a term $nL + \frac{n^2}{2}b + O(1)$ that accounts for the $n - 1$ communications;
- a term $\frac{n^2}{2}w' + O(1)$ for the column preparations; and
- a term $\frac{n^3}{3p}w + O(n^2)$ for the column updates.

Note that to be precise we use w' to denote the time for a basic column preparation operation (one division and one negation), and w to denote the time for a basic column update operation (one multiplication and one addition). As n grows, the overall execution time becomes asymptotically equivalent to $\frac{n^3}{3p}w$. Note that the total number of update operations to be performed is given by

$$\sum_{i=1}^{i=n} ops(i) ,$$

which can be computed by replacing $ops(i)$ by its expression. Simple calculations show that the above sum is asymptotically equal to $\frac{n^3}{3}w$ when n becomes large. The execution time of our algorithm is asymptotically p times shorter than this amount, and consequently our algorithm has an asymptotic parallel efficiency of 1. In spite of this good asymptotic result, the communication term above can still be significant in practical situations. In the next two sections we describe techniques to reduce communication costs.

```
1  var A:  array[0..n − 1,0..r − 1] of real
2  var buffer:  array[0..n − 1] of real
3  q ←My_Num()
4  p ←Num_Procs()
5  l ← 0
6  for k = 0 to n − 2 do
7      if k = q mod p then
8          Prep(k) :
9              for i = k + 1 to n − 1 do
10                 buffer[i − k − 1] ← A[i, l] ← −A[i, l]/A[k, l]
11         l ← l + 1
12         Send(buffer, n − k − 1)
13     else
14         Recv(buffer, n − k − 1)
15         if q ≠ k − 1 mod p then
16             Send(buffer, n − k − 1)
17     for j = l to r − 1 do
18         Update(k, j) :
19             for i = k + 1 to n − 1 do
20                 A[i, j] ← A[i, j] + buffer[i − k − 1] × A[k, j]
```

ALGORITHM 4.8: Pipelined LU factorization algorithm on a ring of processors.

4.4.2 Pipelining on the Ring

The algorithm we developed in the previous section does not rely on the fact that the processors are arranged in a logical ring. The algorithm just uses a broadcast primitive that could be implemented on any logical topology. This makes our algorithm portable, which is good, but, on the other hand, it does not exploit the ring topology to the best of its potential. In particular, the $n-1$ broadcasts are not overlapped with any computations. On a ring topology and with a cyclic allocation of the matrix columns to the processors, realizing a good overlap is, however, natural. In fact, one can almost directly insert the source code of the broadcast algorithm into the LU algorithm!

We modify the algorithm so that as soon as a processor receives column k (at step k), it sends it to its successor. The processor performs this communication before starting to update the columns it holds in memory. The algorithm is shown in Algorithm 4.8. An application execution is depicted in Figure 4.8, where each column corresponds to one of four processors and time progresses from top to bottom. For the sake of illustration, we depict column preparation, sending, receiving, and column updating as taking the same

P_0	P_1	P_2	P_3
Prep(0)			
Send(0)	Receive(0)		
U(0,4)	Send(0)	Receive(0)	
U(0,8)	U(0,1)	Send(0)	Receive(0)
U(0,12)	U(0,5)	U(0,2)	U(0,3)
	U(0,9)	U(0,6)	U(0,7)
	U(0,13)	U(0,10)	U(0,11)
	Prep(1)	U(0,14)	U(0,15)
	Send(1)	Receive(1)	
	U(1,5)	Send(1)	Receive(1)
Receive(1)	U(1,9)	U(1,2)	Send(1)
U(1,4)	U(1,13)	U(1,6)	U(1,3)
U(1,8)		U(1,10)	U(1,7)
U(1,12)		U(1,14)	U(1,11)
		Prep(2)	U(1,15)
		Send(2)	Receive(2)
Receive(2)		U(2,6)	Send(2)
Send(2)	Receive(2)	U(2,10)	U(2,3)
U(2,4)	U(2,5)	U(2,14)	U(2,7)
U(2,8)	U(2,9)		U(2,11)
U(2,12)	U(2,13)		U(2,15)
			Prep(3)
Receive(3)			Send(3)
Send(3)	Receive(3)		U(3,7)
U(3,4)	Send(3)	Receive(3)	U(3,11)
U(3,8)	U(3,5)	U(3,6)	U(3,15)
U(3,12)	U(3,9)	U(3,10)	
	U(3,13)	U(3,14)	

FIGURE 4.8: Example execution of the pipelined LU algorithm on a ring.

amount of time. This figure makes the pessimistic assumption that communication cannot be overlapped with computation at a processor, even though the two may be independent from each other. Even with this assumption one can see that some communication phases take place concurrently with certain computation phases. While column k travels on the ring, the first processors have started their column updates while the last processors are still in the process of receiving column k. As a result, processor P_q is ahead of processor P_{q+1}, which makes it possible to pipeline the communication and computation phases as seen in the figure. After each p steps there is a gap, also called a pipeline bubble. These bubbles happen when the processor that is the most behind holds the column that must be sent to all the other processors.

Assume now, like we have done throughout this chapter, that our processors can perform concurrent computation and communication. This assumption makes it possible for the column for step $k+1$ to be communicated along the ring completely asynchronously while the computations for step k are taking place. In this case, communication is, for the most part, fully overlapped with

P_0	P_1	P_2	P_3
PREP(0)			
SEND(0) U(0,4)	RECEIVE(0)		
U(0,8)	SEND(0) U(0,1)	RECEIVE(0)	
U(0,12)	U(0,5)	SEND(0) U(0,2)	RECEIVE(0)
	U(0,9)	U(0,6)	U(0,3)
	U(0,13)	U(0,10)	U(0,7)
	PREP(1)	U(0,14)	U(0,11)
	SEND(1) U(1,5)	RECEIVE(1)	U(0,15)
	U(1,9)	SEND(1) U(1,2)	RECEIVE(1)
RECEIVE(1)	U(1,13)	U(1,6)	SEND(1) U(1,3)
U(1,4)		U(1,10)	U(1,7)
U(1,8)		U(1,14)	U(1,11)
U(1,12)		PREP(2)	U(1,15)
		SEND(2) U(2,6)	RECEIVE(2)
RECEIVE(2)		U(2,10)	SEND(2) U(2,3)
SEND(2) U(2,4)	RECEIVE(2)	U(2,14)	U(2,7)
U(2,8)	U(2,5)		U(2,11)
U(2,12)	U(2,9)		U(2,15)
	U(2,13)		PREP(3)
RECEIVE(3)			SEND(3) U(3,7)
SEND(3) U(3,4)	RECEIVE(3)		U(3,11)
U(3,8)	SEND(3) U(3,5)	RECEIVE(3)	U(3,15)
U(3,12)	U(3,9)	U(3,6)	
PREP(4)	U(3,13)	U(3,10)	
SEND(4) U(4,8)	RECEIVE(4)	U(3,14)	

FIGURE 4.9: Example execution of the pipelined LU algorithm on a ring, with concurrent communication and computation at each processor.

computation and pipeline bubbles can be reduced. Application execution, with this assumption, is depicted in Figure 4.9. Note, however, that while the idle periods are somewhat reduced, they are not completely eliminated.

4.4.3 Look-Ahead Algorithm

Is it possible to improve on the previous algorithm? It turns out that it is, based on a rather clever way of interleaving communication and computation phases. Consider processor P_1:

1. At step $k = 0$, processor P_1 receives column 0 from processor P_0.
2. It immediately sends it to its successor, processor P_2.
3. Then, it updates its columns by executing UPDATE$(0, j)$ for all j such that $j = 1 \bmod p$.
4. It then moves on to step $k = 1$ and prepares column $k = 1$ by executing PREP(1).

5. It sends column $k = 1$ to processor P_2.

6. It updates its own columns via executions of $\textsf{Update}(1, j)$.

We can improve the above sequence. The key insight is that, instead of executing all the $\textsf{Update}(0, j)$ followed by $\textsf{Prep}(1)$, and then sending column $k = 1$, processor P_1 can execute $\textsf{Update}(0, 1)$, then $\textsf{Prep}(1)$, then send column $k = 1$, and then execute all remaining $\textsf{Update}(0, j)$ for $j = p+1, 2p+1, 3p+1$, etc. From the perspective of the other processors, both sequences of operations are equivalent. But in the second sequence, column $k = 1$ is sent earlier, which ends up greatly reducing and sometimes removing the pipeline bubbles mentioned earlier. The basic principle here is, again, to perform communications as early as possible. The pseudo-code of the look-ahead algorithm is shown in Algorithm 4.9, omitting the code for the Prep() and Update() inlined functions. Consequently, we have added an extra argument to these functions to specify the buffer array that is to be used. Note the use of two different buffers in this version of the algorithm, so that a processor can call Prep() in step k of the algorithm and then place Update() calls that are "left over" from step $k - 1$.

Application execution when using the look-ahead algorithm described above is depicted in Figure 4.10. By comparing this figure with Figure 4.9 we can see that the new algorithm is indeed more efficient than the simple pipelined algorithm from the previous section because it removes most of the pipeline bubbles. Remember that our depiction of the execution makes the unrealistic assumption that column preparation, column update, and column communication all take the same time. Nevertheless, the ability of this look-ahead algorithm to reduce idle time is well observed in practice.

4.5 A Second Look at Stencil Applications

In Section 4.3 we studied a stencil application in which a cell value is updated using the *updated* value of its North and West neighbor cells. Many other stencils are possible. For instance, stencils that use the *non-updated* values of a cell's West, East, North, and South neighbors are commonly employed in the Jacobi numerical method [97]:

$$c_{new} \leftarrow \textsf{Update}(c_{old}, W_{old}, E_{old}, N_{old}, S_{old}) \; .$$

With this stencil, which is applied on a 2-D $n \times n$ domain, it is not possible to update the domain in place without overwriting cell values prematurely. Instead, two domains must be stored in memory: the original one, which we call A, and an empty one into which updated values will be stored, which we

```
1  var A: array[0..n-1,0..r-1] of real
2  var received_buffer: array[0..n-1] of real
3  var sent_buffer: array[0..n-1] of real
4  q ←My_Num()
5  p ←Num_Procs()
6  for k = 0 to n-2 do
7      if k = q mod p then
8          Prep(k, sent_buffer)
9          Send(sent_buffer, n-k-1)
10         if k > 1 then
11             forall j such that j mod p = q, j > k do
12                 Update(k-1, j, received_buffer)
13         forall j such that j mod p = q, j > k do
14             Update(k, j, sent_buffer)
15     else
16         Recv(received_buffer, n-k-1)
17         if q ≠ k-1 mod p then
18             Send(received_buffer, n-k-1)
19         if q = k+1 mod p then
20             Update(k, k+1, received_buffer)
21         else
22             forall j such that j mod p = q, j > k do
23                 Update(k, j, received_buffer)
```

ALGORITHM 4.9: Look-ahead LU factorization algorithm on a ring of processors.

call B. As a result, the execution of the algorithm is in fact simpler than that for the stencil in Section 4.3, and so is the data distribution.

4.5.1 Parallelization on a Unidirectional Ring

We consider our usual unidirectional ring of p processors. As done previously, we assume that p divides n to simplify the performance analysis. The most natural data distribution is to allocate $r = n/p$ consecutive rows of the domain to each processor, exactly like we did for the matrix in Section 4.1. Therefore, each processor has the following declarations:

```
var A, B: array[0..r-1,0..n-1] of real;
```

P_0	P_1	P_2	P_3
PREP(0)			
SEND(0) U(0,4)	RECEIVE(0)		
U(0,8)	SEND(0) U(0,1)	RECEIVE(0)	
U(0,12)	PREP(1)	SEND(0) U(0,2)	RECEIVE(0)
	SEND(1) U(0,5)	RECV(1) U(0,6)	U(0,3)
	U(0,9)	SEND(1) U(0,10)	RECV(1) U(0,7)
RECEIVE(1)	U(0,13)	U(0,14)	SEND(1) U(0,11)
U(1,4)	U(1,5)	U(1,2)	U(0,15)
U(1,8)	U(1,9)	PREP(2)	U(1,3)
U(1,12)	U(1,13)	SEND(2) U(1,6)	RECV(2) U(1,7)
RECEIVE(2)		U(1,10)	SEND(2) U(1,11)
SEND(2) U(2,4)	RECEIVE(2)	U(1,14)	U(1,15)
U(2,8)	U(2,5)	U(2,6)	U(2,3)
U(2,12)	U(2,9)	U(2,10)	PREP(3)
RECEIVE(3)	U(2,13)	U(2,14)	SEND(3) U(2,7)
SEND(3) U(3,4)	RECEIVE(3)		U(2,11)
PREP(4)	SEND(3) U(3,5)	RECEIVE(3)	U(2,15)
SEND(4) U(3,8)	RECV(4) U(3,9)	U(3,6)	U(3,7)
U(3,12)	SEND(4) U(3,13)	RECV(4) U(3,10)	U(3,11)
U(4,8)	U(4,5)	SEND(4) U(3,14)	RECV(4) U(3,15)

FIGURE 4.10: Example execution of the look-ahead LU algorithm on a ring, with concurrent communication and computation at each processor.

where array A contains initial cell values, and array B is to be filled with updated cell values. Each processor can update rows 1 to $r-2$ of B using the values stored in A. However, the updating of the first row of array B at processor P_q requires values stored in the last row of array A at processor P_{q-1}. Similarly, the updating of the last row of array B at processor P_q requires values stored in the first row of array A at processor P_{q+1}. For the sake of simplicity, we assume that the domain "wraps around" in the sense that processors P_0 and P_{q-1} exchange rows. This assumption removes the need for the $q=0$ or $q=p-1$ tests that cluttered the algorithms in Section 4.3. Furthermore, this assumption does not decrease performance. Indeed, since the execution goes only as fast as the slowest processor, communication savings on only two out of p processors does not reduce execution time. In practice, it is actually common to avoid special cases and have all the processors perform the exact same communications. This leads to simpler algorithms, as we will see, and data that have been communicated superfluously can be explicitly ignored by some processors when they perform their computations.

With the above assumption, each processor must send data to both its successor and its predecessor on the ring. Given that we use a unidirectional ring, sending data to a predecessor amounts to sending those data all around the ring with $p-1$ hops, with intermediate processors merely relaying messages. Our parallel stencil algorithm on the unidirectional ring is shown in Algorithm 4.10.

```
1  var A, B: array[0..r-1,0..n-1] of real
2  var fromP, fromS: array[0..n-1] of real
3  p ←NUM_PROCS()

4  begin                                                  { Phase 1: communications }
5  |   tempS ←ADDR(A[0,0])
6  |   for k = 1 to p-2 do
7  |   |   SEND(tempS, n) || RECV(tempR, n)
8  |   |_  tempS ↔tempR
9  |   SEND(tempS, n) || RECV(fromS, n)
10 |   SEND(ADDR(A[r-1,0]), n) || RECV(fromP, n)
11 end
12 ||
13 begin                                                  { Phase 1: computations }
14 |   for i = 1 to r-2 do
15 |   |   B[i,0] ←UPDATE(A[i,0], Nil, A[i,1], A[i-1,0], A[i+1,0])
16 |   |   for j = 1 to n-2 do
17 |   |   |_ B[i,j] ←UPDATE(A[i,j], A[i,j-1], A[i,j+1], A[i-1,j], A[i+1,j])
18 |   |   B[i,n-1] ←
19 |   |_     UPDATE(A[i,n-1], A[i,n-2], Nil, A[i-1,n-1], A[i+1,n-1])
20 end

21 begin                                                  { Phase 2 }
22 |   B[0,0] ←UPDATE(A[0,0], Nil, A[0,1], fromP[0], A[1,0])
23 |   B[r-1,0] ←UPDATE(A[r-1,0], Nil, A[r-1,1], A[r-2,0], fromS[0])
24 |   for j = 1 to n-2 do
25 |   |   B[0,j] ←UPDATE(A[0,j], A[0,j-1], A[0,j+1], fromP[j], A[1,j])
26 |   |   B[r-1,j] ←
27 |   |_     UPDATE(A[r-1,j], A[r-1,j-1], A[r-1,j+1], A[r-2,j], fromS[j])
28 end

29 B[0,n-1] ←
       UPDATE(A[0,n-1], A[0,n-2], Nil, fromP[n-1], A[1,n-1])
30 B[r-1,n-1] ←
       UPDATE(A[r-1,n-1], A[r-1,n-2], Nil, A[r-2,n-1], fromS[n-1])
```

ALGORITHM 4.10: Algorithm for our second stencil application on a unidirectional ring of processors.

This algorithm proceeds in two phases: a first phase from statements 4 to 20 and a second phase from statements 21 to 28. The first phase consists of two concurrent sub-phases. In the first sub-phase, statements 4 to 11, each processor communicates the top and bottom rows of array A to its predecessor and its successor, respectively. In the second sub-phase, statements 13 to 20, each processor computes all rows of B but for the top and bottom rows. As done previously in this chapter, we assume that processors can compute and communicate concurrently, and can send and receive concurrently. Removing these assumptions would simply mean removing the || sign in the program and changing our performance analysis. By the beginning of the second phase of the algorithm each processor has received the bottom row from its predecessor and the top row from its successor. Therefore, in the second phase, each processor can finally compute the top and bottom rows of array B. We detail the critical steps of this algorithm below.

Statement 2 in the program declares array *fromP* for storing the bottom row sent by the predecessor processor, and array *fromS* for storing the top row sent by the successor processor. Statement 5 sets the pointer *tempS* to the beginning of the first row of array A, this processor's top row. Note our use of a function ADDR to access the address of an array element (equivalent to the & operator in C, for instance), and our reliance on the fact that array A is stored in row-major fashion as explained in Section 4.4.1. At the first iteration of the for loop at statement 6 the processor sends its top row to its successor and receives its predecessor's top row. At every following iteration the processor forwards what it receives from its predecessor to its successor. After $p-2$ iterations, the final step in statement 9 consists in doing a last forwarding and finally receiving the top row of the processor's successor. In statement 10 the processor sends its bottom row to its successor, and receives its predecessor's bottom row. As usual, we assume that the SEND is non-blocking while the RECV is blocking.

After completion of statements 4 to 11, each processor has received the two rows needed from its predecessor and its successor. While all these communications are happening we assume that computation can take place at each processor. Statements 13 to 20 just apply the stencil to compute all rows of B that can be computed without the bottom row of A from the predecessor processor and without the top row of A from the successor processor, that is, all rows of B but row 0 and row $r-1$. This is done with the for loop in statement 14. Each iteration of this loop deals with the leftmost and the rightmost elements of row i of B separately as the former has no West neighbor and the latter has no East neighbor.

Once both the previously described communication and computation sub-phases have completed, with statements 21 to 28 the processor computes row 0 and row $r-1$ of array B using the received values stored in arrays $fromP$ and $fromS$, respectively. The computations are similar to those in the for loop in statement 14.

One can easily compute the execution time of the algorithm. As usual, let w be the computation time for one basic cell update computation, b the time to communicate one cell value in steady state, and L the network startup cost. The communication sub-phase of the first phase consists of a sequence of p concurrent sends and receives of a row of n cell values, and thus takes time $pL + pnb$. The computation sub-phase of the first phases consists in computing $r - 2$ rows, and thus takes time $(r - 2)nw$. Therefore, given that $r = n/p$, the first phase of the algorithm takes time

$$\max\{pL + pnb, (n^2/p - 2n)w\} .$$

The second phase of the algorithm consists in computing two rows, and thus takes time $2nw$. The overall execution time of the algorithm, $T(n,p)$, is

$$T(n,p) = \max\{pL + pnb, (n^2/p - 2n)w\} + 2nw . \quad (4.1)$$

When n becomes large, $T(n,p) \sim wn^2/p$. Since the sequential execution time is wn^2, the parallel algorithm's asymptotic efficiency is 1.

4.5.2 Parallelization on a Bidirectional Ring

The previous algorithm is overly complicated. Indeed, because we used a unidirectional ring, we had to send messages from a processor to its predecessor around the ring with $p - 1$ hops. In spite of its optimal asymptotic parallel efficiency, in practice the first term of the maximum in Equation (4.1) may be large (for instance, if w and n are relatively small). Furthermore, and perhaps more importantly, using a unidirectional ring makes the code for the communication phase rather cumbersome.

Since as parallel algorithm designers we can choose which logical topology to use, we can just decide to use a bidirectional ring if we feel it is more appropriate. Let us then redefine the SEND and RECV functions to allow direct communication to both the predecessor and the successor processor. We just add an argument, with value *pred* or *succ*, to both calls to specify the direction of the communication. For instance, SEND(*pred*, *addr*, 1) is used by a processor to send one data element stored at address *addr* to its predecessor. The predecessor must place a matching call such as RECV(*succ*, *addr*, 1). With this new logical topology, the communication phase of the algorithm is rewritten as

```
...
SEND(pred, ADDR(A[0,0]), n) || RECV(succ, fromS, n)
SEND(succ, ADDR(A[r-1,0]), n) || RECV(pred, fromP, n)
...
```

With this new communication phase, the algorithm's execution time becomes

$$T(n,p) = \max\{2(L + nb), (n^2/p - 2n)w\} + 2nw .$$

The program is much simpler. Furthermore, the communication time is lower (it no longer depends on p as messages only use one hop), and thus the maximum in the above equation is likely lower in practice. We conclude that the bidirectional ring is a very natural match for our stencil, and in fact it is the one used in practice for this type of stencil.

4.6 Implementing Logical Topologies

In this chapter, we have seen our first example of changing a logical topology to better fit the requirements of an algorithm. We even made the statement that as designers of parallel algorithms we can choose the logical topology. Some readers may wonder how this can be done in practice. It turns out that message passing libraries, for instance, ones implemented according to the MPI standard [109], assign an integer identifier to each processor and allow communications between any pair of processors. This is done via the SEND and RECV functions that take processor identifiers as arguments for specifying communication sources and destinations. As a result, the developer has complete control over which logical communication paths are used by the algorithm. Using a logical topology restricts communications to only a few paths, which makes algorithm design simpler (provided an appropriate logical topology is chosen). When defining a logical topology, one just arbitrarily defines for each processor which processors are its neighbors. One then decides which way communication can flow on each logical link between two neighboring processors. The logical topology is then implemented via a set of functions by which processors identify their neighbors. For instance, one could implement a function LEFTNEIGHBOR(q) that, given a processor's identifier q, returns the identifier of the processor that is the left neighbor of processor q in the logical topology, for some definition of "left." A call such as SEND(LEFTNEIGHBOR(MY_NUM()), $addr$, 1) then implements communication on the logical topology. While in this chapter we have only used a ring topology, many others are used in practice. In the next chapter, for instance, we will see logical 2-D grid and torus topologies. Other common topologies include single-level and multi-level trees, or topologies that combine any of the above.

A difficult question is that of matching the logical topology to the physical topology. Indeed, it is often the case that an algorithm can be implemented using different logical topologies. The SCALAPACK library [36], for instance, allows the user to choose among several logical topologies for executing parallel linear algebra algorithms. The common wisdom is that a logical topology that resembles the underlying physical topology should lead to good performance. If the physical network topology is, say, a tree, implementing a logical ring topology would cause several performance difficulties. For instance, the communication times between neighboring processors on the logical ring would vary among pairs of processors, which could lead to unexpected idle time in the execution of a parallel algorithm. Paradoxically, the point of using a logical topology is often to hide the complexity of the underlying physical topology. Choosing a logical topology that closely resembles the physical topology could make algorithm design very difficult. The goal is then to choose a logical topology that at the same time makes algorithm design natural and is reasonably compatible with the underlying physical topology. Achieving this goal can be straightforward or challenging, depending on the application and on the platform. Note that some modern supercomputers provide multiple physical networks to better support various communication patterns and thus various logical topologies. Also, platforms such as commodity clusters often use switched networks that are akin to fully connected (but possibly hierarchical) topologies. Such physical topologies often support various logical topologies reasonably well, but often extensive benchmarking is required to determine the best logical topology for a given algorithm on a given platform. In this book we do not study this question of matching logical and physical topology in depth because it depends on the physical characteristics of the parallel platforms at hand, which vary across platforms and across platform generations. However, the logical topologies in this chapter and the ones we study in the next chapter are known to be useful in the majority of practical scenarios.

4.7 Distributed vs. Centralized Implementations

When designing the parallel algorithms in this chapter we have always assumed that data (i.e., arrays) were already distributed among processors at the onset of the algorithm execution. For instance, in our matrix-vector multiplication algorithm in Section 4.1, we assumed that at the onset of the parallel execution each processor magically holds a particular subset of the rows of matrix A. The reader may wonder how this distribution of data occurs in the first place and whether the algorithm should distribute the data directly. One can distinguish two approaches: *distributed* and *centralized*. In this book we have opted for always using the distributed approach. As a result, our algorithm implementations do not perform any data distribution and we assume that the data are already distributed. In the centralized approach one assumes instead that data reside at a single "master" location (a processor or, if the data are large, a file on disk). The algorithm must then distribute the data among available processors, apply the desired parallel algorithm, and "un-distribute" (gather) the results of the computation back to the master location. While data distribution has to occur one way or another, the question of which approach is best arises when developing a general-purpose library of routines that implement parallel, distributed-memory algorithms. In other terms, should the library user or the library itself be responsible for distributing data?

One advantage of the centralized approach is that each library routine can enforce whichever data distribution is deemed best by the library developer. As a result, the library user does not have to worry about data distribution, which is another advantage. Unfortunately, for best performance, the choice of data distribution for a given algorithm must account for the particular underlying physical topology. Since this topology is not known at library development time, the library developer may wish to implement multiple versions of each routine for different possible data distributions. The user then would have to choose which routine to call depending on the underlying platform. As discussed in Section 4.6, this choice may be difficult without extensive benchmarking. The main drawback of the centralized approach, however, emerges when the user calls multiple library routines in sequence with the same data. For instance, imagine that a user wishes to compute a matrix product $C = A \times B$, followed by an LU factorization of matrix C, followed by a sequence of $(LU)x_i = b_i$ linear system resolutions. Data will be distributed and un-distributed at each library call, which typically leads to prohibitive overhead. Unfortunately, such use of a library is common and, as a result, most library developers opt for a distributed implementation.

For the same reasons as those given in the case of a centralized approach, routines in a distributed library implementation should support multiple data distribution schemes for best performance. This leads to an interesting trade-off. Indeed, the user may have called a library routine and used a given distribution for a given matrix. When calling another routine for which a different distribution would be best, a trade-off arises: Is it better to incur the overhead of re-distributing the matrix from the old distribution to the new distribution, which can be significant, or is it better to call the next routine with a sub-optimal data distribution? This trade-off is difficult to resolve, especially at library development time. Consequently, the typical approach is to provide routines that can execute the necessary algorithms using a variety of data distribution schemes, and routines that can redistribute data from any distribution scheme to any other. The user is then given maximum flexibility for data distributions and data re-distributions. By contrast, a centralized implementation offers at best marginal flexibility. Some users argue that this flexibility only leads to endless dilemmas, but it is the price to pay for the hope of achieving best performance.

4.8 Summary of Algorithmic Principles

Throughout this chapter we have highlighted several principles that are commonly employed when designing and implementing distributed memory algorithms to reduce processor idle time and/or to reduce communication overhead. While our focus was the ring topology, these principles are generally applicable. Unfortunately, as seen in previous sections, these principles often conflict with each other and force the algorithm developer to understand and choose among multiple trade-offs. We summarize the principles below:

- *Sending data in bulk* – Aggregating communication operations reduces communication overhead due to network latencies.
- *Sending data early* – Sending data as early as possible allows processors to start computing as early as possible, which reduces processor idle time.
- *Overlapping communication and computation* – It is always a good idea to overlap communication and computation in order to hide communication time.
- *Block data distribution* – Using a block distribution by which processors are assigned blocks of contiguous data elements reduces the amount of communication (and can often improve cache reuse).
- *Cyclic data distribution* – Using a cyclic distribution by which data elements are interleaved among processors makes it possible to achieve better load balancing between processors, which reduces processor idle time.

Bibliographical Notes

In terms of algorithms, a seminal reference is the book by Kumar et al. [76]. The content of this chapter's section on matrix-vector and matrix-matrix multiplications belongs to popular parallel computing knowledge. The discussion and performance modeling of stencil applications are inspired by the articles by Miguet and Robert [90, 91]. Finally, the parallel Gaussian elimination algorithm used in our LU factorization is a classic (see [101] and other cited references).

4.9 Exercises

We start with two classical linear algebra sequential algorithms and their parallelization on a ring of processors. The third exercise revisits the definition of parallel speedup and introduces the important notion of scaled speedup. The fourth exercise is a hands-on implementation of a matrix multiplication algorithm on a parallel platform using the MPI message-passing standard.

◇ Exercise 4.1 : Solving a Triangular System

Consider the problem of solving linear system $Ax = b$, where A is a lower triangular matrix of rank n and b is a vector with n components. We assume the computing platform is a unidirectional ring of processors $P_0, P_1, \ldots, P_{p-1}$. For simplicity we assume that p divides n.

1. We wish to distribute columns of matrix A among the processors. What is likely the best strategy? Give the pseudo-code for the corresponding parallel algorithm.

2. Now we wish to distribute rows of A among processors. What is likely the best strategy? Give the pseudo-code for the corresponding parallel algorithm.

◇ Exercise 4.2 : Givens Rotation

To triangularize a matrix A of rank n in a way that is numerically stable, one can use Givens rotations. The basic operation, which we denote $\textsf{Rot}(i, j, k)$, consists in combining rows i and j of the matrix. These two rows must both have $k-1$ leading zeros and the goal of the combination is to zero out element $a_{j,k}$:

$$\begin{pmatrix} 0 & \ldots & 0 & \mathbf{a'_{i,k}} & a'_{i,k+1} & \cdots & a'_{i,n} \\ 0 & \ldots & 0 & \mathbf{0} & a'_{j,k+1} & \cdots & a'_{j,n} \end{pmatrix} \leftarrow \begin{pmatrix} \cos\theta & -\sin\theta \\ \sin\theta & \cos\theta \end{pmatrix} \begin{pmatrix} 0 & \ldots & 0 & \mathbf{a_{i,k}} & a_{i,k+1} & \cdots & a_{i,n} \\ 0 & \ldots & 0 & \mathbf{a_{j,k}} & a_{j,k+1} & \cdots & a_{j,n} \end{pmatrix}.$$

The reader can easily compute the value of θ so that element $a_{j,k}$ is zeroed out. The sequential algorithm can be written as follows:

```
1 GIVENS(A)
2   for k = 1 to n − 1 do
3     for i = n DownTo k + 1 do
4       ROT(i − 1, i, k)
```

Consider that one rotation executes in one unit of time, independently of the value of k.

1. Give the pseudo-code for this algorithm on a ring of $p = n$ processors. Give a performance model.

2. Write the pseudo-code for this algorithm on a (bidirectional) ring of only $\lfloor \frac{n}{2} \rfloor$ processors.

◇ Exercise 4.3 : Parallel Speedup

This exercise has the reader "discover" several fundamental laws of parallel computing (Amdahl, Gustafson).

1. Consider a sequential computation that needs to be parallelized, but with a fraction f of the sequential execution time that cannot be parallelized. Show that the parallel speedup is bounded by $1/f$ no matter how many processors are used. Assume that the part of the execution that can be parallelized can be parallelized perfectly.

2. Consider a matrix computation for $n \times n$ matrices so that:

- the number of (arithmetic) operations to execute is n^α, where α is a constant;
- the number of matrix elements to store in memory is βn^2, where β is a constant;
- the number of non-parallelizable operations (in this exercise we assume they are simply I/O operations) is γn^2, where γ is a constant.

Let M be the maximum memory size on a processor. What is the parallel speedup with p processors for a "large problem," i.e., one that uses all available memory? What are the implications?

3. Give examples of scenarios in which one may observe a super-linear speedup, i.e., a parallel efficiency strictly superior to 1.

Exercise 4.4 : MPI Matrix-Matrix Multiplication

1. Implement the matrix multiplication algorithm described in Section 4.2 using MPI, for $n \times n$ matrices, on a ring of p processors. For simplicity, assume that p divides n. Each processor initially holds a piece of matrix A and a piece of matrix B in two arrays. Initialize these arrays with matrix elements defined as $A_{i,j} = i$ and $B_{i,j} = i + j$ (these are global indices starting at 0). At the end of the execution, each processor holds a piece of matrix C stored in a third array. Your program should check the validity of the results (indeed, they can be computed analytically as $c_{i,j} = i \times n \times (n-1)/2 + i \times j \times n$, using global indices).

2. Modify your program from question 1 so that the "p divides n" assumption is no longer necessary.

3. Run your program on a parallel platform, e.g., a cluster, and plot the parallel speedup and efficiency for two, four, and eight processors as functions of the matrix size, n. Each data point should be obtained as an average over ten runs.

4. If you have not done so, experiment with non-blocking MPI communication and see whether there is any impact on your program's performance when you overlap communication and computation.

4.10 Answers

Exercise 4.1 (Solving a Triangular System)

▷ **Question 1.** If we distribute the columns of A among the processors, we must parallelize operations within the rows of A. (Parallelizing the operations within a column would lead to sequential execution.) For each row, each processor should contribute some computations for the columns it holds. Consider the typical sequential algorithm

```
1 for i = 0 to n − 1 do
2     s ← 0
3     for j = 0 to i − 1 do
4         s ← s − a_{i,j} × x_j
5     x_i ← (b_i − s)/a_{i,i}
```

This algorithm is well suited to parallelizing row operations. At step i, one computes the scalar product of the elements of the i-th row of A that are before the diagonal and of the elements of vector x that have already been computed. To balance the load at each step, one must distribute the columns of A among the processors in a cyclic fashion. More precisely, one allocates column i of matrix A and component b_i of vector b to processor $\text{ALLOC}(i) = i \bmod p$. This processor is responsible for the computation of the x_i components of vector x. With this allocation, we obtain the parallel algorithm

```
1 for i = 0 to n − 1 do
2     t ← 0
3     forall j ∈ MyCols, j < i do
4         t ← t + a_{i,j} × x_j
5     s = GATHER(ALLOC(i), t, 1)
6     if i ∈ MyCols then
7         x_i ← (b_i − s)/a_{i,i}
```

In this algorithm, we use the variable *MyCols* to denote the set of the indices of the columns allocated to each processor. The GATHER operation is used so that the sum of the partial scalar products, computed locally at each processor, is computed and stored in the memory of processor ALLOC(i). The above pseudo-code is written using only global indices, and we let the reader write it using local array indices. This can be done using the same technique as for the LU factorization algorithm in Section 4.4.

▷ **Question 2.** We only sketch the main idea of the solution, which is similar to that for Question 1. If matrix rows are distributed to processors, we need to operate on a matrix column at each step, so that each processor can contribute by updating the fraction of the column corresponding to the processor's local rows. This implies swapping the two loops in the sequential version:

```
1 for j = 0 to n − 1 do
2     x[j] ← b[j]/a[j, j]
3     for i = j + 1 to n − 1 do
4         b[i] ← b[i] − a[i, j] × x[j]
```

As before, a cyclic allocation will nicely balance the work among the processors. One allocates row i of matrix A and component b_i to processor $\mathsf{ALLOC}(i) = i \bmod p$, which is responsible for the computation of x_i. With this allocation, we obtain the parallel algorithm

```
1 for j = 0 to n − 1 do
2     if j ∈ MyRows then
3         x[j] ← b[j]/a[j, j]
4     BROADCAST(()alloc(j),x[j],1)
5     for i ∈ MyRows, i > j do
6         b[i] ← b[i] − a[i, j] × x[j]
```

Exercise 4.2 (Givens Rotation)

▷ **Question 1.** Implementing the algorithm with n processors, which we denote $P_i, i = 1, \ldots, n$, is straightforward: Processor P_k is responsible for computing all $\textsc{Rot}(i-1, i, k)$ rotations. One can then view the execution as feeding the rows of the matrix to the ring starting with the last row as seen below ($\boxed{i}$ represents row i of matrix A):

$$\boxed{1}\,\boxed{2}\,\boxed{3}\,\boxed{4}\,\boxed{5}\,\boxed{6}\,\boxed{7}\,\boxed{8} \rightarrow P_1 \rightarrow P_2 \rightarrow P_3 \rightarrow P_4 \rightarrow P_5 \rightarrow P_6 \rightarrow P_7 \rightarrow P_8.$$

When a row arrives at a processor, it stays there for one step. When a second row arrives at that processor, it is combined with the first row. The first row is then sent to the next processor and the combined row takes its place. We obtain the following execution, shown as an example for $n = 8$ (each table element shows which matrix row each processor contains at each algorithm step):

Step	P_1	P_2	P_3	P_4	P_5	P_6	P_7	P_8
$t = 1$	$\boxed{8}$							
$t = 2$	Rot$(7, 8, 1)$							
$t = 3$	Rot$(6, 7, 1)$	$\boxed{8}$						
$t = 4$	Rot$(5, 6, 1)$	Rot$(7, 8, 2)$						
$t = 5$	Rot$(4, 5, 1)$	Rot$(6, 7, 2)$	$\boxed{8}$					
$t = 6$	Rot$(3, 4, 1)$	Rot$(5, 6, 2)$	Rot$(7, 8, 3)$					
$t = 7$	Rot$(2, 3, 1)$	Rot$(4, 5, 2)$	Rot$(6, 7, 3)$	$\boxed{8}$				
$t = 8$	Rot$(1, 2, 1)$	Rot$(3, 4, 2)$	Rot$(5, 6, 3)$	Rot$(7, 8, 4)$				
$t = 9$	$\boxed{1}$	Rot$(2, 3, 2)$	Rot$(4, 5, 3)$	Rot$(6, 7, 4)$	$\boxed{8}$			
$t = 10$	$\boxed{1}$	$\boxed{2}$	Rot$(3, 4, 3)$	Rot$(5, 6, 4)$	Rot$(7, 8, 5)$			
$t = 11$	$\boxed{1}$	$\boxed{2}$	$\boxed{3}$	Rot$(4, 5, 4)$	Rot$(6, 7, 5)$	$\boxed{8}$		
$t = 12$	$\boxed{1}$	$\boxed{2}$	$\boxed{3}$	$\boxed{4}$	Rot$(5, 6, 5)$	Rot$(7, 8, 6)$		
$t = 13$	$\boxed{1}$	$\boxed{2}$	$\boxed{3}$	$\boxed{4}$	$\boxed{5}$	Rot$(6, 7, 6)$	$\boxed{8}$	
$t = 14$	$\boxed{1}$	$\boxed{2}$	$\boxed{3}$	$\boxed{4}$	$\boxed{5}$	$\boxed{6}$	Rot$(7, 8, 7)$	
$t = 15$	$\boxed{1}$	$\boxed{2}$	$\boxed{3}$	$\boxed{4}$	$\boxed{5}$	$\boxed{6}$	$\boxed{7}$	$\boxed{8}$

At step $2n - 1$, each processor contains a row of the triangularized matrix. We let the reader write the actual pseudo-code and the performance model.

▷ **Question 2.** The idea is to "fold in half" the ring used in the previous question so that rows travel first to the right and then to the left.

Exercise 4.3 (Parallel Speedup)

▷ **Question 1.** Let T_1 be the execution time with one processor, which we decompose into $T_1 = T_{\text{seq}} + T_{\text{par}}$, where T_{seq} is the sequential part and T_{par} is the parallelizable part. By hypothesis, $T_{\text{seq}} = f \cdot T_1$. With p processors, the best we can do is divide T_{par} by a factor p (perfect parallelism, no overhead). The total execution time T_p with p processors is thus

$$T_p \geqslant T_{\text{seq}} + \frac{T_{\text{par}}}{p} = f \cdot T_1 + \frac{(1-f)T_1}{p},$$

and the speedup is bounded by

$$S_p = \frac{T_1}{T_p} \leqslant \frac{1}{f + \frac{1-f}{p}} \leqslant \frac{1}{f}.$$

As a consequence, if $f = 5\ \%$, the potential speedup is limited to 20, even with $1,000$ processors. For a problem of fixed size, the degree of parallelism that can be achieved is bounded independently of the number of available processors. This result is known as Amdahl's law [6], which seems to preclude massive parallelism.

▷ **Question 2.** The sequential execution time for a problem of size n is $T_1(n) = n^\alpha w + \gamma n^2 \tau_{\text{i/o}}$, where w and $\tau_{\text{i/o}}$ denote the processor's compute speed and I/O speed. With one processor, the maximum feasible problem size, $n_{\max}(1)$, is defined by $\beta(n_{\max}(1))^2 = M$.

With p processors, one can compute a larger problem since one can aggregate all the processors' memories: $n_{\max}(p) = \sqrt{p} \cdot n_{\max}(1)$. How do we then compute the parallel speedup for a problem of size $n_{\max}(p)$, that is, for a problem too large to be executed on a single processor? The simple idea is to *scale* the speedup: We compute $A(p)$, the mean time to perform an arithmetic operation with p processors for a problem of maximum size $n_{\max}(p)$. The new definition of the speedup,

proposed by Gustafson [65], is $S_p = \frac{A(1)}{A(p)}$.

For our matrix computation, assuming perfect parallelization of the arithmetic computations, we have

$$A(1) = \frac{n^\alpha w + \gamma n^2 \tau_{\mathrm{i/o}}}{n^\alpha} \qquad \text{with } \beta n^2 = M, \text{and}$$

$$A(p) = \frac{\frac{n^\alpha w}{p} + \gamma n^2 \tau_{\mathrm{i/o}}}{n^\alpha} \qquad \text{with } \beta n^2 = pM\ .$$

If $\alpha = 2$, $A(1) = w + \gamma\tau_{\mathrm{i/o}}$, and $A(p) = \frac{w}{p} + \gamma\tau_{\mathrm{i/o}}$, we obtain the traditional parallel speedup that is bounded by a value that does not depend on p. But for $\alpha \geqslant 3$, we have

$$A(1) = w + \frac{\gamma\tau_{\mathrm{i/o}}}{\left(\frac{M}{\beta}\right)^{\alpha-2}}\ , \text{and}$$

$$A(p) = \frac{w}{p} + \frac{\gamma\tau_{\mathrm{i/o}}}{\left(\frac{M}{\beta}\right)^{\alpha-2} p^{\alpha-2}}\ .$$

The speedup then increases roughly linearly with p.

The conclusion from this development is that massive parallelism is useful, but only as the problem size becomes large and the fraction of the sequential execution that is non-parallelizable becomes negligible. Fortunately, most matrix operations meet these two criteria.

▷ **Question 3.** There are several kinds of answers to the question:

- **Cheating** – Here is a simple example: the execution of a loop to add two vectors of size 100. The sequential execution requires computing 100 additions, plus some overhead to control the loop (i.e., to increment the loop index, to test whether the loop is finished). With 100 processors, we have a single addition per processor but no more control. Consequently, we compute (slightly) more than 100 times faster. Of course we could have unrolled the loop with a single processor, thereby removing the control overhead. This is why we say that this solution is "cheating."

- **Algorithms** – Sometimes the algorithm used with p processors may converge surprisingly fast. Consider a *branch-and-bound* algorithm to solve some optimization problem. Processors will search independent sub-trees in parallel. One processor may then (luckily) immediately find the optimal solution in its sub-tree, independently of the time spent in the sequential execution (hence a super-linear speedup).

Here is another example, in which the algorithm used with p processor is different from the sequential algorithm in some subtle way. Consider the resolution of a sparse linear system by some iterative method. Typically, a pre-conditioning matrix is used. With two processors, a natural idea is to zero out non-diagonal blocks of the pre-conditioning matrix so that the execution can go fully parallel. It may well be the case that this ad-hoc preconditioning leads to convergence in fewer iterations than the sequential one, hence a super-linear speedup again. It is true that the speedup should be computed as the ratio of the time of the best sequential algorithm over the time of the parallel algorithm, and a posteriori we could have simulated the parallel algorithm with a single processor. However, in general we do not know the best sequential algorithm, and parallelism leads us to invent new algorithms that would never have been considered (as in this example) for a purely sequential execution.

- **Hardware** – This is the classic answer. Never forget that p processors have more memory than one. When computing a matrix-vector product with two processors, there always exists a problem size such that (i) distributed data (half the matrix and the vector) fit in the main memory of each processor, but (ii) accessing the whole data (full matrix and vector) requires disk reads and writes with a single processor. The dramatic difference in access speed between memory and disk guarantees super-linear speedup!

Let us conclude with some parallel computing humor: A single mover takes infinite time to move a piano up a few floors, but two movers only take a few minutes: infinite speedup!

Chapter 5

Algorithms on Grids of Processors

5.1 Logical Two-Dimensional Grid Topologies

In Chapter 4 we have developed algorithms on a ring of processors. In this chapter we motivate and demonstrate the use of a more complex logical topology: a two-dimensional (2-D) grid, or *grid* for short. Figure 5.1(a) shows an example square grid with $p = q^2$ processors. Processors in the corners have only two neighbors, processors on the edges have three, and processors in the middle have four. Processors are indexed by their row and column in the grid, $P_{i,j}$, $0 \leqslant i, j < q$. One popular variation of the grid topology is obtained by adding loopbacks from edge to edge, forming what is commonly called a 2-D torus, or *torus* for short, as shown in Figure 5.1(b). In this case every processor belongs to two different rings. As in the case of a ring topology, links can be either unidirectional or bidirectional. Choosing an appropriate flavor of the grid topology is up to the parallel algorithm developer, but we will see that a bidirectional torus proves very convenient and this is the topology we will use by default. For the sake of simplicity we will always assume that the number of processors is a perfect square, leading to square grids. The algorithms presented in this chapter can be adapted to rectangular grids, which often require only more cumbersome index calculations.

Just as with a ring, an assumption here is that communication can occur in parallel on different links. So, using the grid in Figure 5.1(b), $P_{0,0}$ can send data to $P_{1,0}$ while $P_{3,2}$ sends data to $P_{0,2}$. Additionally, as with a ring, the three standard assumptions about concurrent sending, receiving, and computing apply (see Section 3.3). For all the algorithms developed in this chapter, we will assume that each processor can be engaged in computing, sending, and receiving concurrently as long as there are no race conditions. Two new issues arise with a bidirectional grid topology.

The first issue is that of a processor concurrently sending and receiving on the same bidirectional link. We were careful that such concurrent communication on a single link did not arise in our use of a bidirectional link for the stencil application described in Section 4.5.2. In this chapter, we will make the assumption that links are *full-duplex*, meaning that communications can flow both ways on each link without contention. In other terms, a processor can concurrently send and receive a given amount of data on a link in the same

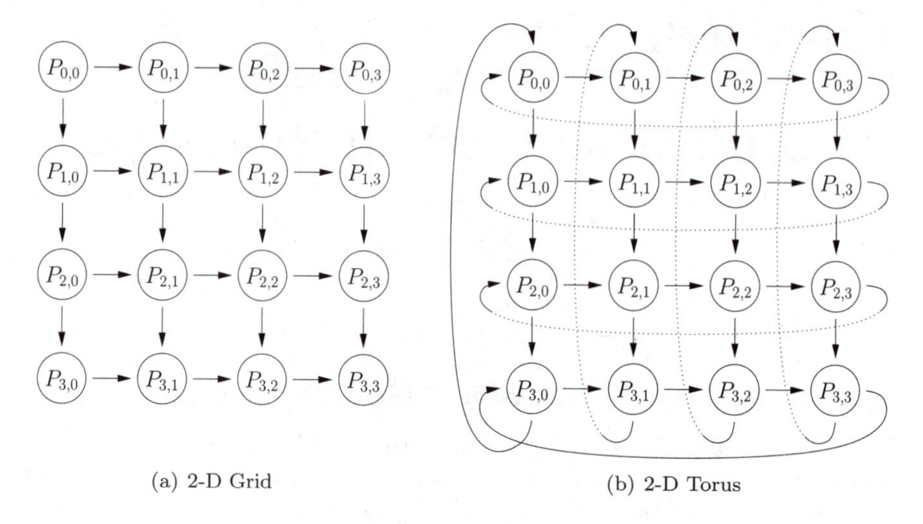

FIGURE 5.1: Two options for a logical unidirectional grid topology of $p = 16$ processors.

time as it can send (or receive) that amount of data. This assumption may or may not be realistic for the underlying physical platform. It is straightforward to modify the programs presented in this chapter and, more importantly, their performance analyses, in case the full-duplex assumption does not hold.

The second issue is that of the number of communications in which a single processor can be engaged simultaneously. With four bidirectional links, conceivably a processor can be involved in one send and one receive on all its network links, all concurrently. The assumption that such concurrent communications are allowed at each processor with no decrease in communication speed when compared to a single communication is termed the *multi-port model.* In the case of our grid topology, we talk of a *4-port model.* If instead at most two concurrent communications are allowed, one of them being a send and one of them being a receive, then one talks of a *1-port model.* Going back to our stencil algorithm on a bidirectional ring (Section 4.5.2), the reader will see that we had implicitly used the 1-port model. In this chapter, we will show performance analyses for both the 1-port and the 4-port model. Once again, it is typically straightforward to adapt these analyses to other assumptions regarding concurrency of communications.

As discussed at the end of Chapter 4, an important issue is impedance matching between logical and physical topologies: How do the grid and ring logical topologies compare in terms of realism for a given physical platform? It turns out that there are platforms whose physical topologies are or include grids and/or rings. A famous example of a supercomputer using a grid is the defunct Intel Paragon. A more recent example, at least at the time this book is being written, is IBM's Blue Gene/L supercomputer and its three-dimensional torus topology, which contains grids and rings. When both a ring and a grid map well to the physical platform, the grid is preferable for many algorithms. For a given number of processors p, a torus topology uses $2p$ network links (and a grid topology $2(p - \sqrt{p})$), twice more than a ring topology, which uses only p network links. As a result, more communications can occur in parallel and there are more opportunities for developing parallel algorithms with lower communication costs. Interestingly, even on platforms whose physical topology does not contain a grid (e.g., on a platform using a switched interconnect), using a logical grid topology can allow for more concurrent communications. In this chapter, we will see that the opportunity for concurrent communications is the key advantage of logical grid topologies for implementing popular algorithms. But we will also see that even if the underlying platform offers no possibility of concurrent communications, writing some algorithms assuming a grid topology is inherently beneficial!

5.2 Communication on a Grid of Processors

In this section, we define communication primitives and convenient functions that we will use in the upcoming sections to write parallel algorithms on grids of processors. Processor $P_{i,j}$, $0 \leqslant i, j < q$, is said to be in *processor row* i and to be in *processor column* j. A processor can obtain the indices of its processor row and column via the two following functions:

$$\textsc{My\_Proc\_Row}() \text{ and } \textsc{My\_Proc\_Col}().$$

A processor can determine the total number of processors, $p = q^2$, by calling function NUM_PROCS(). Since we assume square grids, this function makes it possible to know the number of processors in each processor row or column. Two separate functions may be needed in the case of rectangular grids.

A processor can send a message of L data items stored at address *addr* to one of its neighbor processors by calling the following primitive:

$$\textsc{Send}(dest, addr, L),$$

where *dest* has value $North$, $South$, $West$, or $East$. For a torus topology, which is the topology we will use for the majority of the algorithms in this chapter, the North, South, West, and East neighbors of processor $P_{i,j}$ are $P_{i-1,j \bmod q}$, $P_{i+1,j \bmod q}$, $P_{i,j-1 \bmod q}$, and $P_{i,j+1 \bmod q}$, respectively. We often omit the modulo and assume that all processor indices are taken modulo q implicitly. If the topology is a grid, then some *dest* values are not allowed for some source processors. Each SEND call has a matching RECV call:

$$\textsc{Recv}(src, addr, L).$$

As in Chapter 4, we assume asynchronous communications, i.e., non-blocking sends and blocking or non-blocking receives. We remind the reader that it is typically straightforward to adapt algorithms and performance analyses to more restrictive assumptions.

In addition to the above point-to-point communication primitives we also assume two additional primitives that implement a broadcast within a processor row or processor column (which is technically a multi-cast). Broadcasting from processor $P_{i,j}$ to all the processors in processor row i is done as follows:

$$\text{BroadcastRow}(i, j, \mathit{srcaddr}, \mathit{dstaddr}, L),$$

where *srcaddr* is the address, in the memory of processor $P_{i,j}$, of a message of length L that is to be sent. Once received, the message is stored at address *dstaddr* in the memory of all involved processors. In the same spirit as for the broadcast function described in Chapter 3, all processors in the processor row place the same call and the source processor sends a message to itself. We assume a similar function for broadcasting within processor column j:

$$\text{BroadcastCol}(i, j, \mathit{srcaddr}, \mathit{dstaddr}, L).$$

Note that in the case of a torus, each processor row and processor column is a ring embedded in the processor grid. Therefore, the above two functions can use the pipelined implementation of the broadcast on a ring developed in Section 3.3.4. If, in addition, links are bidirectional and one assumes a 4-port model (or in fact just a 2-port model in this case), then the broadcast can be done faster by sending data from the source processors in both directions simultaneously. We will see later in this chapter that this does not change the asymptotic performance of the broadcast. If the topology is not a torus but links are bidirectional, then this broadcast can be implemented by sending messages both ways from the source processor. If the topology is not a torus and links are unidirectional, then these functions cannot be implemented. We assume that a processor that calls these functions and is not in the relevant processor row or column returns immediately. This assumption will simplify the pseudo-code of our algorithms by removing the need for processor row and column indices before calling BroadcastRow() or BroadcastCol().

5.3 Matrix Multiplication on a Grid of Processors

In this section, we present four popular algorithms for computing the matrix product $C = A \times B$ on a square grid of processors. The first question that arises is that of the distribution of the matrices among the processors. Let us assume that the grid is square and contains $p = q^2$ processors, that matrices are also square of dimension $n \times n$, and that q divides n. In the case of a one-dimensional (1-D) topology, i.e., a ring, we had used a natural 1-D data distribution. Here, our 2-D topology naturally induces a 2-D data distribution. We define $m = n/q$. The standard approach is to assign an $m \times m$ block of each matrix to each processor according to the grid topology. More precisely, processor $P_{i,j}$, $0 \leqslant i, j < q$, holds matrix elements $A_{k,l}$, $B_{k,l}$, and $C_{k,l}$ with $i.m \leqslant k < (i+1).m$ and $j.m \leqslant l < (j+1).m$. We denote the three matrix blocks assigned to processor $P_{i,j}$ by $\widehat{A_{i,j}}$, $\widehat{B_{i,j}}$, and $\widehat{C_{i,j}}$, as depicted in Figure 5.2 for matrix A. All algorithms hereafter use this distribution scheme.

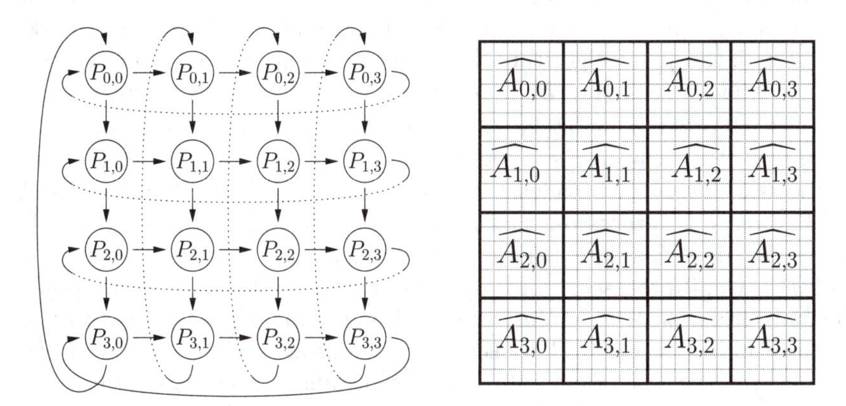

FIGURE 5.2: 2-D block distribution of an $n \times n$ matrix ($n = 24$) on a unidirectional grid/torus of p processors ($p = 4^2 = 16$).

5.3.1 The Outer-Product Algorithm

While the standard algorithm for matrix multiplication is often written as a sequence of inner product computations (as in Section 4.2), operations can be ordered in many different ways. One option is to write the algorithm as a series of outer products, which amounts to simply switching the order of the loops. Assuming that all elements of matrix C are initialized to zero, the sequential algorithm is written as

for $k = 0$ **to** $n-1$ **do**
  for $i = 0$ **to** $n-1$ **do**
    for $j = 0$ **to** $n-1$ **do**
      $C_{i,j} \leftarrow C_{i,j} + A_{i,k} \times B_{k,j}$

It turns out that this so-called "outer-product algorithm" [1, 55, 76] leads to a particularly simple and elegant parallelization on a torus of processors. The algorithm proceeds in n steps, that is, n iterations of the outer loop. At each step k, $C_{i,j}$ is updated using $A_{i,k}$ and $B_{k,j}$. Recall that all three matrices are partitioned in q^2 blocks of size $m \times m$, as on the right side of Figure 5.2.

The algorithm above can be written in terms of matrix blocks and of matrix multiplications, and it proceeds in q steps as follows:

for $k = 0$ **to** $q - 1$ **do**
  for $i = 0$ **to** $q - 1$ **do**
    for $j = 0$ **to** $q - 1$ **do**
      $\widehat{C_{i,j}} \leftarrow \widehat{C_{i,j}} + \widehat{A_{i,k}} \times \widehat{B_{k,j}}$

Now consider the execution of this algorithm on a torus of $p = q^2$ processors. Processor $P_{i,j}$ holds block $\widehat{C_{i,j}}$ and is responsible for updating this block at each step of the above algorithm. To perform this update at step k, processor $P_{i,j}$ needs blocks $\widehat{A_{i,k}}$ and $\widehat{B_{k,j}}$. At step $k = j$, $P_{i,j}$ already happens to hold $\widehat{A_{i,k}}$. For all other steps, $P_{i,j}$ must receive $\widehat{A_{i,k}}$ from the processor that holds it, that is, $P_{i,k}$. This is true for all $P_{i,j}$, $j \neq k$, processors. Therefore, at step k processor $P_{i,k}$ must broadcast its block of matrix A to all processors $P_{i,j}$, $j \neq k$, that is, all processors that are on processor $P_{i,k}$'s processor row. This is true for all i. Similarly, blocks of matrix B must be broadcasted at step k by $P_{k,j}$ to all processors on its processor column, for all j. The resulting communication pattern is illustrated in Figure 5.3. The figure shows which blocks of matrices A and B are sent to which processors at step $k = 1$ of the algorithm in the case of a 4×4 torus. For instance, block $\widehat{A_{2,1}}$, which is held by processor $P_{2,1}$, is sent to processors $P_{2,0}$, $P_{2,2}$, and $P_{2,3}$.

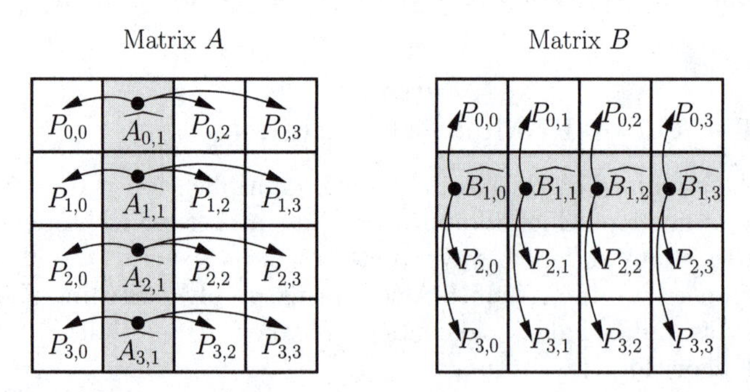

FIGURE 5.3: Communications of blocks of matrices A and B at step $k = 1$ of the outer-product matrix multiplication algorithm on a 4×4 torus of processors.

The outer-product algorithm on a torus of processors is given in Algorithm 5.1. Statement 1 in the algorithm declares the square blocks of the three matrices stored by each processor, assuming that elements of array C are initialized to zero, and that arrays A and B contain the blocks of matrices A and B according to the data distribution shown in Figure 5.2. Statement 2 declares two helper buffers that will be used by the processors to store received

```
1  var A, B, C:  array[0..m−1,0..m−1] of real
2  var bufferA, bufferB:  array[0..m−1,0..m−1] of real
3  q ←SRQT(NUM_PROCS())
4  myrow ←MY_PROC_ROW()
5  mycol ←MY_PROC_COL()
6  for k = 0 to q − 1 do
7      for i = 0 to m − 1 do                      { Broadcast A along rows }
8          BROADCASTROW(i, k, A, bufferA, m × m)
9      for j = 0 to m − 1 do                      { Broadcast B along columns }
10         BROADCASTCOL(k, j, B, bufferB, m × m)
11                                                { Multiply matrix blocks }
12     if (myrow = k) And (mycol = k) then
13         MATRIXMULTIPLYADD(C, A, B, m)
14     else if (myrow = k) then
15         MATRIXMULTIPLYADD(C, bufferA, B, m)
16     else if (mycol = k) then
17         MATRIXMULTIPLYADD(C, A, bufferB, m)
18     else
19         MATRIXMULTIPLYADD(C, bufferA, bufferB, m)
```

ALGORITHM 5.1: The outer-product algorithm on a grid of processors.

blocks of matrices A and B. In statement 3, each processor determines the square root of the number of processors. In statements 4 and 5, each processor obtains its location on the processor torus. The q steps of the algorithm are implemented by the loop in statement 6. At iteration k, three things happen. First, in statements 7 and 8, the q processors in processor column k broadcast their block of A in their respective processor rows. Note that these statements are correct because we assume that processors that should not participate in a broadcast return immediately from the BROADCASTROW() call. Then, statements 9 and 10 implement similar broadcasts for blocks of matrix B along processor columns. Once all these broadcasts have completed, each processor holds all the necessary blocks: Each processor multiplies a block of A by a block of B and adds the result to the block of C for which it is responsible. If the processor is both on processor row k and on processor column k, then this processor can just multiply the two blocks of A and B it holds (statements 12 and 13). Note that for brevity we assume the existence of a function MATRIXMULTIPLYADD() defined as MATRIXMULTIPLYADD(C, A, B, n), which multiplies two $n \times n$ matrices stored in arrays A and B and adds the result to an $n \times n$ matrix stored in array C. If the processor is on processor row k and not on processor column k, then it must multiply the block of A it has just received with the block of B it holds (statements 14 and 15). Similarly, a processor on processor column k but not on processor row k multiplies the block of A it holds with the block of B it has just received (statements 16 and 17). Finally, in statements 18 and 19, a processor that is neither on processor row k nor on processor column k multiplies the two blocks of A and B it has received. Note that it is possible to adapt this algorithm to the case of non-square matrices and/or non-square grids by allocating rectangular blocks of the matrix to processors, which we leave as an exercise for the reader.

Performance analysis of the algorithm is straightforward. At each step, each processor is involved in two broadcasts of messages containing m^2 matrix elements to $q-1$ processors. Using the pipelined broadcast implementation on a ring described in Section 3.3.4, the time for each broadcast is

$$T_{bcast} = \left(\sqrt{(q-2)L} + \sqrt{m^2 b}\right)^2,$$

where L is the communication startup cost, and b is the time to communicate a matrix element in steady state. After the two broadcasts, each processor performs an $m \times m$ matrix multiplication, which takes time $m^3 w$, where w is the computation time for a basic operation (multiplying two matrix elements and adding the result to another matrix element). After the communication phase in step 0, communication at step k can always occur in parallel with computation at step $k-1$. The computation phase in the last step does not occur in parallel with any computation. Since there are q steps, the overall execution time $T(n,p)$ is

$$T(m,q) = 2T_{bcast} + (q-1)\max\left(2T_{bcast}, m^3 w\right) + m^3 w. \qquad (5.1)$$

The above analysis is for the *1-port* model, with the horizontal and the vertical broadcasts happening in sequence at each step. With the *4-port* model, both broadcasts can occur concurrently, and the execution time of the algorithm is obtained by removing the factor 2 in front of each T_{bcast} in the above equation. When n becomes large, $T_{bcast} \sim n^2 b/p$, and thus $T(m,q) \sim n^3 w/p$, which shows that the algorithm achieves an asymptotic efficiency of 1. This algorithm is used by the ScaLAPACK [36] library, albeit often using a block-cyclic data distribution (see Section 5.4).

5.3.2 Grid vs. Ring?

The reader will remember that in Section 4.2 we have already given a distributed memory matrix multiplication algorithm with optimal asymptotic efficiency. Furthermore, that algorithm was for a ring of processors, which is a simpler topology. Therefore, one may wonder why there is any advantage to the grid topology. It is true that asymptotically there is no advantage, due to the fact that matrix multiplication has an $O(n^3)$ computational complexity and an $O(n^2)$ data size. However, communication terms that become negligible asymptotically do matter for practical values of n, and in fact many algorithm designers have striven to reduce communication costs for parallel matrix multiplication. We discuss below in what way a grid topology is advantageous compared to a ring topology.

In practice, for large values of n, up to a point, the algorithm's execution time can be dominated by communication time. This happens, for instance, when the ratio w/b is low. The communication time (which is then approximately equal to the execution time) on a grid, using the outer-product algorithm described in the previous section, would be $2n^2b/\sqrt{p}$, assuming a 1-port model. The communication time of the matrix multiplication algorithm on a ring, developed and analyzed in Section 4.2, is $pn^2/pb = n^2b$. The time spent in communication when using a grid topology is thus a factor $\frac{1}{2}\sqrt{p}$ smaller than when using a ring topology. This is easily seen when examining the communication patterns of both algorithms. When using a ring topology, at each step the communication time is equal to that needed for sending n^2/p elements between neighbor processors, and there are p such steps. By contrast, for the algorithm on a grid, the communication at each step involves two broadcasts of n^2/p matrix elements, and there are $\sqrt{p}$ such steps. With the pipelined implementation of the broadcast, broadcasting n^2/p matrix elements in a processor row or column can be done in approximately the same time as sending n^2/p matrix elements from one processor to another on the ring (provided that n is not too small). Since there are two broadcasts, at each step the algorithm on the grid spends twice as much time communicating as on the ring. But it performs a factor $\sqrt{p}$ fewer steps! So the algorithm on a grid spends a factor $\frac{1}{2}\sqrt{p}$ less time communicating than the algorithm on a ring. With a 4-port model, this factor is $\sqrt{p}$.

The above advantage of the grid topology can be attributed to the presence of more network links and to the fact that many of these links can be used concurrently. In fact, for matrix multiplication, the 2-D data distribution induced by a grid topology is inherently better than the 1-D data distribution induced by a ring topology, regardless of the underlying physical topology! To see this, let us just compute the total amount of data that needs to be communicated in both versions of the algorithm.

The algorithm on a ring communicates p matrix stripes, each containing n^2/p elements at each step, for p steps, amounting to a total of $p.n^2$ matrix elements sent on the network. The algorithm on a grid proceeds in $\sqrt{p}$ steps. At each step $2 \times \sqrt{p}$ blocks of n^2/p elements are sent, each to $\sqrt{p} - 1$ processors, for a total of $2\sqrt{p}.\sqrt{p-1}.n^2/p \leqslant 2.n^2$ elements. Since there are $\sqrt{p}$ steps, the total number of matrix elements sent on the network is lower than $2\sqrt{p}.n^2$, i.e., at least a factor $2\sqrt{p}$ lower than in the case of the algorithm on a ring! We conclude that when using a 2-D data distribution one inherently sends less data then when using a 1-D distribution, by a factor that increases with the number of processors. Although we do not show it here formally, this result is general and holds for any (reasonable) matrix multiplication algorithm. The implication of this result is that, for the purpose of matrix multiplication, using a grid topology (and the induced 2-D data distribution) is at least as good as using a ring topology, and possibly better. For instance, when implementing a parallel matrix multiplication in a physical topology on which all communications are serialized (e.g., on a bus architecture like a non-switched Ethernet network), one should opt for a logical grid topology with a 2-D data distribution to reduce the amount of transferred data. Recall, however, that for n sufficiently large, the two logical topologies become equivalent, with execution time dominated by computation time.

5.3.3 Three Matrix Multiplication Algorithms

In this section, we review three classic algorithms for multiplying square matrices, named after their designers: Cannon, Fox, and Snyder. These algorithms are more complex than the outer-product algorithm, but they are interesting to study and should be part of the culture of all parallel algorithm designers. We assume a $q \times q$ torus of $p = q^2$ processors as shown on the left side of Figure 5.2, with the corresponding 2-D data distribution for matrices A, B, and C as shown on the right side of Figure 5.2. In this section, we write algorithms using high-level pseudo-code for the sake of simplicity. Indeed, although these algorithms are easy to understand intuitively and visually with examples, the codes for the full algorithms can be rather lengthy. At this point in this book the reader should be able to develop full-detailed implementations of these algorithms, and we leave such implementations as exercises.

The Cannon Algorithm

The Cannon algorithm [38] requires an initial re-distribution of matrices A and B as follows. Each block row of matrix A is shifted (by zero, one, or more positions) so that each processor in the first processor column holds a diagonal block of the matrix. Similarly, each block column of matrix B is shifted so that each processor in the first processor row holds a diagonal block of the matrix. After these shifts, block $\widehat{A_{i,j}}$ of matrix A is stored on processor $P_{i,(i-j) \bmod q}$ and block $\widehat{B_{i,j}}$ of matrix B is stored on processor $P_{(i-j) \bmod q,j}$. The resulting distributions are shown in Figure 5.4 for a 4×4 torus. At the end of the algorithm similar shifts are necessary to restore matrices A and B to their initial distributions. These initial and final steps are typically called *preskewing* and *postskewing*.

The Cannon algorithm only requires point-to-point communication between neighboring processors. Once the preskewing is done, the algorithm proceeds in q steps. At each step, each processor computes the product of the blocks of A and B it holds and adds the result to its block of C (whose elements were initialized to zero). Then, blocks of matrix A are shifted by one position toward the left (horizontal shifts within processor rows) and blocks of matrix B are shifted by one position toward the top (vertical shifts within processor columns). Blocks of matrix C never move. Communication and

$$\begin{bmatrix} \widehat{A_{00}} & \widehat{A_{01}} & \widehat{A_{02}} & \widehat{A_{03}} \\ \widehat{A_{11}} & \widehat{A_{12}} & \widehat{A_{13}} & \widehat{A_{10}} \\ \widehat{A_{22}} & \widehat{A_{23}} & \widehat{A_{20}} & \widehat{A_{21}} \\ \widehat{A_{33}} & \widehat{A_{30}} & \widehat{A_{31}} & \widehat{A_{32}} \end{bmatrix} \qquad \begin{bmatrix} \widehat{B_{00}} & \widehat{B_{11}} & \widehat{B_{22}} & \widehat{B_{33}} \\ \widehat{B_{10}} & \widehat{B_{21}} & \widehat{B_{32}} & \widehat{B_{03}} \\ \widehat{B_{20}} & \widehat{B_{31}} & \widehat{B_{02}} & \widehat{B_{13}} \\ \widehat{B_{30}} & \widehat{B_{01}} & \widehat{B_{12}} & \widehat{B_{23}} \end{bmatrix}$$

FIGURE 5.4: Block data distribution of matrices A and B after the preskewing phase of the Cannon algorithm (on a 4×4 processor grid).

computation can be overlapped, provided that incoming data can be buffered by processors, as seen for many of the algorithms we have discussed so far in this book. The Cannon algorithm, written in high-level pseudo-code, is given in Algorithm 5.2.

```
1 var A, B, C:  array[0..m − 1,0..m − 1] of real
    { Preskewing }
2 Horizontal preskewing of A
3 Vertical preskewing of B
    { Computation phase }
4 for k = 1 to q do
5 |   MatrixMultiplyAdd(C, A, B, m)
6 |   Horizontal shift of A
7 |   Vertical shift of B
    { Postskewing }
8 Horizontal postskewing of A
9 Vertical postskewing of B
```

ALGORITHM 5.2: The Cannon algorithm.

We depict the first two steps of the algorithm in Figure 5.5, which shows which block multiplications are performed by each processor. The symbol $\circledast$ indicates block-wise matrix multiplications. For instance, in the first step, processor $P_{2,3}$ updates its block of matrix C, $\widehat{C_{2,3}}$, by adding to it the result of the $\widehat{A_{2,1}} \times \widehat{B_{1,3}}$ product, while processor $P_{1,0}$ adds $\widehat{A_{1,1}} \times \widehat{B_{1,0}}$ to $\widehat{C_{1,0}}$. In the second step, blocks of A and B have been shifted horizontally and vertically, as seen in the figure. So, during this step, processor $P_{2,3}$ adds $\widehat{A_{2,2}} \times \widehat{B_{2,3}}$ to $\widehat{C_{2,3}}$, while processor $P_{1,0}$ adds $\widehat{A_{1,2}} \times \widehat{B_{2,0}}$ to $\widehat{C_{1,0}}$. Intuitively one can see that eventually each processor $P_{i,j}$ will have computed all $\widehat{A_{i,l}} \times \widehat{B_{l,j}}$ products, $l = 0, \ldots, q-1$, needed to obtain the final value of $\widehat{C_{i,j}}$.

$$\left[\begin{array}{c|c|c|c} \widehat{C_{00}} & \widehat{C_{01}} & \widehat{C_{02}} & \widehat{C_{03}} \\ \hline \widehat{C_{10}} & \widehat{C_{11}} & \widehat{C_{12}} & \widehat{C_{13}} \\ \hline \widehat{C_{20}} & \widehat{C_{21}} & \widehat{C_{22}} & \widehat{C_{23}} \\ \hline \widehat{C_{30}} & \widehat{C_{31}} & \widehat{C_{32}} & \widehat{C_{33}} \end{array}\right] += \left[\begin{array}{c|c|c|c} \widehat{A_{00}} & \widehat{A_{01}} & \widehat{A_{02}} & \widehat{A_{03}} \\ \hline \widehat{A_{11}} & \widehat{A_{12}} & \widehat{A_{13}} & \widehat{A_{10}} \\ \hline \widehat{A_{22}} & \widehat{A_{23}} & \widehat{A_{20}} & \widehat{A_{21}} \\ \hline \widehat{A_{33}} & \widehat{A_{30}} & \widehat{A_{31}} & \widehat{A_{32}} \end{array}\right] \circledast \left[\begin{array}{c|c|c|c} \widehat{B_{00}} & \widehat{B_{11}} & \widehat{B_{22}} & \widehat{B_{33}} \\ \hline \widehat{B_{10}} & \widehat{B_{21}} & \widehat{B_{32}} & \widehat{B_{03}} \\ \hline \widehat{B_{20}} & \widehat{B_{31}} & \widehat{B_{02}} & \widehat{B_{13}} \\ \hline \widehat{B_{30}} & \widehat{B_{01}} & \widehat{B_{12}} & \widehat{B_{23}} \end{array}\right]$$

$$\left[\begin{array}{c|c|c|c} \widehat{C_{00}} & \widehat{C_{01}} & \widehat{C_{02}} & \widehat{C_{03}} \\ \hline \widehat{C_{10}} & \widehat{C_{11}} & \widehat{C_{12}} & \widehat{C_{13}} \\ \hline \widehat{C_{20}} & \widehat{C_{21}} & \widehat{C_{22}} & \widehat{C_{23}} \\ \hline \widehat{C_{30}} & \widehat{C_{31}} & \widehat{C_{32}} & \widehat{C_{33}} \end{array}\right] += \left[\begin{array}{c|c|c|c} \widehat{A_{01}} & \widehat{A_{02}} & \widehat{A_{03}} & \widehat{A_{00}} \\ \hline \widehat{A_{12}} & \widehat{A_{13}} & \widehat{A_{10}} & \widehat{A_{11}} \\ \hline \widehat{A_{23}} & \widehat{A_{20}} & \widehat{A_{21}} & \widehat{A_{22}} \\ \hline \widehat{A_{30}} & \widehat{A_{31}} & \widehat{A_{32}} & \widehat{A_{33}} \end{array}\right] \circledast \left[\begin{array}{c|c|c|c} \widehat{B_{10}} & \widehat{B_{21}} & \widehat{B_{32}} & \widehat{B_{03}} \\ \hline \widehat{B_{20}} & \widehat{B_{31}} & \widehat{B_{02}} & \widehat{B_{13}} \\ \hline \widehat{B_{30}} & \widehat{B_{01}} & \widehat{B_{12}} & \widehat{B_{23}} \\ \hline \widehat{B_{00}} & \widehat{B_{11}} & \widehat{B_{22}} & \widehat{B_{33}} \end{array}\right]$$

FIGURE 5.5: The first two steps of the Cannon algorithm on a 4×4 grid of processors, where at each step each processor multiplies one block of A and one block of B and adds this product to a block of C.

The Fox Algorithm

This algorithm developed by Fox in [55] was originally designed for CalTech's hypercube platform, but it uses a torus logical topology. The algorithm performs broadcasts of blocks of matrix A and is also known as the *broadcast-multiply-roll* algorithm. Unlike Cannon's algorithm, the Fox algorithm does not require any preskewing or postskewing of the matrices. The algorithm proceeds in q steps and at each step it performs a vertical shift of blocks of matrix B. At step k, $1 \leqslant k \leqslant q$, the algorithm also performs horizontal broadcasts of all blocks of the k-th block diagonal of matrix A within processor rows. A block $\widehat{A_{i,j}}$ is on the k-th block diagonal, $k \geqslant 1$, if $j = i + k - 1 \bmod q$. Therefore, at step k, processor $P_{i,i+k-1 \bmod q}$ sends its block of matrix A to all processors $P_{i,j}$, $j \neq i + k - 1 \bmod q$. The Fox algorithm, written in high-level pseudo-code, is shown in Algorithm 5.3.

1 var A, B, C: array[0..$m-1$,0..$m-1$] of real
2 **for** $k = 1$ **to** q **do**
3   Horizontal broadcasts of the blocks of the k-th diagonal of A
4   MatrixMultiplyAdd(C, A, B, m)
5   Vertical shift of B

ALGORITHM 5.3: The Fox algorithm.

As for the Cannon algorithm, we illustrate the first two steps of this algorithm in Figure 5.6. The figure shows which matrix block multiplications are performed by each processor at each step. During the first step relevant blocks of A are blocks $\widehat{A_{i,i}}$, $0 \leqslant i \leqslant q$, that is, blocks of the first block diagonal of matrix A. For instance, in the first step, processor $P_{2,3}$ updates its block of matrix C, $\widehat{C_{2,3}}$, by adding to it the results of the $\widehat{A_{2,2}} \times \widehat{B_{2,3}}$ product, while processor $P_{1,0}$ adds $\widehat{A_{1,1}} \times \widehat{B_{1,0}}$ to $\widehat{C_{1,0}}$. In the second step, relevant blocks of A are blocks $\widehat{A_{i,i+1 \bmod q}}$, that is, blocks on the second block diagonal. During this step, processor $P_{2,3}$ adds $\widehat{A_{2,3}} \times \widehat{B_{3,3}}$ to $\widehat{C_{2,3}}$, and processor $P_{1,0}$ adds $\widehat{A_{1,2}} \times \widehat{B_{2,0}}$ to $\widehat{C_{1,0}}$. Here again it is easy to see that eventually processor $P_{i,j}$ will have computed all block products necessary to obtain the final value of $\widehat{C_{i,j}}$.

$$\left[\begin{array}{c|c|c|c} \widehat{C_{00}} & \widehat{C_{01}} & \widehat{C_{02}} & \widehat{C_{03}} \\ \hline \widehat{C_{10}} & \widehat{C_{11}} & \widehat{C_{12}} & \widehat{C_{13}} \\ \hline \widehat{C_{20}} & \widehat{C_{21}} & \widehat{C_{22}} & \widehat{C_{23}} \\ \hline \widehat{C_{30}} & \widehat{C_{31}} & \widehat{C_{32}} & \widehat{C_{33}} \end{array}\right] += \left[\begin{array}{c|c|c|c} \widehat{A_{00}} & \widehat{A_{00}} & \widehat{A_{00}} & \widehat{A_{00}} \\ \hline \widehat{A_{11}} & \widehat{A_{11}} & \widehat{A_{11}} & \widehat{A_{11}} \\ \hline \widehat{A_{22}} & \widehat{A_{22}} & \widehat{A_{22}} & \widehat{A_{22}} \\ \hline \widehat{A_{33}} & \widehat{A_{33}} & \widehat{A_{33}} & \widehat{A_{33}} \end{array}\right] \circledast \left[\begin{array}{c|c|c|c} \widehat{B_{00}} & \widehat{B_{01}} & \widehat{B_{02}} & \widehat{B_{03}} \\ \hline \widehat{B_{10}} & \widehat{B_{11}} & \widehat{B_{12}} & \widehat{B_{13}} \\ \hline \widehat{B_{20}} & \widehat{B_{21}} & \widehat{B_{22}} & \widehat{B_{23}} \\ \hline \widehat{B_{30}} & \widehat{B_{31}} & \widehat{B_{32}} & \widehat{B_{33}} \end{array}\right]$$

$$\left[\begin{array}{c|c|c|c} \widehat{C_{00}} & \widehat{C_{01}} & \widehat{C_{02}} & \widehat{C_{03}} \\ \hline \widehat{C_{10}} & \widehat{C_{11}} & \widehat{C_{12}} & \widehat{C_{13}} \\ \hline \widehat{C_{20}} & \widehat{C_{21}} & \widehat{C_{22}} & \widehat{C_{23}} \\ \hline \widehat{C_{30}} & \widehat{C_{31}} & \widehat{C_{32}} & \widehat{C_{33}} \end{array}\right] += \left[\begin{array}{c|c|c|c} \widehat{A_{01}} & \widehat{A_{01}} & \widehat{A_{01}} & \widehat{A_{01}} \\ \hline \widehat{A_{12}} & \widehat{A_{12}} & \widehat{A_{12}} & \widehat{A_{12}} \\ \hline \widehat{A_{23}} & \widehat{A_{23}} & \widehat{A_{23}} & \widehat{A_{23}} \\ \hline \widehat{A_{30}} & \widehat{A_{30}} & \widehat{A_{30}} & \widehat{A_{30}} \end{array}\right] \circledast \left[\begin{array}{c|c|c|c} \widehat{B_{10}} & \widehat{B_{11}} & \widehat{B_{12}} & \widehat{B_{13}} \\ \hline \widehat{B_{20}} & \widehat{B_{21}} & \widehat{B_{22}} & \widehat{B_{23}} \\ \hline \widehat{B_{30}} & \widehat{B_{31}} & \widehat{B_{32}} & \widehat{B_{33}} \\ \hline \widehat{B_{00}} & \widehat{B_{01}} & \widehat{B_{02}} & \widehat{B_{03}} \end{array}\right]$$

FIGURE 5.6: The first two steps of the Fox algorithm on a 4×4 grid of processors, where at each step each processor multiplies one block of A and one block of B and adds this product to a block of C.

The Snyder Algorithm

Our third algorithm, proposed by Snyder in [83], uses both a preskewing and postskewing phase, as well as vertical shifts of matrix B and global sum operations. The preskewing step consists in simply transposing the blocks of matrix B (so that block $\widehat{B_{i,j}}$ is stored on processor $P_{j,i}$). Like the previous ones, this algorithm proceeds in q steps. At each step k, $1 \leqslant k \leqslant q$, each processor multiplies the blocks of matrices A and B it holds and performs a vertical shift of B (by one position upward). Then, processor $P_{i,i+k-1 \bmod q}$ receives all block products computed by the processors on its processor row, including itself, which are summed together and stored into the block of matrix C this processor holds, i.e., $C_{i,i+k-1 \bmod q}$. Note that these blocks of C are those of the k-th block diagonal of matrix C. The Snyder algorithm, written in high-level pseudo-code, is shown in Algorithm 5.4.

```
1  var A, B, C:  array[0..m − 1,0..m − 1] of real
2  var bufferC:  array[0..m − 1,0..m − 1] of real
      { Preskewing }
3  Transpose B
      { Computation Phase }
4  MatrixMultiplyAdd(bufferC, A, B, m)
5  Vertical shift of B
6  for k = 1 to q − 1 do
7  |   Global sum of bufferC on proc. rows into C_{i,(i+k−1) mod q}
8  |   MatrixMultiplyAdd(bufferC, A, B, m)
9  |_  Vertical shift of B
10 Global sum of bufferC on proc. rows into C_{i,(i+q−1) mod q}
      { Postskewing }
11 Transpose B
```

ALGORITHM 5.4: The Snyder algorithm.

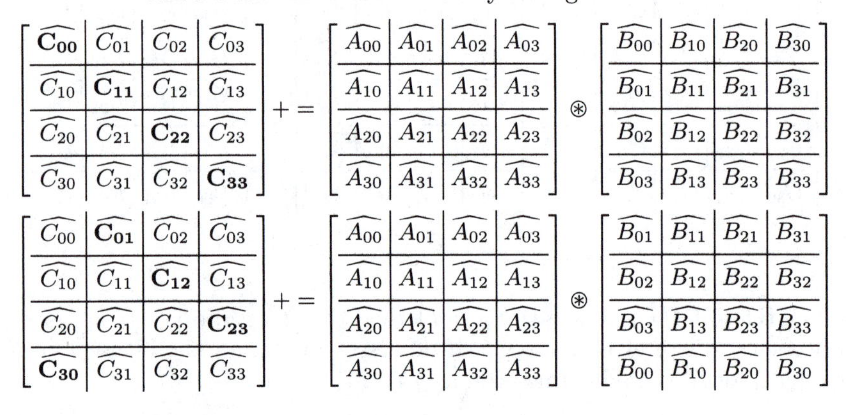

FIGURE 5.7: The first two steps of the Snyder algorithm on a 4×4 grid of processors, where at each step each processor multiplies one block of A and one block of B. All such products are added together and added to the blocks of C shown in boldface, within each processor row.

Figure 5.7 shows the first two steps of the algorithm. The blocks of C that are updated by the global sum operations are shown in boldface, along the first diagonal for the first step, and the second diagonal for the second step. In this sense the meaning of the $+=$ sign in this figure is different from that in Figures 5.5 and 5.6. Indeed, only q blocks of matrix C are updated at each step, as opposed to q^2 blocks. But each block is updated only once during the execution of the algorithm. For instance, in the second step, processor $P_{2,3}$ updates block $\widehat{C_{2,3}}$ by adding to it the three products $\widehat{A_{2,0}} \times \widehat{B_{0,3}}$, $\widehat{A_{2,1}} \times \widehat{B_{1,3}}$, and $\widehat{A_{2,2}} \times \widehat{B_{2,3}}$ received from processors $P_{2,0}$, $P_{2,1}$, and $P_{2,2}$ respectively, and

TABLE 5.1: Features of the Cannon, Fox, and Snyder algorithms.

Algorithm	**Cannon**	**Fox**	**Snyder**
preskewing	shifts of A and B	none	transposition of B
matrix products	in place	in place	sums on proc. rows
A movements	horizontal shifts	horizontal broadcasts	none
B movements	vertical shifts	vertical shifts	vertical shifts

the locally computed product $\widehat{A_{2,3}} \times \widehat{B_{3,3}}$.

5.3.4 Performance Analysis of the Three Algorithms

Table 5.1 summarizes the essential features of the three algorithms. In this section, we analyze their performance on a $q \times q$ torus, both under the *1-port* assumption and the less stringent *4-port* assumption. In both cases we assume full-duplex bidirectional network links. For completeness, we also mention results obtained in the literature assuming that the underlying platform implements wormhole routing. Note that a typical approach is to consider a hypercube topology to develop global communication algorithms. One then "casts" these algorithms to a grid topology, which is straightforward as a grid topology is easily embeddable in an hypercube topology. Essentially, the wormhole routing assumption makes it possible to ignore issues of processor proximity when accounting for communication time. More details on the hypercube, on the notion of topology embedding, and on wormhole routing are presented in Chapter 3.

We use the following notations, with which the reader should be familiar by now: $m = n/q$, L is the communication startup cost, b is the time necessary to send a matrix element over a network link in steady state, and w is the time necessary to multiply an element of matrix A by an element of matrix B and add the result to an element of matrix C.

Cannon Algorithm

Let us start with the 4-port model. The algorithm's execution time is the sum of two terms, T^{4p}_{skew}, the time for preskewing and postskewing the matrices, and T^{4p}_{comp}, the time to perform the computation ("$4p$" stands for 4-port).

For the preskewing and postskewing steps, one can limit the number of shifts of blocks of matrices A and B to $\lfloor q/2 \rfloor$. To understand this, consider a processor row and the shifts of blocks of matrix A that must be performed by processors in that row. It should be clear to the reader that performing $q-1$ left shifts is equivalent to performing one right shift. More generally, performing x left shifts is equivalent to performing $q-x$ right shifts. Therefore, in the worst case, a processor row only needs to perform $\lfloor q/2 \rfloor$ shifts, this maximum number of shifts being performed by the middle processor row(s). Therefore, the preskewing of matrix A takes time $\lfloor \frac{q}{2} \rfloor (L + m^2 b)$. The time to preskew matrix B is identical. In the 4-port model, horizontal and vertical communications can occur concurrently, and thus the total time for preskewing is simply $\left\lfloor \frac{q}{2} \right\rfloor (L + m^2 b)$. The time for postskewing is identical to the time for preskewing. Therefore, the total time spent for preskewing and postskewing is

$$T^{4p}_{skew} = 2 \left\lfloor \frac{q}{2} \right\rfloor (L + m^2 b).$$

At each step, each processor computes an $m \times m$ matrix multiplication, which is done in time $m^3 w$, and sends an $m \times m$ block of A and an $m \times m$ block of B to its neighbors, which both take time $L + m^2 b$. These two communications can occur concurrently in the 4-port model because one is horizontal and the other is vertical. Assuming that computation can occur concurrently with communication, and accounting for the fact that there are q steps to the computation, we obtain the overall execution time as

$$T^{4p}_{comp} = q. \max\left(m^3 w, L + m^2 b\right).$$

Using the same basic algorithm with a 1-port model, the added constraint is that horizontal and vertical communications cannot happen concurrently. Consequently, preskewing (and postskewing) for matrices A and B must happen in sequence and we obtain $T^{1p}_{skew} = 2T^{4p}_{skew}$ ("$1p$" stands for 1-port). Similarly, at each step the communications of blocks of A and blocks of B cannot happen concurrently, leading to

$$T^{1p}_{comp} = q. \max\left(m^3 w, 2L + 2m^2 b\right).$$

In the end we obtain the overall execution times for both models:

$$T^{4p} = 2 \left\lfloor \frac{q}{2} \right\rfloor (L + m^2 b) + q. \max\left(m^3 w, L + m^2 b\right),$$

$$T^{1p} = 4 \left\lfloor \frac{q}{2} \right\rfloor (L + m^2 b) + q. \max\left(m^3 w, 2L + 2m^2 b\right).$$

Fox Algorithm

The Fox algorithm proceeds in q steps, with no preskewing or postskewing, and thus the execution time is simply the time taken by a single step multiplied by q. The computation time at each step is $m^3 w$, just like for the Cannon algorithm.

At each step, there are q concurrent broadcasts of blocks of matrix A, one broadcast in each processor row. Using the pipelined broadcast presented in Section 3.3.4 with the optimal packet size, the time for the broadcast, T_{bcast}, is

$$T_{bcast} = \left(\sqrt{(q-2)L} + \sqrt{m^2 b}\right)^2.$$

Due to the fact that links are bidirectional, with the 4-port model the broadcast time above can be reduced by having the source processor simultaneously send data in both directions on the ring. With this technique, the execution time of the broadcast is obtained by replacing q in the above equation by $\lceil q/2 \rceil$. This is because the first packets sent in both directions go through at most $\lceil q/2 \rceil$ hops. Note that the asymptotic performance of the broadcast when m gets large is unchanged by this modification.

The shift of the blocks of matrix B can be done in time $L + m^2 b$. In the 4-port model, this shift can occur concurrently with the broadcasts of the blocks of matrix A at each step. As a result, the time to perform the shift is completely hidden ($T_{bcast} \geqslant L + m^2 b$, for $q > 1$). Computation at each step occurs concurrently with these communications. However, processors must all wait for the first broadcast to complete before proceeding. Then, at each step, the computation for that step, the shift for that step, and the broadcast for the next step can occur concurrently. In the last step, only the computation and the shift occur. We obtain the overall execution time in the 4-port model, T^{4p}, as

$$
\begin{aligned}
T^{4p} = \lceil \frac{q}{2} \rceil \left(\sqrt{(q-2)L} + \sqrt{m^2 b} \right)^2 + \\
(q-1) \max \left(m^3 w, \lceil \frac{q}{2} \rceil \left(\sqrt{(q-2)L} + \sqrt{m^2 b} \right)^2 \right) + \\
\max \left(m^3 w, L + m^2 b \right).
\end{aligned}
$$

With the 1-port model, horizontal and vertical communications cannot occur concurrently and during the broadcast the source can only send data in one direction at a time. Therefore, the execution time is simply

$$
\begin{aligned}
T^{1p} = & \left(\sqrt{(q-2)L} + \sqrt{m^2 b} \right)^2 + \\
& (q-1) \max \left(m^3 w, (\sqrt{(q-2)L} + \sqrt{m^2 b})^2 + L + m^2 b \right) + \\
& \max \left(m^3 w, L + m^2 b \right).
\end{aligned}
$$

If the underlying platform implements wormhole routing, the execution time for the broadcasts in the 1-port model can be reduced approximately to being only logarithmic in q due to the use of a (binary) broadcast tree, while in the above equation it is linear in q. The broadcast starts with one processor sending its matrix block to another processor. Then, both these processors send their blocks to another two processors, and so on, for a total of $\log q$ steps. See Section 3.4.2 for a complete description of the broadcast algorithm with wormhole routing.

Snyder Algorithm

The major differences between the Snyder algorithm and the previous ones are the use of a pre- and post-transposition of matrix B and the use of global sums to compute blocks of matrix C by accumulation. Let us discuss both these operations.

There are various ways to implement a matrix transposition on a grid of processors. A simple approach is discussed in Exercise 3.4. We discuss here another simple approach (note that we transpose the blocks, but not the content of the blocks). Consider a block of matrix B held initially by processor $P_{i,j}$, where $i > j$ (i.e., a block in the lower left part of the matrix). One way for this block to reach its destination, processor $P_{j,i}$, is for it to travel to the right, all the way to processor $P_{i,i}$ (which is on the diagonal of the processor grid), and then upward all the way to processor $P_{j,i}$. Blocks in the upper right half of the matrix travel similarly in the other direction. In the 4-port model, all blocks can be communicated concurrently and the blocks that require the largest number of hops are the blocks owned by processors $P_{q,1}$ and $P_{1,q}$. These blocks are communicated between neighboring processors $2(q-1)$ times, giving us the time for the entire transposition as

$$T_{transpose}^{4p} = 2(q-1)(L + m^2 b).$$

If our logical topology is a torus, this time can be cut roughly in half. Indeed, with a torus, transposing the block initially held by processor $P_{q,1}$ takes only two communication steps, using the loopback links. The factor $2(q-1)$ above then becomes $2\lfloor q/2 \rfloor$. The 1-port model precludes concurrent communications of the blocks originally in the upper half of the matrix and of the blocks originally in the lower half of the matrix. Therefore, the execution time is multiplied by a factor 2.

Another approach could consist in shifting the blocks in each row so that they are in their destination columns. As soon as a block has reached its destination column it can be shifted vertically to reach its destination row provided there is no link contention. While in a 4-port model this algorithm has the same execution time as the algorithm above, in the 1-port model it uses a smaller number of steps. Writing the algorithm to satisfy the "there is no link contention" constraint requires care, and we leave it as an exercise for the reader. Note that enforcing this constraint is not necessary to have a correct algorithm, but it may be difficult to analyze its performance.

Finally, note that if wormhole routing is implemented on the underlying platform, then a clever recursive transposition algorithm described in [45] leads to a shorter execution time:

$$T_{transpose}^{1p} = \lceil \log_2(q) \rceil (L + m^2 b).$$

Let us first consider the 4-port model when developing the performance analysis of the rest of the algorithm. The execution consists of a sequence of q steps, where each step consists in a product of matrix blocks, a shift of blocks of matrix B along processor columns, and a global sum of blocks of matrix C along processor rows (see Algorithm 5.4). The first global sum can only be done after the matrix products have been performed. These matrix products take time $m^3 w$, and can be done concurrently with the shift of matrix B, which takes time $L + m^2b$. As usual, with appropriate use of buffers and pointers, all three operations can be done in parallel for $q-1$ iterations of the for loop in the algorithm. By now, it should be clear to the reader that the shift of matrix B is completely hidden by the global sum of matrix C as it entails strictly fewer communication steps. The algorithm completes with a final global sum of matrix C blocks. Using Γ to denote the execution time of this global sum, which is to be determined, the execution time of the q steps is

$$\max\left(m^3w, L + m^2b\right) + (q-1)\max\left(m^3w, \Gamma\right) + \Gamma$$

The global sum of blocks of matrix C can be implemented in various ways. A complex but likely efficient approach would be to have partial sums computed recursively and in parallel by increasingly large groups of processors within a row. Instead, let us design a simple algorithm. At each step, in each processor row, the two processors the farthest away from the destination processor (i.e., the one on the k-th diagonal of the processor grid at step $k = 1, \ldots, q$) send a matrix block toward this processor in opposite directions on the ring. Each processor receiving this block adds it to its own block and forwards the result toward the destination processor, and so on. The destination processor will receive two blocks, which will be added to its own block of matrix C. If q is odd, then these two blocks will arrive at the same time and thus will be added one after the other. It follows that this algorithm requires $\lceil q/2 \rceil$ communication steps and $\lceil (q+1)/2 \rceil$ computation steps, for an overall execution time of

$$\Gamma = \left\lceil \frac{q}{2} \right\rceil (L + m^2b) + \left\lceil \frac{q+1}{2} \right\rceil m^2w'.$$

Note our use of w' in this equation to denote the time needed to add two matrix elements together, which is lower than w, the time needed to multiply two matrix elements together and add them to a third matrix element.

Alternatively, one can design a pipelined algorithm on a unidirectional ring in the same fashion as the pipelined broadcast developed in Section 3.3.4. Note that it is possible to implement a faster broadcast using the fact that our rings are bidirectional, as for the Fox algorithm. However, note that here we have both communications and computations (for the summing of blocks of C), so this technique is only useful if communication time is larger than computation time. We leave the development of a bidirectional global sum of blocks of C as an exercise for the reader. We split the m^2 matrix elements to be sent and added by each processor into r individual chunks. Let us consider a given processor row i. Without loss of generality, let us also assume that $P_{i,q-1}$ is the destination processor and that communication occurs in the direction of increasing processor column indices. In the first step of the algorithm, processor P_0 sends a chunk of m^2/r matrix elements to P_1. In the second step, processor P_1 adds these matrix elements to the corresponding matrix elements it owns (to perform the global sum) and receives the next chunk from P_0. In the third step, processor P_1 is engaged in three activities: receiving m^2/r elements, adding m^2/r elements, and sending m^2/r elements, which can all occur concurrently and take time $\max(L+m^2b/r, m^2w'/r)$. With this scheme, processor P_{q-1} finishes computing the first chunk of its block of matrix C after $2(q-1) = 2q-2$ steps. Since the first step entails only communication and all the others entail both communication and computation, processor P_q finishes computing its first chunk by time

$$L + m^2b/r + (2q-3)\max\left(L + m^2b/r, m^2w'/r\right).$$

Afterwards, a new chunk is computed at every step. Since there are $r-1$ remaining chunks, the overall execution time for the global sum, $\Gamma(r)$, is

$$\Gamma(r) = L + m^2b/r + (2q-4+r)\max\left(L + m^2b/r, m^2w'/r\right).$$

As in the case of the simple pipelined broadcast, one goal here is to find the optimal value for r. Fortunately, we can apply the same technique. For instance, assuming that $b \geqslant w'$, the maximum in the above equation becomes simply $L+m^2b/r$. The optimal execution time is obtained for $r = r_{opt}$, where

$$r_{opt} = \sqrt{\frac{(2q-3)m^2b}{L}}.$$

Recall that the above can be obtained by directly applying the "goat in a pen" theorem, or by simply computing the derivative of $\Gamma(r)$ with respect to r (see Sections 3.2.2 and 3.3.4). The optimal execution time for the global sum when $r = r_{opt}$, Γ_{opt}, is then

$$\Gamma_{opt} = (2q-2)L + m^2b + 2m\sqrt{(2q-3)bL}.$$

The case $b < w'$ is more involved, and is left as an exercise for the interested reader. We finally have the overall execution time for the algorithm, T^{4p}, in the case $b \geqslant w'$, as

$$\begin{aligned} T^{4p} = {} & 2T^{4p}_{transpose} + \max\left(m^3w, L + m^2b\right) + \\ & (q-1)\max\left(m^3w, (2q-2)L + m^2b + 2m\sqrt{(2q-3)bL}\right) + \\ & (2q-2)L + m^2b + 2m\sqrt{(2q-3)bL}. \end{aligned}$$

With the 1-port model, the shifts of blocks of matrix B cannot occur concurrently with the global sums of blocks of matrix C. We obtain the overall execution time as

$$\begin{aligned} T^{1p} = {} & 2T^{1p}_{transpose} + \max\left(m^3w, L + m^2b\right) + \\ & (q-1)\max\left(m^3w, L + m^2b + (2q-2)L + m^2b + 2m\sqrt{(2q-3)bL}\right) + \\ & L + m^2b + (2q-2)L + m^2b + 2m\sqrt{(2q-3)bL}. \end{aligned}$$

If the underlying platform implements wormhole routing, the global sum can be implemented following an approach similar to the one for the broadcast in the Fox algorithm, that is, using a binary broadcast tree. The only difference is that some computation is involved at each step to compute the global sum. Essentially, one processor sends its block of matrix C to another processor, which adds the received block to the block it holds. Then, both these processors send their blocks to another two processors, and so on, for a total of $\log q$ steps. Remembering that with wormhole routing the data transfer times do not depend on the distance between the two communicating processors (or only in a way that is typically negligible), the execution time of the global sum is then approximately equal to $\lceil \log(q) \rceil (L + m^2 w + m^2 b))$.

Conclusion

When n gets large, all these algorithms achieve an asymptotic parallel efficiency of 1. By now it should be clear to the reader that this is not terribly difficult to achieve for matrix multiplication, given that the computation is $O(n^3)$ and the communication is $O(n^2)$. The above performance analyses make it possible to compare the three algorithms for particular values of n and q and of the characteristics of the platform. More importantly, the main merit of going through these admittedly lengthy performance analyses is to expose the reader to several typical algorithms such as matrix transposition or global sums, to put principles such as pipelining to use, and to better understand the impact of 1-port and 4-port models on algorithm design and performance.

5.4 Two-Dimensional Block Cyclic Data Distribution

On a 2-D grid we have used the natural 2-D data distribution of matrices by which square blocks of a matrix are assigned to processors. This distribution was well suited to, for instance, the multiplication of two matrices. In Chapter 4, we have seen applications that strongly benefit from cyclic 1-D data distributions, namely, stencil applications and the LU factorization. We demonstrated that cyclic distributions can reduce idle time via latency reduction, improved pipelining of communication and computation, and/or improved load balancing. It is therefore natural to combine the advantages of cyclic data distributions and of 2-D data distributions, in so-called *2-D block cyclic* data distributions.

Figure 5.8 depicts an example 2-D block cyclic distribution of a square matrix on a square processor grid, in which processors are allocated 2×2 blocks of the matrix. While we have seen 1-D distributions that were cyclic across rows or across columns, this 2-D distribution is cyclic across both rows and columns. More formally, for a matrix A of dimension $n \times n$ distributed in a 2-D block cyclic fashion on a processor grid with $p = q^2$ processors using $b \times b$ blocks, block $\widehat{A_{k,l}}$ is stored on processor $P_{k \bmod q, l \bmod q}$. Note that for

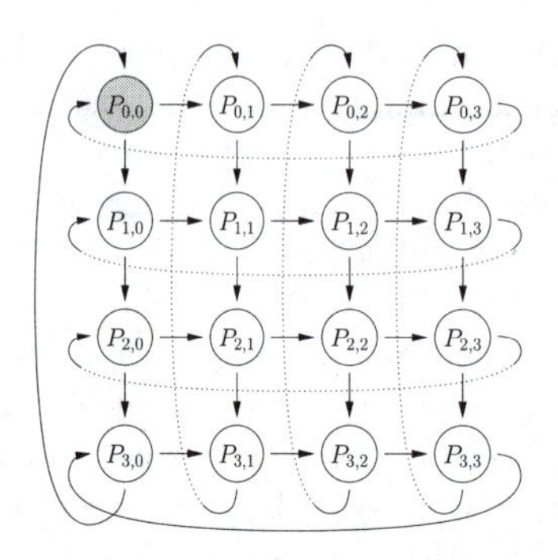

0,0	0,1	0,2	0,3	0,0	0,1	0,2	0,3	0,0	0,1
1,0	1,1	1,2	1,3	1,0	1,1	1,2	1,3	1,0	1,1
2,0	2,1	2,2	2,3	2,0	2,1	2,2	2,3	2,0	2,1
3,0	3,1	3,2	3,3	3,0	3,1	3,2	3,3	3,0	3,1
0,0	0,1	0,2	0,3	0,0	0,1	0,2	0,3	0,0	0,1
1,0	1,1	1,2	1,3	1,0	1,1	1,2	1,3	1,0	1,1
2,0	2,1	2,2	2,3	2,0	2,1	2,2	2,3	2,0	2,1
3,0	3,1	3,2	3,3	3,0	3,1	3,2	3,3	3,0	3,1
0,0	0,1	0,2	0,3	0,0	0,1	0,2	0,3	0,0	0,1
1,0	1,1	1,2	1,3	1,0	1,1	1,2	1,3	1,0	1,1

FIGURE 5.8: 2-D block cyclic distribution of a $n \times n$ matrix ($n = 20$) on a grid/torus of p processors ($p = 4^2 = 16$), with block size $r = 2$. Each 2×2 block is labeled by the indices of the processor holding that block. Blocks held by processor $P_{0,0}$ are highlighted in gray.

simplicity we assume that b divides n. Depending on the values of n, q, and b, some processors may hold more blocks of A than some others. For the example shown in Figure 5.8, processor $P_{0,0}$ holds nine blocks of the matrix while processor $P_{2,3}$ holds only four blocks.

All the algorithms we have seen in this chapter can be implemented using a 2-D block cyclic distribution. Another advantage of the 2-D bock cyclic distribution is that block sizes can be chosen arbitrarily, rather than be limited to being $n/q \times n/q$. This provides more flexibility for performance tuning and more opportunities to utilize each processor's memory hierarchy to the best of its potential. Both the ScaLAPACK [36] and PLAPACK [116] libraries provide 2-D block cyclic implementations of linear algebra functions. Such implementations are usually not conceptually difficult, but require the implementation of a software infrastructure to support and implement the 2-D block cyclic abstraction. This is especially true when the algorithms cannot rely on simplifying assumptions and must instead handle, say, rectangular matrices, grids, or blocks, and prime matrix, grid, or block dimensions. Note also that the High-Performance Fortran (HPF) language supports 2-D block cyclic data distributions natively [75]. We do not develop algorithms that use a 2-D block cyclic data distribution in this book and refer the reader to the aforementioned libraries for example implementations.

Bibliographical Notes

In addition to the many references cited in this chapter, a valuable reference for examples of algorithms on 2-D processor grids is, once again, the book by Kumar et al. [76]. Of interest is also the book by Cosnard et al. [45]. Finally, both the ScaLAPACK [36] and PLAPACK [116] libraries contain many implementations of interesting algorithms on 2-D processor grids.

5.5 Exercises

In the first two exercises, we write the pseudo-code for two algorithms that we have seen in this chapter, namely, the matrix block transposition in Snyder's algorithm and a stencil application, on 2-D processor grids. The third exercise deals with the parallelization of the well-known Gauss-Jordan method for solving a linear system of equations. Finally, the fourth exercise is a hands-on implementation of the outer-product matrix multiplication algorithm on a parallel platform using the MPI message-passing standard.

◇ Exercise 5.1 : Matrix Transposition

Write the pseudo-code for the 4-port matrix block transposition described as part of the Snyder algorithm in Section 5.3.4, for an $n \times n$ matrix on a non-toric $q \times q$ 2-D processor grid, where q divides n.

◇ Exercise 5.2 : Stencil Application

1. Write the pseudo-code for the stencil application described in Section 4.5 for a $n \times n$ domain, but this time on a $q \times q$ 2-D processor grid, where q divides n.

2. Give a performance model for your algorithm, assuming a 4-port bidirectionel model.

Exercise 5.3 : Gauss-Jordan Method

The Gauss-Jordan method is an algorithm for solving a linear system $Ax = b$, where A is a matrix of rank n, b is a right-hand side vector with n components, and x is a vector of unknowns with n components. The method consists in eliminating all the unknowns but x_i from equation i. The sequential algorithm, assuming that indices start at 0, is easily written as in Algorithm 5.5.

1. Write the pseudo-code for the above algorithm on a $q \times q$ 2-D processor torus, where q divides n.

2. Give a performance model for your algorithm with the 4-port assumption and the 1-port assumption.

Exercise 5.4 : Outer-Product Algorithm in MPI

1. Implement the outer-product matrix multiplication algorithm described in Section 5.3.1 using MPI, for $n \times n$ matrices, on a $q \times q$ processor grid. Assume that q divides n. Each processor initially holds a piece of matrix

```
1 GaussJordan(A, b, x)
2   for i = 0 to n − 1 do
3     for j = 0 to n − 1 do
4       if i ≠ j then
5         for k = j to n − 1 do
6           A_{i,k} ← A_{i,k} − (A_{i,j}/A_{j,j}) × A_{j,k}
7         b_i ← b_i − (A_{i,j}/A_{j,j}) × b_j
8   for i = 0 to n − 1 do
9     x_i ← b_i/A_{i,i}
```

ALGORITHM 5.5: The (sequential) Gauss-Jordan algorithm.

A and matrix B in two arrays. Initialize these arrays with matrix elements defined as $a_{i,j} = i$ and $b_{i,j} = i + j$ (these are global indices starting at 0). At the end of the execution each processor holds a piece of matrix C stored in a third array. Your program should check the validity of the results (indeed, they can be computed analytically as $c_{i,j} = i.n.(n-1)/2 + i.j.n$, using global indices).
Hint: It may be a good idea to use multiple MPI communicators, one for each processor row, and one for each processor column, as a convenient way to implement the BroadcastRow() and BroadcastColumn() functions.

2. Run your program on a parallel platform, e.g., a cluster, and plot the parallel speedup and efficiency for two, four, and eight processors as functions of the matrix size, n. Each data point should be obtained as the average over ten runs.

3. If you have not done so, experiment with non-blocking MPI communication and see whether there is any impact on your program's performance when you overlap communication and computation.

5.6 Answers

Exercise 5.1 (Matrix Transposition)

According to the description of the algorithm, a block stored on a processor in the lower part of the processor grid travels to the right on that processor's processor row until the processor on the diagonal of the processor grid is reached. Then, the block travels upward in that processor's column until it reaches its destination. Blocks stored on processors in the upper part of the processor grids travel similarly but first downward and then to the left.

Given the above algorithm, processor $P_{i,i}$ ($0 \leqslant i \leqslant q-1$) is involved in forwarding a total of $2i$ blocks. Each processor $P_{i,j}$ in the lower part of the matrix sends $\min(i,j)$ blocks to its West neighbor and $\min(i+1,j+1)$ blocks to its East neighbor, with corresponding block receptions by these neighbors. Each processor $P_{i,j}$ in the upper part of the matrix sends $\min(i,j)$ blocks to its North neighbor and $\min(i+1,j+1)$ blocks to its South neighbor, with corresponding block receptions by these neighbors. The figure below shows an example for a 5×5 processor grid; for each processor it depicts how many send/receive operations this processor performs throughout the overall execution.

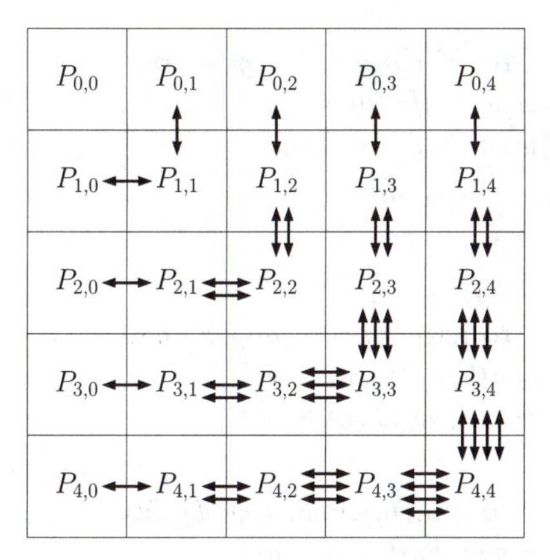

Assuming as usual that sends are non-blocking and that receives are blocking, we can now write the pseudo-code as shown in Algorithm 5.6, with $m = n/q$. The algorithm is written so that the clockwise and the counterclockwise communications happen in parallel. Note that this algorithm can be further improved by allowing a message reception for step k_1+1 (resp. k_2+1) to occur in parallel with a message sending for step k_1 (resp. k_2), as, for instance, in Algorithm 4.1. This improvement leads to the performance model given in Section 5.3.4.

```
1  var A, buff1, buff2: array[0..m-1,0..m-1] of real
2  myrow ←My_Proc_Row()
3  mycol ←My_Proc_Col()
4  if myrow == mycol then
5      for k1 = 1 to myrow do
6          Recv(north, buff1, m × m)
7          Send(west, buff1, m × m)
8      ||
9      for k2 = 1 to myrow do
10         Recv(west, buff2, m × m)
11         Send(north, buff2, m × m)
12 else if myrow > mycol then
13     Send(east, A, m × m)
14     for k1 = 1 to min(myrow, mycol) do
15         Recv(west, buff1, m × m)
16         Send(east, buff1, m × m)
17     ||
18     for k2 = 1 to min(myrow, mycol) do
19         Recv(east, buff2, m × m)
20         Send(west, buff2, m × m)
21     Recv(east, A, m × m)
22 else
23     Send(south, A, m × m)
24     for k1 = 1 to min(myrow, mycol) do
25         Recv(south, buff1, m × m)
26         Send(north, buff1, m × m)
27     ||
28     for k2 = 1 to min(myrow, mycol) do
29         Recv(north, buff2, m × m)
30         Send(south, buff2, m × m)
31     Recv(south, A, m × m)
```

ALGORITHM 5.6: Matrix block transposition algorithm.

Exercise 5.2 (Stencil Application)

▷ **Question 1.** The stencil requires that each cell be updated using its North, South, East, and West neighbors. Let us consider processor $P_{i,j}$, which holds a square block of dimension $\frac{n}{q} \times \frac{n}{q}$. $P_{i,j}$ can update all "internal" cells of its block without need for communication with its neighboring processors. Cells on an edge of the block require values that are stored on the edge of a block held by a neighbor processor. Therefore, each processor must exchange $\frac{n}{q}$ cell values with each of its neighbors. This is depicted in Figure 5.9. If the domain is assumed not to wrap around, there are special cases for the processors on the edges and in the corners of the processor grid as they are involved in fewer communications. We let the reader write the pseudo-code based on this high-level description of the algorithm.

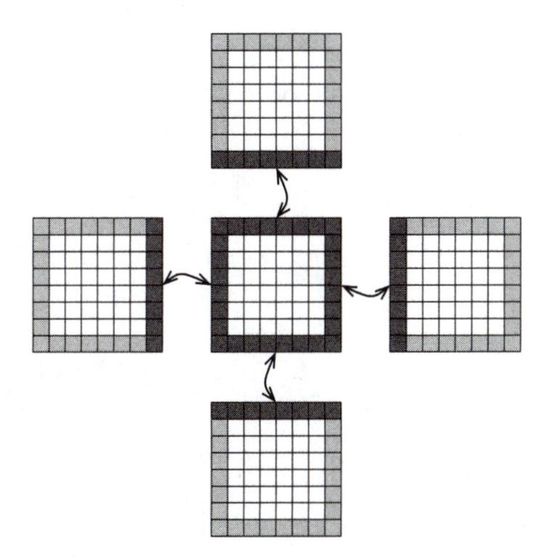

FIGURE 5.9: Depiction of the block held by a processor, in the center, and of the blocks held by its four neighbors. Each processor can update the inside cells of its block (in white) independently of its neighbors. The edge cells (in gray) must be exchanged between neighbors so that they can be updated by all the processors.

▷ **Question 2.** Each processor can compute internal cells of its block while exchanging edge cells with its neighbors. Let us consider a processor not on the edge of the processor grid. Using the same notation as in Section 4.5, the time for this processor to update its internal cells is

$$\left(\frac{n}{q} - 1\right)^2 w \,,$$

and the time for this processor for exchanging its edge cells with its neighbors in the 4-port, bidirectional model is

$$L + \frac{n}{q} b \,.$$

Once both these (simultaneous) operations have completed, each processor can then update all its edge cells in time

$$4 \left(\frac{n}{q} - 1 \right) w \,.$$

We obtain the overall execution time $T(n, q)$ as

$$T(n, q) = \max \left[\left(\frac{n}{q} - 1 \right)^2 w, L + \frac{n}{q} b \right] + 4 \left(\frac{n}{q} - 1 \right) w \,.$$

As n gets large, the execution time becomes equivalent to $\left(\frac{n}{q}\right)^2 w$, denoting an asymptotic efficiency of 1.

Chapter 6

Load Balancing on Heterogeneous Platforms

In Chapters 4 and 5 we discussed and developed parallel algorithms for various logical topologies, various data distributions, and various assumptions regarding the concurrency of communication and computation. But we made one assumption that was common to all our algorithms, namely, that the underlying computing platform is *homogeneous*. In this context homogeneity means that all processors have the same computing capabilities, and that all network links have the same physical characteristics. The homogeneous assumption is relevant as the most popular platforms for running distributed-memory applications today are commodity clusters, which are typically purchased and installed as homogeneous systems. Originally, the notion of heterogeneous parallel computing stemmed from the desire to build cheap parallel computers by using individual workstations already available in, say, an academic department. Such workstations are typically heterogeneous as they are purchased at different times, for different users, and possibly for different purposes. Such ad-hoc parallel computers were initially the main target of the PVM [58] message-passing library, which contributed to their popularity.

The use of such ensembles of heterogeneous workstations has since decreased due to the availability of affordable commodity clusters. However, the need for parallel algorithms that run on heterogeneous platforms has not disappeared. Well-known examples of highly heterogeneous systems that are used every day to run parallel applications are Internet-wide volunteer computing systems that harvest idle CPU cycles of donated home and office computers, such as SETI@home [7]. Heterogeneity also arises in the context of commodity clusters. During their life cycles clusters are often upgraded with additional compute nodes. These compute nodes use newer technology and typically have faster compute speeds than the original nodes, resulting in a heterogeneous cluster. An application that attempts to use all the available compute power of the cluster would then have to use heterogeneous nodes effectively. Perhaps the most compelling case for parallel algorithms that support heterogeneous platforms comes from the notion of *grid computing* [54, 25] (formerly known, to some extent, as "metacomputing"). The computational requirements of many scientific applications make it difficult to run these applications on a single cluster. Given the increasing availability of high-speed wide-area networks, it has become possible to run applications on increasingly distributed collections of individual clusters. These clusters typically belong to different institutions, were deployed at different times, and the overall system thus comprises heterogeneous compute nodes.

This chapter provides an introduction to parallel algorithms on heterogeneous platforms. The primary focus is on *load balancing*, which ensures that processors are assigned amounts of data and computation commensurate with their computing capabilities. We first address a simple load balancing problem (Section 6.1): Given M identical and independent tasks and p heterogeneous processors, how many tasks should be assigned to each processor to minimize execution time? While the solution is straightforward, it provides the basis for solving useful load balancing problems, as we will see in the context of a stencil application (Section 6.1.3) and of the LU factorization (Section 6.1.4). Both applications were discussed in previous chapters, but we revisit them here in the context of heterogeneous platforms. Although the load balancing problem is often straightforward in the case of one-dimensional (1-D) data distributions, it becomes difficult (i.e., NP-complete) for two-dimensional (2-D) data distributions. This chapter highlights this difficulty in the case of the outer-product algorithm for multiplying matrices (Sections 6.2 and 6.3).

While this chapter focuses solely on load balancing, the issue of heterogeneity is also explored in Chapters 7 and 8 in the context of application scheduling.

6.1 Load Balancing for One-Dimensional Data Distributions

The problem of allocating tasks to heterogeneous resources in a way that optimizes application performance (e.g., that minimizes execution time) has been studied extensively. Two broad variants of the problem have been defined, depending on whether processors are *non-dedicated* or whether they are *dedicated*. In the non-dedicated case, the compute speed that a processor delivers to the application varies through time and is in general not perfectly predictable, if at all. In this case, one must resort to *dynamic* strategies to allocate work to processors. Such strategies range from the simple workqueue scheme, by which a master processor maintains a list of tasks that are yet to be executed and a worker processor picks the first task in this list whenever idle, to more sophisticated approaches, which typically make work allocation decisions based on the past performance of the processors. In this book we only consider the dedicated case and refer the reader to [26] for a discussion of dynamic strategies for non-dedicated processors.

Even when assuming dedicated resources, the notion of dynamic resource allocation is attractive. Indeed, letting each processor request tasks when idle should lead the system to automatically converge to a resource allocation that is load balanced. Unfortunately, in many cases, due to data dependences, a dynamic strategy can lead to poor load balancing, as we will see later in this chapter. In such cases, one must resort to *static* task allocation schemes, which are the focus of this chapter.

6.1.1 Basic Static Task Allocation Algorithm

The first step for implementing a sound task allocation on a heterogeneous set of processors is to assess task execution times on each processor. These assessments can be based on off-line benchmark measurements of the actual tasks, or on performance models that estimate task execution times based on task and processor characteristics, or on a combination of both. Alternately, task execution times can be assessed during the initial stages of application execution, and a static task allocation based on these assessments can be computed and implemented for the remainder of application execution. Let us assume that our application consists of M tasks that are identical, independent, and atomic. That is, tasks do not require communications, can be completed in any order, and are indivisible. The computing platform consists of p processors, $P_1, \ldots, P_p$, and we denote the task execution time on each processor by $t_1, \ldots, t_p$. These times are often called *processor cycle times.* We thus have a problem with homogeneous tasks and heterogeneous processors.

Intuitively, the number of tasks to allocate to each processor P_i, which we denote by c_i, should be inversely proportional to t_i, so that the processors are assigned a number of tasks that is commensurate to their processor speeds. More formally,

$$\forall i, 1 \leqslant i \leqslant 1, \quad c_i \times t_i = K ,$$

where K is a constant. One can then compute the number of tasks allocated to each processor as

$$c_i = \frac{\frac{1}{t_i}}{\sum_{k=1}^{p} \frac{1}{t_k}} \times M .$$

If M is a multiple of $C = \operatorname{lcm}(t_1, t_2, \ldots, t_p) \sum_{i=1}^{p} \frac{1}{t_i}$, then load balancing is perfect, that is, $c_i \times t_i$ is a constant that does not depend on i. If M is not a multiple of C, the above formula does not produce an integral task allocation, which is problematic because tasks are atomic. One may then wonder how to round off the obtained solution. For instance, should one allocate a task to processor P_i if $c_i = 0.5$?

It turns out that the optimal allocation is easily obtained with Algorithm 6.1, which was originally used in [87]. This algorithm simply computes an integral lower bound on the allocation of each processor, and then iteratively increases the allocations. At each step, the allocation of the processor that can complete a new task the earliest is increased by one.

PROPOSITION 6.1. *Algorithm 6.1 produces an optimal task allocation.*

```
1  ALLOCATION((t_1, ..., t_p), M)
     { Initialization: compute c_i values such that
       c_i × t_i ≈ Constant and c_1 + c_2 + ... + c_p ⩽ M }
2    for i = 1 to p do
3      c_i = ⌊ (1/t_i) / (Σ_{k=1}^{p} 1/t_k) × M ⌋
     { Iteratively increment the c_i value so that execution
       time is minimized, until Σ_{i=1}^{p} c_i = M }
4    while Σ_{i=1}^{p} c_i < M do
5      k = argmin_{1⩽i⩽p} {t_i × (c_i + 1)}
6      c_k = c_k + 1
7    return (c_1, c_2, ..., c_k)
```

ALGORITHM 6.1: Optimal static allocation of M identical, independent, atomic tasks on p processors with cycle times $t_1, \ldots, t_p$.

Proof. Let us consider an optimal allocation, which we denote by $o_1, o_2, \ldots, o_p$ (o_i is the number of tasks allocated to processor P_i). Let j be the index of a processor with the maximum execution time: $\forall i \in \{1, \ldots, p\}$, $T = o_j t_j \geqslant o_i t_i$, where T is the overall execution time. To prove that the algorithm produces the optimal allocation we prove the following invariant:

$$\forall i \in \{1, \ldots, p\},\ c_i t_i \leqslant o_j t_j \ . \qquad (I)$$

After the initialization phase of the algorithm, for all i, $c_i \leqslant \frac{\frac{1}{t_i}}{\sum_{k=1}^{p} \frac{1}{t_k}} \times M$. Since $M = \sum_{k=1}^{p} o_k \leqslant o_j t_j \times \sum_{k=1}^{p} \frac{1}{t_k}$, we have $c_i t_i \leqslant \frac{M}{\sum_{k=1}^{p} \frac{1}{t_k}} \leqslant o_j t_j$, and invariant (I) is satisfied.

We now prove by induction that (I) is satisfied after each incrementation step in the algorithm. Assume that at a given step c_k is incremented. Before this step, $\sum_{i=1}^{p} c_i < M$, and there exists $k' \in \{1, \ldots, p\}$ such that $c_{k'} < o_{k'}$. We have $t_{k'}(c_{k'} + 1) \leqslant t_{k'} o_{k'} \leqslant t_j o_j$, and the choice of k' implies that $t_k(c_k + 1) \leqslant t_{k'}(c_{k'} + 1)$. Therefore, invariant (I) is satisfied after this step. Finally, the obtained allocation, $(c_1, c_2, \ldots, c_p)$, is optimal because for all i $\max\{c_i \times t_i\} \leqslant o_j t_j = \max(o_i t_i)$. □

After the initialization phase we have $c_1 + c_2 + \ldots + c_p \geqslant M - p$. The algorithm proceeds in at most p steps and its complexity is thus $O(p^2)$. This complexity can be further reduced, for instance, by using sorted data structures to speed up the computation for the argmin in statement 5 of the algorithm.

6.1.2 Incremental Algorithm

The algorithm presented in the previous section makes it possible to compute the optimal allocation on a set of processors for a given number of independent tasks M. However, if one wishes to compute the optimal allocations for all numbers of tasks between 1 and M, a simple dynamic programming algorithm is more efficient. We will see later in this chapter why knowing all these allocations is useful and leads to elegant load balancing solutions.

We develop an incremental algorithm that returns the optimal task allocation for 1 to M tasks. We illustrate the principle of the algorithm in an example with three processors with cycle times $t_1 = 3$, $t_2 = 5$, and $t_3 = 8$. The table on the left-hand side of Figure 6.1 summarizes the task allocation computed by the algorithm up to $M = 10$. In the table, the column "Chosen proc." shows the index of the processor that computes the next task. At each step, the chosen processor minimizes the overall execution time. As before, let c_i be the number of tasks allocated to processor P_i. The execution time of allocation $\mathcal{C} = (c_1, c_2, \ldots, c_p)$ is $\max_{1 \leqslant i \leqslant p} c_i t_i$. Therefore, the mean task execution time is $t_{mean} = \frac{\max_{1 \leqslant i \leqslant p} c_i t_i}{\sum_{i=1}^{p} c_i}$, which is shown in the table at each step.

Cycle time	$t_1 = 3$	$t_2 = 5$	$t_3 = 8$		
Num. of tasks	c_1	c_2	c_3	t_{mean}	Chosen proc.
0	0	0	0		P_1
1	1	0	0	3	P_2
2	1	1	0	2.5	P_1
3	2	1	0	2	P_3
4	2	1	1	2	P_1
5	3	1	1	1.8	P_2
6	3	2	1	1.67	P_1
7	4	2	1	1.71	P_1
8	5	2	1	1.87	P_2
9	5	3	1	1.67	P_3
10	5	3	2	1.6	

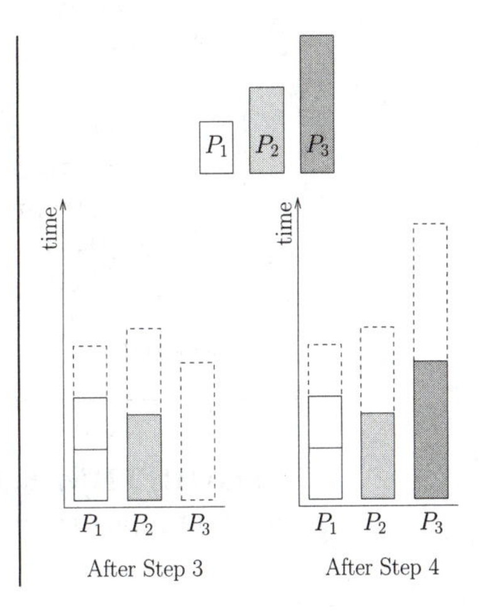

FIGURE 6.1: Steps of the incremental algorithm with three processors with cycle times $t_1 = 3$, $t_2 = 5$, and $t_3 = 8$.

Consider step 4 as an example, that is, the step in which the fourth task is allocated to a processor. The algorithm bases the allocation decision on the allocation obtained so far, that is, $(c_1, c_2, c_3) = (2, 1, 0)$. This allocation is depicted on the Gantt chart labeled as "After Step 3" on the right-hand side of Figure 6.1. This diagram depicts the three tasks already allocated, two to processor P_1 and one to processor P_2, with time along the vertical axis. The allocation decision boils down to choosing which c_i should be incremented. There are three possibilities: $(c_1+1, c_2, c_3) = (3, 1, 0)$, $(c_1, c_2+1, c_3) = (2, 2, 0)$, and $(c_1, c_2, c_3 + 1) = (2, 1, 1)$, with execution times $\frac{9}{4}$ (P_1 finishes last), $\frac{10}{4}$ (P_2 finishes last), and $\frac{8}{4}$ (P_3 finishes last), respectively. These three possible allocations are depicted as three dashed-line boxes on the "After Step 3" Gantt chart. The third allocation, $(c_1, c_2, c_3) = (2, 1, 1)$, is picked because it leads to the lowest execution time. This allocation is depicted on the "After Step 4" Gantt chart, which also depicts the next three possible choices for resource allocation. One can see that the best choice will be to allocate the next task to processor P_1 as it leads to the lowest overall execution time, as seen on the fifth row of the table in Figure 6.1.

Algorithm 6.2 shows the algorithm more formally. This algorithm produces an allocation that is encoded as a list of processor indices, $\mathcal{A}$, in the order of the "Chosen proc." column of the table in Figure 6.1. The list is initially empty, and at each step one processor index is added to it. The complexity of the algorithm is simply $O(pM)$.

```
1  IncrementalDistribution(t_1, ..., t_p, M)
       { Initialization }
2      C = (c_1, ..., c_p) = (0, ..., 0)
3      A = {}
       { Iterative computation of allocation A }
4      for n = 1 to M do
5          i = argmin_{1≤j≤p}(t_j × (c_j + 1))
6          A[n] = i
7          c_i ← c_i + 1
8      return A
```

ALGORITHM 6.2: Incremental allocation of M tasks on p processors with cycle times $t_1, \ldots t_p$.

PROPOSITION 6.2. *Every prefix of length m of the allocation computed by the incremental algorithm for M tasks on p processors corresponds to an optimal allocation of m tasks onto the p processors, $m \leqslant M$.*

Proof. The proof is very similar to that of the previous proposition. One can simply modify the invariant as

$$\forall i \in \{1, \ldots, p\},\ c_i t_i \leqslant o_{j_m}^{(m)} t_{j_m} = \max_{1 \leqslant j \leqslant p} o_j^{(m)} t_j, \qquad (I_m)$$

where $\mathcal{O}^{(m)} = (o_1^{(m)}, o_2^{(m)}, \ldots, o_p^{(m)})$ is an optimal allocation with m tasks, and j_m is the index of the processor that finishes last for this allocation. By induction, at step m, invariant (I_m) is satisfied. Since the overall execution time can only increase when one executes an extra task, we have $o_{j_m}^{(m)} t_{j_m} \leqslant o_{j_{m+1}}^{(m+1)} t_{j_{m+1}}$. Thus, before step $m+1$, invariant (I_{m+1}) is satisfied. Since step $m+1$ is identical to all steps of Algorithm 6.1, the proof of the previous proposition shows that (I_{m+1}) holds after step $m+1$. □

6.1.3 Application to a Stencil Application

Let us revisit the stencil application discussed in Section 4.3. Recall that the goal is to use the simple stencil shown in Figure 4.4 to update all cells of a 2-D domain. This stencil requires that the West and North neighbors of a given cell be updated before it. We have seen that, in the case of homogeneous processors, a cyclic allocation of rows, or blocks of rows, to the processors is efficient since it leads to (asymptotically) optimal load balancing. Remember also that each row is partitioned in several chunks, so as to strike a good compromise between communication overhead and parallelism. In this section, we consider the execution of the same stencil application on a set of heterogeneous processors.

In the case of heterogeneous processors, one could think of employing a dynamic allocation strategy in the hope that processors self-regulate and converge to a load balanced execution. Unfortunately, due to data dependences, a dynamic allocation is not effective for our stencil application. For instance, consider the natural approach in which a master processor maintains a list of rows to be computed, and each worker processor, whenever idle at the onset of the application or after computing a row, is allocated the next row in the list. The problem with this approach from a load balancing perspective stems from the fact that a chunk of cells cannot be computed before the chunk above it has been computed. Consider an example with two worker processors, a slow one and a fast one. Let us assume for instance that the first row of the domain is assigned to the fast processor and that the second row is assigned to the slow processor. Once the fast processor finishes computing the first row, the slow processor is still computing the second row. The fast processor starts processing the third row and may catch up to the slow processor. Once the slow processor has finished computing the last chunk of the second row, the fast processor can start computing the last chunk of the third row. While this computation is taking place the slow processor starts processing the fourth row, and so on, with each processor computing every other row. As a result, each processor is allocated an equal amount of work as opposed to an amount of work commensurate to its processing speed, and poor load balance ensues. In fact, in spite of being attractive at first glance, any dynamic allocation approach, whether it allocates entire rows (as in our example), chunks of cells, or individual cells, is doomed to produce an allocation that gives equal amounts of work to the processors.

Let us consider a platform with three processors with respective cycle times $t_1 = 3$, $t_2 = 5$, and $t_3 = 8$, where the cycle time denotes the time to compute one row of the domain. Let us consider a domain that consists of N rows. The sequential execution time is obtained by allocating all rows to the fastest processor, P_1, and is thus equal to $3N$. A simple cyclic allocation like the one we used in the homogeneous case, by which row j is assigned to processor $P_{j \bmod 3}$, leads to an execution time approximately equal to $\frac{8}{3}N$. Indeed, each processor computes in locked step with the slowest processor, with cycle time 8, and computes one third of the rows. This is also what a dynamic row allocation strategy would achieve. By contrast, Algorithm 6.1 leads to a much better allocation, as seen hereafter.

For the same reasons as in the homogeneous case, a cyclic distribution is required. However, we have seen that a simplistic cyclic distribution leads to poor load balancing. The idea is to use Algorithm 6.1 to compute a load balanced assignment of a small and fixed number of rows, and then use this assignment in a periodic fashion. For instance, let us consider only 10 rows. For 10 rows, the allocation computed by the algorithm for our three processors is $c_1 = 5$, $c_2 = 3$, and $c_3 = 2$. Based on this allocation we allocate the first five rows to P_1, the next three rows to P_2, the next two rows to P_3, the next five rows to P_1, and so on, until all rows have been assigned. This heterogeneous cyclic allocation is depicted in Figure 6.2. On the left-hand side of the figure we see the rows allocated to the processors, where each row consists of six chunks of cells in our example. On the right-hand side we depict which sets of chunks are computed at which steps of the application execution.

The main thing to note is that it takes almost the same time for P_1 to compute five chunks, for P_2 to compute three chunks, and for P_3 to compute two chunks, as ensured by the static allocation algorithm. Indeed, we have $c_1 \times t_1 = c_2 \times t_2 = 15$ and $c_3 \times t_3 = 16$. As a result, the processor idle time due to load imbalance is reduced. Furthermore, the execution time of the algorithm is approximately equal to the time it takes for the most loaded processor, in this case P_3, to compute its allocated rows. P_3 computes one fifth of the rows, and takes 8 time units to compute a row. Therefore, the overall execution time of the application is $\frac{8}{5}N$, which is a factor $\frac{5}{3}$ lower than with a simplistic cyclic allocation.

In our example we picked $\pi = 10$ rows arbitrarily as the period of our cyclic allocation, with the same pattern being repeated for each group of 10 rows. Let us see what would have happened if we had picked, say, $\pi = 6$. In this case Algorithm 6.1 would have returned $c_1 = 3$, $c_2 = 2$, and $c_3 = 1$. We compute $c_1 \times t_1 = 9$, $c_2 \times t_2 = 10$, and $c_3 \times t_3 = 8$. The load imbalance appears larger, since both P_1 and P_3 have to wait for P_2. To check this, we compute the execution time of the most loaded processor, P_2. We derive the value $\frac{5}{3}N$, since P_2 executes one third of the rows.

Can we determine a better period π? Only π values that make $c_i \times t_i$

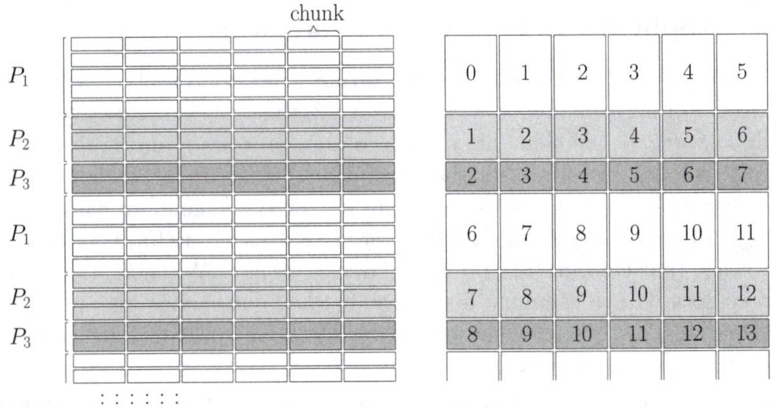

FIGURE 6.2: Static cyclic allocation of rows to three heterogeneous processors with cycle times $t_1 = 3$, $t_2 = 5$, and $t_3 = 8$, for the stencil application.

products all identical lead to perfect load balancing. In our case, the smallest such value is actually $\pi = 79$. Indeed, we obtain $c_1 = 40$, $c_2 = 24$, and $c_3 = 15$, and $c_i \times t_i = 120$ for all i. The execution time is $\frac{120}{79}N$, and it is the optimal value that can be achieved.

One can thus increase the granularity of the application (to trade off parallelism for lower communication overhead) by choosing π values that are multiples of the smallest π value that achieves perfect load balancing. This is exactly the same process as when using larger blocks of rows in the cyclic allocation used in the homogeneous case. Note, however, that the smallest π value that achieves perfect load balancing may be large. With p processors of cycle time t_i, $1 \leqslant i \leqslant p$, the smallest π value is $\pi = L\sum_{i=1}^{p}\frac{1}{t_i}$, where $L = \mathrm{lcm}(t_1, \ldots, t_p)$ is the least common multiple of the cycle times. In our previous example we had $L = 120$. But, for instance, with cycle times $t_1 = 11$, $t_2 = 23$, and $t_3 = 31$, we obtain the least common multiple $L = 7,843$ and the smallest π value is then $\pi = 1,307$. With such a large period, unless the total number of rows N is orders of magnitude larger than π, the reduced parallelism for the first and second steps of the application execution would have a prohibitive impact on performance. In this case, one may opt for a value of π that does not lead to perfect load balancing but that allows processors P_2 and P_3 to start computing earlier. For instance, using $\pi = 88$ leads to an execution in which the load imbalance between the processors at each step is lower than 1% (i.e., the relative differences between all processor execution times at each step are all below 1%). Indeed, we obtain $c_1 \times t_1 = 48 \times 11 = 528$, $c_2 \times t_2 = 23 \times 23 = 529$, and $c_3 \times t_3 = 17 \times 31 = 527$.

6.1.4 Application to the LU Factorization

Recall the principle of the LU factorization of an $n \times n$ matrix, as described in Section 4.4: At each step k, $k = 0, \ldots, n-2$, the processor that holds matrix column k *prepares* the elements of this column on rows k and higher through a simple scaling; this processor then broadcasts these updated elements to all other processors; all processors can then *update* all elements of columns $k+1$ and higher on rows $k+1$ and higher. Note that this algorithm is non-blocked, in the sense that columns are processed one at a time. While this is inefficient on processors with a memory hierarchy, it does not change the spirit of the algorithm. Let us consider the execution of this algorithm on p processors, $P_i, i = 0, \ldots, p-1$, with cycle times $t_1, \ldots t_p$.

The performance analysis of the LU factorization algorithm in Section 4.4 shows that the bulk of the computation time is due to the column updates, while the time for performing column preparation is asymptotically negligible. Our goal here then is to determine an allocation of columns to the processors that leads to a well-balanced execution of the column updates. Let us consider the first step of the algorithm. Once the first column has been prepared, all columns that need to be updated can be updated independently. Therefore, if the matrix is of size n, $n-1$ update tasks must be allocated to the processors.

A simple idea is to use Algorithm 6.1 to determine the column allocations. While this would lead to good load balance during the first step of the algorithm, unfortunately the number of columns to be updated decreases as the algorithm makes progress. In the second step, only $n-2$ columns need to be updated, and thus the allocation of columns to processors needs to be recomputed to lead to good load balancing. One then faces a conundrum. On the one hand, as the algorithm makes progress the column allocation likely leads to worsening load balance. On the other hand, redistributing the columns among the processors at each step according to the optimal column allocation at this step makes the algorithm much more complicated and would likely cause large overhead. One way to strike a compromise would be to only re-allocate columns periodically every x iteration, where x is chosen appropriately given n and the characteristics of the underlying platform. However, it is difficult to determine the best choice for x and one must resort to some empirical method based on previous runs and benchmarks.

It turns out that there is an elegant solution to the above conundrum, which does not require column re-allocations among the processors and which achieves optimal load balance at each step of the algorithm. Let us denote the $(n-k+1)$ update tasks at step $k = 0, \ldots, n-1$ of the algorithm as $u_{k+1}, \ldots, u_{n-1}$. This is a slight abuse of notation as we do not use a subscript to indicate the algorithm step, meaning that, for instance, u_{n-1} denotes the last update task at all steps. We seek an allocation of $u_1, \ldots, u_{n-1}$ onto the p processors, with the following constraint: For each $i \in \{1, \ldots, n-1\}$, the number of tasks allocated to processor P_j among tasks $u_i, \ldots, u_{n-1}$ should be (approximately) proportional to its cycle time t_j.

The attentive reader will recognize that the above constraint corresponds almost exactly to the task allocation computed by our incremental algorithm from Section 6.1.2. The only difference is that we need to run the algorithm *in reverse*. Indeed, the algorithm produces an optimal task allocation for all subsets $[1, i]$ of $[1, n-1]$, while we need an optimal task allocation for all subsets $[i, n-1]$ of $[1, n-1]$. Let us go back to our example of three processors with cycle times $t_1 = 3$, $t_2 = 5$, and $t_3 = 8$ for $\pi = 10$ column update tasks. To obtain our desired allocation for the LU factorization algorithm, we can just read the table on the left-hand side of Figure 6.1 from bottom to top, thus allocating column 0 to P_3, column 1 to P_2, and so on. The entire pattern is $(P_3P_2P_1P_1P_2P_1P_3P_1P_2P_1)$. For illustration purposes, Figure 6.3 depicts two allocations. The left-hand side figure shows the "reversed" allocation computed by the above algorithm, which is optimally load balanced at each step of the LU factorization. On the right-hand side, the figure shows the "non-reversed" allocation, which is optimally load balanced only for the first step of the LU factorization. Above each allocation we plot the compute time at each algorithm step, thus showing that using the reversed heterogeneous allocation leads to better performance.

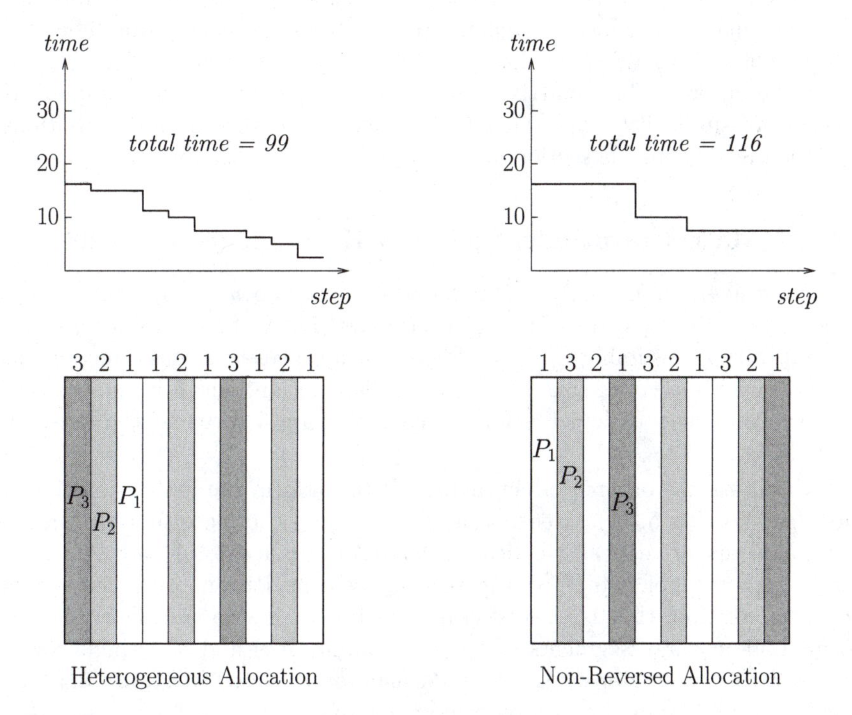

FIGURE 6.3: Comparison of the heterogeneous allocation that is optimal at all steps of the LU factorization algorithm (on the left) and of the heterogeneous allocation that is optimal only for the first step (on the right), with three processors ($t_1 = 3$, $t_2 = 5$, $t_3 = 8$) and ten column update tasks.

When n is large, it may be time consuming to compute the full heterogeneous allocation via the incremental algorithm. Furthermore, with a more irregular (i.e., less periodic) allocation of columns to processors, it is more difficult to overlap communication and computation. Therefore, on the one hand one desires the optimal allocation computed by the incremental algorithm over the n columns, and on the other hand one needs to have a periodic allocation. The typical approach is to pick a certain number of columns, say B, for which one computes the optimal allocation. This allocation pattern is then repeated every B columns.

6.2 Load Balancing for Two-Dimensional Data Distributions

In the previous section we have seen a simple algorithm for balancing computational load across a set of heterogeneous processors. We have seen two straightforward applications of this algorithm, both for 1-D data distributions. In turns out that the load balancing problem becomes much more difficult in the case of 2-D data distributions. In this section, we highlight this difficulty using the outer-product matrix multiplication algorithm as an example. For the sake of simplicity we consider non-cyclic, non-blocked data distributions, but the message in this section is unchanged for such distributions.

6.2.1 Matrix Multiplication on a Heterogeneous Grid

We consider the $C = A \times B$ multiplication of two $n \times n$ matrices, to be executed on a $q \times q$ torus of $p = q^2$ processors. Let us first consider a homogeneous torus. Assuming that q divides n, each processor stores a rectangular block of each matrix of size $\frac{n}{q} \times \frac{n}{q}$. This distribution, which is the same for all three matrices, is depicted for the case of a 3×3 torus on the left-hand side of Figure 6.4.

We will use the outer-product algorithm to perform the matrix multiplication (see Section 5.3.1). Recall that at each step the algorithm performs q horizontal and q vertical broadcasts of blocks of matrices A and B, respectively, for a total of q steps. In this section, we consider a non-blocked version of this algorithm, that is, a version in which at each step the algorithm performs broadcasts of segments of a single column of A and of a single row of B, for a total of n steps. These broadcasts are depicted in Figure 6.4 for a particular column of A and a particular row of B.

As seen on the right-hand side of the figure, this non-blocked outer-product algorithm is easily adaptable to the heterogeneous case. The only difference is that processors are assigned rectangular blocks of different sizes, with each size commensurate to each processor's computing speed.

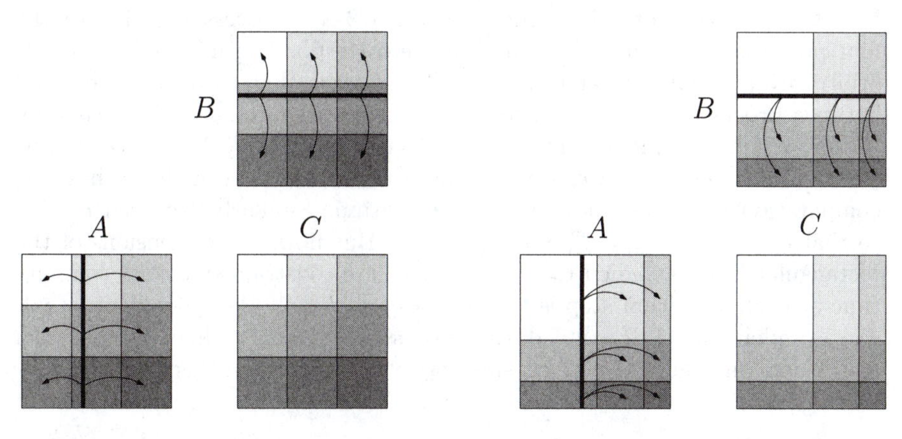

FIGURE 6.4: Example 2-D matrix distributions for matrix multiplication using the outer-product algorithm on a homogeneous 3×3 torus of processors (on the left) and on a heterogeneous 3×3 torus (on the right).

We denote the processors by $P_{i,j}$, with $1 \leqslant i, j \leqslant q$. We use $t_{i,j}$ to denote the cycle time of processor $P_{i,j}$, in this case the time to perform the multiplication of two elements of matrices A and B and to add the result to an element of matrix C. Let us assume that processor $P_{i,j}$ is allocated a rectangle of size $r_i \times c_j$, as shown on an example in Figure 6.5. At each step of the algorithm this processor computes for time $t_{i,j} \times r_i \times c_j$.

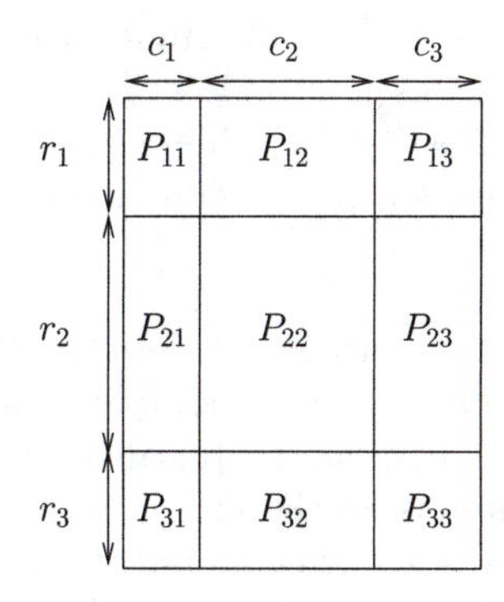

FIGURE 6.5: Example 2-D matrix distribution on a 3×3 heterogeneous processor grid.

Let us note that, depending on the cycle time of the processors and on the size of the matrices, it is not always possible to achieve perfect load balancing. See the example shown in Figure 6.6, for a 2×2 processor grid with the four processor cycle times indicated in the corresponding matrix blocks. We arbitrarily normalize r_1 and c_1 to 1. We observe that to balance the load between processor $P_{1,1}$ and processor $P_{1,2}$, we need $c_2 = \frac{1}{2}$ (which may only be approximated if the matrix dimension is not divisible by 2). Indeed, in this way $t_{1,1} \times r_1 \times c_1 = t_{1,2} \times r_1 \times c_2 = 1$ and thus both processors have the same computation time at each step of the algorithm. Similarly, we need $r_2 = \frac{1}{3}$ so that $t_{1,1} \times r_1 \times c_1 = t_{2,1} \times r_2 \times c_1 = 1$. But now, the dimensions of the rectangular block assigned to processor $P_{2,2}$ are constrained and its compute time at each algorithm step is $t_{2,2} \times r_2 \times c_2 = \frac{5}{6} < 1$. As a result processor $P_{2,2}$ is underutilized. If, instead, this processor had a cycle time $t_{2,2} = 6$, then load balancing could be perfect, since $t_{2,2} \times r_2 \times c_2 = 1$ in this case.

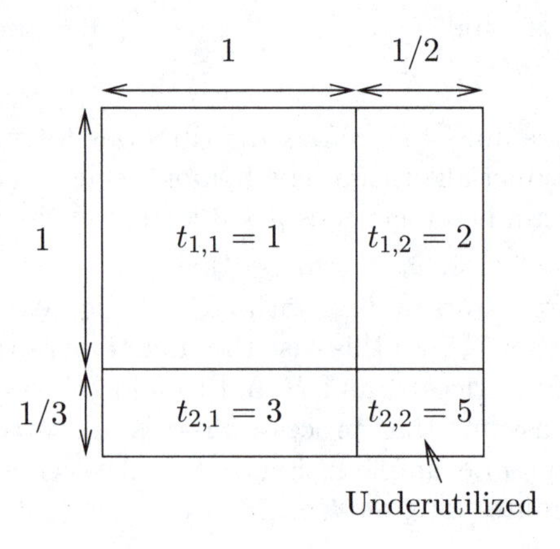

FIGURE 6.6: An example matrix distribution on a 2×2 heterogeneous processor grid with cycle times $t_{1,1} = 1$, $t_{1,2} = 2$, $t_{2,1} = 3$ and $t_{2,2} = 5$, which shows that it is not possible to achieve perfect load balance when the matrix of processor cycle times is not of rank 1. In this case processor $P_{2,2}$ is underutilized.

The above result, which we just showed on an example, is very easy to generalize: Perfect load balancing is achievable if and only if the matrix of the processor cycle times is of rank 1. Recall that the rank of a matrix is the number of rows, or columns, of the matrix that are linearly independent. Let us call T this matrix, whose elements are $(t_{i,j})$, $i, j = 1, \ldots, q$. If T is of rank 1, then we set $r_1 = 1$, $c_1 = 1/t_{1,1}$, $r_i = 1/(t_{i,1} \times c_1)$ (for $i \geqslant 2$), and $c_j = 1/t_{1,j}$ (for $j \geqslant 2$). We can then see that $t_{i,j} r_i c_j = 1$ for all $i, j \geqslant 2$. Indeed, $t_{i,j} \times r_i \times c_j = t_{i,j}/(t_{i,1} \times c_1 \times t_{1,j})$. But since the matrix is of rank 1, the determinant $\begin{vmatrix} t_{11} & t_{1j} \\ t_{i1} & t_{ij} \end{vmatrix}$ is equal to zero. Therefore, $t_{i,1} \times t_{1,j} = t_{1,1} \times t_{i,j}$. One then obtains $t_{i,j} \times r_i \times c_j = 1/(t_{1,1} \times c_1)$, which is equal to 1 by definition of c_1. Conversely, if load balancing is perfect, then $t_{i,j} \times r_i \times c_j = 1$ for all $i, j = 1, \ldots, q$. In this case matrix T has to be of rank 1 (one can easily show that the previous determinant is null for all i and j).

Although it may be impossible to achieve perfect load balancing given the values of the processor cycle times, we are still interested in finding the best possible allocation of rectangular blocks to the processors. Since at each step each processor $P_{i,j}$ computes for $r_i \times c_j \times t_{i,j}$ time units, the overall execution time of a step of the algorithm is $\max_{i,j}\{r_i \times c_j \times t_{i,j}\}$.

For the remainder of this section we normalize the matrix dimension to 1 so that the r_i and c_j values are rational numbers between 0 and 1. One can normalize and obtain the following optimization problem with unknowns r_i, $i = 1, \ldots, q$ and c_j, $j = 1, \ldots, q$:

$$\min_{(\sum_i r_i = 1; \sum_j c_j = 1)} \max_{i,j} \{r_i \times c_j \times t_{i,j}\} .$$

Given a solution to this problem, one computes the actual dimensions of the rectangular blocks assigned to each processor by multiplying the r_i and c_j values by n, the original matrix dimension. One then rounds all values to integers so that the sums of the r_i values and the sum of the c_j values are both equal to n.

The above formulation of the load balancing problem as an optimization problem does not lend itself to an easy solution. But, at any rate, the problem that we need to solve is much more complex. Indeed, the above optimization problem is for a given arrangement of the processors in a 2-D grid. However, there are $p!$ such arrangements! So we must compute the optimal solution to the optimization problem for all possible arrangements, and then pick the optimal solution for the optimal arrangement. As expected, this problem is NP-complete, which is shown in the next section.

6.2.2 On the Hardness of the Two-Dimensional Data Partitioning Problem

In this section we consider the data partitioning problem in more abstract terms. Consider $p = q^2$ processors, $P_1, \ldots, P_p$, with cycle times $t_1, \ldots, t_p$. The problem consists in placing these processors in a square grid of dimension $q \times q$, so that the resulting 2-D data distribution leads to the minimum execution time of the outer-product algorithm. This problem, which we call MAXGRID(s), can be formulated purely algebraically as

DEFINITION 6.1. MAXGRID(s): Consider q^2 real positive numbers $s_1, \ldots,$ s_{q^2}. Find $2q$ real positive numbers $(r_1, \ldots, r_q, c_1, \ldots, c_q)$ and a one-to-one mapping, or bijection, σ from $[1, q] \times [1, q]$ to $[1, q^2]$ such that

$$\forall (i, j) \in [1, q] \times [1, q], \quad r_i c_j \leqslant s_{\sigma(i,j)} \text{ and } \left(\sum_{i=1}^{q} r_i\right)\left(\sum_{j=1}^{q} c_j\right) \text{ is maximum.}$$

We simply replaced the cycle times by computing speeds: $s_i = \frac{1}{t_i}$ is the (relative) speed of processor P_i; σ represents the arrangement of the q^2 processors on the $q \times q$ grid; and r_i and c_j are the same variables as the ones used in the previous section. Note that the condition $r_i \times t_{\sigma(i,j)} \times c_j \leqslant 1$ is equivalent to $r_i c_j \leqslant s_{\sigma(i,j)}$. The objective function maximization simply attempts to have each processor perform as much computation as possible.

The MAXGRID(s) problem is relevant to a large class of algorithms on a 2-D grid of processors: all algorithms for which a 2-D domain needs to be partitioned among the processors and for which each processor is responsible for computing/updating the rectangular block of the domain it is allocated. The associated decision problem is as follows:

DEFINITION 6.2. MAXGRID(s, K): Consider q^2 real positive numbers $s_1, \ldots, s_{q^2}$ and a real positive number K. Are there $2q$ numbers ($r_1, \ldots, r_q$, $c_1, \ldots, c_q$) and a one-to-one mapping, or bijection, σ from $[1, q] \times [1, q]$ to $[1, q^2]$ such that

$$\forall (i, j) \in [1, q] \times [1, q], \quad r_i c_j \leqslant s_{\sigma(i,j)} \text{ and } \left(\sum_{i=1}^{q} r_i\right)\left(\sum_{j=1}^{q} c_j\right) \geqslant K \text{ ?}$$

THEOREM 6.1. MAXGRID(s, K) *is NP-complete.*

The proof of this theorem is long and rather intricate, and the interested reader will find it in [16]. The result itself is fundamental and shows the difficulty of balancing the load among heterogeneous processors in 2-D distributions. One must resort to heuristics. For instance, in [98], a (unfortunately) non-guaranteed polynomial heuristic is presented to solve MAXGRID(s). The main idea is to obtain an arrangement of processors in the grid so that the rank of the matrix of the processor cycle times is as close as possible to 1. This is achieved via a decomposition of the problem and via iterative refinement.

6.3 Free Two-Dimensional Partitioning on a Heterogeneous Grid

6.3.1 General Problem

When we introduced the data distribution for a matrix multiplication on a heterogeneous platform we made a seemingly natural but rather arbitrary assumption. We assumed that all processors on the same processor row (resp. column) are allocated matrix blocks of the same height (resp. width), leading to the simple, albeit irregular, checkerboard pattern seen in Figure 6.5. But, in fact, the more general problem is merely to partition matrices in as many rectangles as processors in any possible way. For instance, Figure 6.7 shows a possible data distribution for 13 heterogeneous processors. The goal is the same: Allocate to each processor a rectangular block that is inversely proportional to that processor's cycle time (i.e., proportional to that processor's speed). For this example we picked a prime number of processors to emphasize the fact that for solving the general problem it is not even necessary that the processors be arranged in a grid at all! At each step of the outer-product algorithm, there is a horizontal broadcast of a column of A and a vertical broadcast of a row of B. These broadcasts involve different numbers oW source and destination processors, depending on where the column or row is located. The figure shows an example vertical broadcast of a row of B, which is partially held by four processors. For instance, the processor holding the top right block of the matrix is involved in receiving data from all four source processors, while the processor holding the bottom left block of the matrix is involved in receiving data from only one source processor.

At each step each destination processor receives an amount of data proportional to the half-perimeter of the rectangular block it holds, typically from more than two source processors. This is in contrast with the data distribution seen in Figure 6.4, with which at each step each destination processor is only engaged in two receives (one horizontal and one vertical).

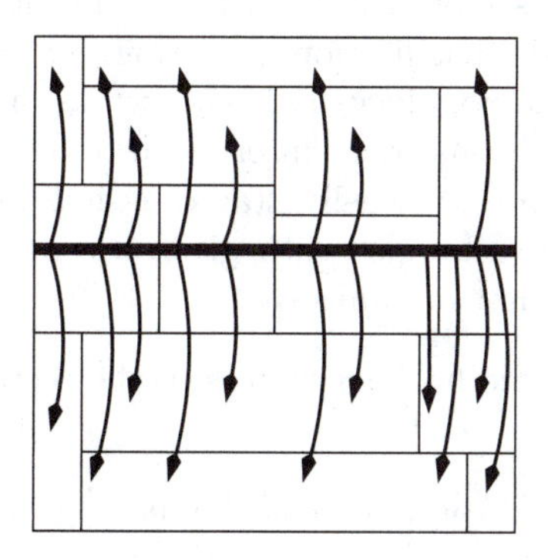

FIGURE 6.7: Example broadcast of a row of matrix B during a step of the outer-product algorithm with 13 heterogeneous processors.

Given p processors with speeds $s_1, \ldots, s_p$, normalized so that $\sum_{i=1}^{p} s_i = 1$, it is always possible to partition the unit square in p rectangles with areas $s_1, s_2, \ldots, s_p$. From now on we always reason on the unit square, knowing that we can then multiply rectangle dimensions by the matrix size n and round to appropriate integer values to obtain actual matrix blocks. One possibility for partitioning the unit square is to simply use a 1-D distribution in block rows. The question is, what is the best partitioning among the ones that achieve perfect load balancing? To formalize this question, one must define an objective function. Given that we only consider partitioning schemes that balance the load perfectly, we can distinguish them by the amount of communication these partitioning schemes induce for our outer-product matrix multiplication algorithm. We said earlier that each processor at each step is involved in communicating an amount of data proportional to the half-perimeter of the rectangular matrix block associated to that processor. This leads us to a very intuitive geometric interpretation of our problem: While the areas of the rectangular blocks are fixed, one can adjust their shapes to lead to the lowest amount of communication.

Depending on the network used by the underlying physical platform, different communication models are possible. For instance, all communications could be sequential if the processors are interconnected by, say, a non-switched Ethernet network. This could be the case if the participating processors are heterogeneous workstations in some laboratory for instance. Probably more typical for modern parallel platforms, such as heterogeneous commodity clusters, the communications can happen concurrently due to the use of a switched interconnect. If communications can happen concurrently, then one wishes to minimize the *maximum* of the half-perimeters of the rectangular blocks. If instead the communications happen sequentially, then one wishes to minimize the *sum* of the half-perimeters.

Given the above, we can now see one of the key points highlighted in Chapter 5: if the platform consists of q^2 homogeneous processors one can achieve lower communication costs by using a 2-D distribution on a grid of processors than by using a 1-D distribution on a ring. On a ring, the sum of the half-perimeters of the matrix blocks is $1 + q^2$, while on a $q \times q$ grid it is $q + q$.

Considering only the geometrical interpretation of the problem, both optimization problems above are easily stated as follows: how can one partition a unit square in p rectangles with given areas $s_1, s_2, \ldots, s_p$, such that $\sum_{i=1}^{p} s_i = 1$, in a ways that minimizes:

- either the sum of the half-perimeters of the rectangles (serialized communications);
- or the maximum half-perimeter of the rectangles (concurrent communications).

Consider an example with $p = 5$ rectangles $R_1, \ldots, R_5$ with areas $s_1 = 0.36$, $s_2 = 0.25$, $s_3 = s_4 = s_5 = 0.13$. One possible partition of the unit square for these five rectangles is shown in Figure 6.8. The areas of the rectangles are as follows: $R_1 = 0.61 \times \frac{36}{61}$, $R_2 = 0.61 \times \frac{25}{61}$, and $R_3 = R_4 = R_5 = 0.39 \times \frac{1}{3}$. The sum of the half-perimeters is 4.39. A lower bound is reached when all rectangles are squares, which is not possible in this example, but which would lead to a sum of the half-perimeters approximately equal to 4.36.

The maximum half-perimeter is for R_1 and approximately equal to 1.2002. Here, a lower bound is achieved when R_1 is a square, with a half-perimeter of 1.2, which is also not possible in this example. We conclude that the depicted partition of the unit square into five rectangles is very satisfactory for both our objective functions.

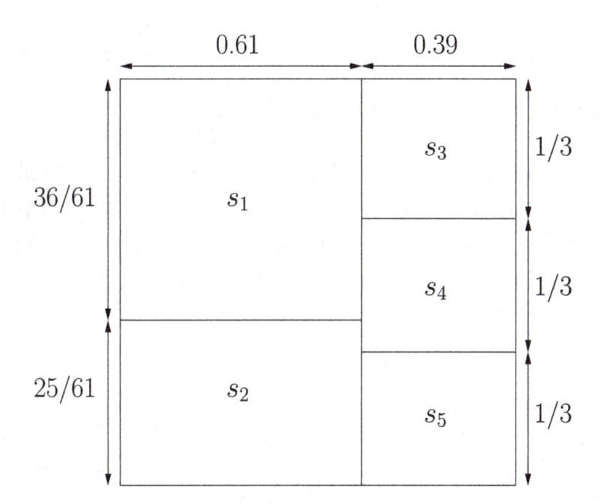

FIGURE 6.8: A partitioning of the unit square using five rectangles.

For all that follows we only consider the first objective function, i.e., the sum of the half-perimeters. The degrees of freedom of the problem are the shapes of the p rectangles used to partition the unit square. We can define the problem formally:

DEFINITION 6.3. PERISUM(s): Given p positive real numbers $s_1, \ldots, s_p$ such that $\sum_{i=1}^{p} s_i = 1$, find a partition of the square unit in p rectangles R_i, $i = 1, \ldots, p$, with areas s_i and of dimensions $h_i \times v_i$, such that $\widehat{C} = \sum_{i=1}^{p}(h_i + v_i)$ is minimum.

There is a trivial lower bound for problem PERISUM(s), which we saw earlier in an example and which can be formalized as

LEMMA 6.1. *For any solution $\widehat{C}$ to* PERISUM(s)*, we have $\widehat{C} \geqslant 2\sum_{i=1}^{p}\sqrt{s_i}$.*

Proof. The half-perimeter of each rectangle R_i is maximized when the rectangle is a square and is then equal to $2\sqrt{s_i}$. As discussed earlier, this lower bound is not always reachable because it is not always possible to partition the square unit using only squares of given areas. □

The decision problem associated to PERISUM(s, K) is as follows:

DEFINITION 6.4. PERISUM(s, K): Given p positive real numbers $s_1, \ldots, s_p$ such that $\sum_{i=1}^{p} s_i = 1$ and a positive real number K, does there exist a partition of the unit square in p rectangles R_i, $i = 1, \ldots, p$, with areas s_i and dimensions $h_i \times v_i$, such that $\sum_{i=1}^{p}(h_i + v_i) \leqslant K$?

The following theorem should be no surprise to the reader:

THEOREM 6.2. PERISUM(s, K) *is NP-complete.*

Like for the 2-D load balancing theorem in Section 6.2.2, the proof is long and involved, and we refer the interested reader to [17].

6.3.2 A Guaranteed Heuristic

It turns out that if one restricts the PERISUM(s, K) problem to consider partitions that consist of columns of rectangles, then one can compute an optimal solution. We describe here the solution and the performance guarantee obtained when compared to the solution of PERISUM(s, K).

The COLPERISUM(s) Problem

Since PERISUM(s) is NP-hard, we consider a more constrained problem, COLPERISUM(s). In this new problem we specify that the partition of the unit square must consist of columns of rectangles, as shown in the example in Figure 6.9. Formally, given p real positive numbers $s_1, \ldots, s_p$ such that $\sum_i s_i = 1$, we want to partition the unit square in $\mathcal{C}$ columns (where $\mathcal{C}$ is to be determined) of widths $c_1, \ldots, c_{\mathcal{C}}$. Each column C_i is itself partitioned in k_i rows (to be determined as well) of areas $s_{\sigma(i,1)}, \ldots, s_{\sigma(i,k_i)}$. The resulting partition of the unit square consists of $\sum_{i=1}^{\mathcal{C}} k_i = p$ rectangles. The goal is to produce such a partition that minimizes the sum of the half-perimeters of the rectangles.

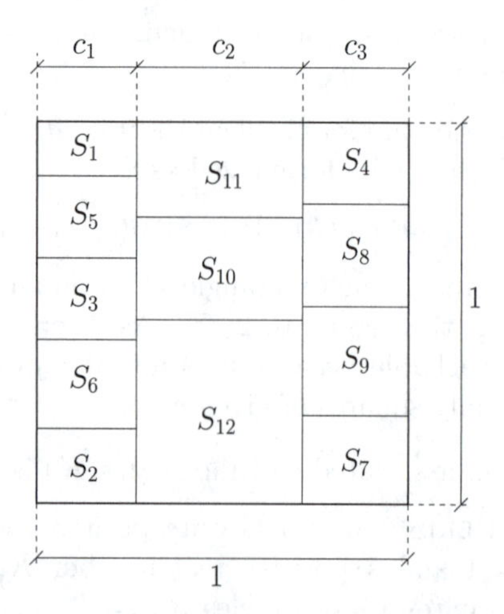

FIGURE 6.9: Column partition of the unit square with $\mathcal{C} = 3$ columns with each $k_1 = 5$, $k_2 = 3$, and $k_3 = 4$ rectangles.

TABLE 6.1: Table showing the values of $f_{\mathcal{C}}(q)$ and of a_0 (separated by a "|") for our eight-rectangle example. The values in boldface indicate the optimal solution.

$\mathcal{C}$ \ q	$q=1$	$q=2$	$q=3$	$q=4$	$q=5$	$q=6$	$q=7$	$q=8$
$\mathcal{C}=1$	1.05 | 0	1.2 | 0	**1.54 | 0**	2.12 | 0	2.9 | 0	4 | 0	5.90 | 0	9 | 0
$\mathcal{C}=2$		2.10 | 1	2.28 | 2	2.56 | 2	2.94 | 3	**3.50 | 3**	4.38 | 4	5.76 | 5
$\mathcal{C}=3$			3.18 | 2	3.38 | 3	3.66 | 4	4 | 4	4.58 | 5	**5.50 | 6**
$\mathcal{C}=4$				4.28 | 3	4.48 | 4	4.78 | 5	5.20 | 6	5.88 | 7
$\mathcal{C}=5$					5.38 | 4	5.60 | 5	5.98 | 6	6.50 | 7
$\mathcal{C}=6$						6.50 | 5	6.80 | 6	7.28 | 7
$\mathcal{C}=7$							7.70 | 6	8.10 | 7
$\mathcal{C}=8$								9 | 7

The algorithm to solve this problem uses dynamic programming and relies on the two following ideas:

1. It renumbers variables $s_1, \ldots, s_p$ so that $s_1 \leqslant s_2 \leqslant \ldots \leqslant s_p$.

2. It iteratively constructs p functions $f_{\mathcal{C}}$ for values of $\mathcal{C}$ going from 1 to p. For $q \in \{1, \ldots, p\}$, $f_{\mathcal{C}}(q)$ is defined as the sum of the half-perimeters in an optimal partition of a rectangle with height 1 and with width $(\sum_{i=1}^{q} s_i)$ in $\mathcal{C}$ columns and q rectangles with areas $s_1, \ldots, s_q$.

The key idea behind the algorithm is that it is straightforward to compute function $f_{\mathcal{C}}$ recursively based on function $f_{\mathcal{C}-1}$ as follows:

$$f_{\mathcal{C}}(q) = \min_{a \in [\mathcal{C}-1, q-1]} \left(1 + (q-a) \sum_{a < i \leqslant q} s_i + f_{\mathcal{C}-1}(a)\right) . \tag{6.1}$$

In the above minimum a corresponds to the number of rectangles in the first $\mathcal{C} - 1$ columns, and thus $q - a$ corresponds to the number of rectangles in the last column. The first term in the minimum, $1 + (q - a) \sum_{a<i\leqslant q} s_i$, is the contribution of the last column to the total sum of the half-perimeters. The intuitive interpretation of Equation (6.1) is that it looks at all possible ways to make use of an additional column to achieve the optimal partition. We denote by a_0 the value of parameter a that achieves the minimum in Equation (6.1). This recursion is initialized by setting

$$f_1(q) = 1 + q \times \sum_{i=1}^{q} s_i ,$$

which simply gives the sum of the half-perimers of a rectangle with height 1 and with width $(\sum_{i=1}^{q} s_i)$ in one column and q rectangles with areas $s_1, \ldots, s_q$.

To better understand how the algorithm works, let us apply it in an example. Consider $p = 8$ rectangles with areas $(0.05; 0.05; 0.08; 0.1; 0.1; 0.12; 0.2; 0.3)$. The algorithm recursively computes all $f_{\mathcal{C}}(q)$ values for $1 \leqslant q, \mathcal{C} \leqslant 8$, as shown in Table 6.1. In each column of the table we show the optimal value, i.e., the one with the smallest $f_C(q)$ value, in boldface. Since we wish to partition the unit square in eight rectangles, we look at column $q = 8$ and find that the optimal $f_C(q)$ value, 5.5, is achieved for $C = 3$, indicating that the optimal partition consists of three columns. Furthermore, the optimal $f_C(q)$ value is achieved for $a_0 = 6$. Therefore, the last column of the optimal partition must contain $8 - 6 = 2$ rectangles. We now look at column $q = 6$ of the table and find out that the optimal $f_C(q)$ value is achieved for $a_0 = 3$. Therefore, the next-to-last column of the optimal partition must contain $6 - 3 = 3$ of the remaining rectangles. Similarly, we determine that in column $q = 3$ of the table the optimal $f_C(q)$ value is achieved for $a_0 = 0$. The first column of the optimal partitioning consists of all remaining $3 - 0 = 3$ rectangles, which makes sense since we know that the optimal partitioning consists of three columns. The widths of the three columns in the optimal partitioning are thus $c_1 = s_1 + s_2 + s_3$, $c_2 = s_4 + s_5 + s_6$, and $c_3 = s_7 + s_8$. This partitioning is shown in Figure 6.10.

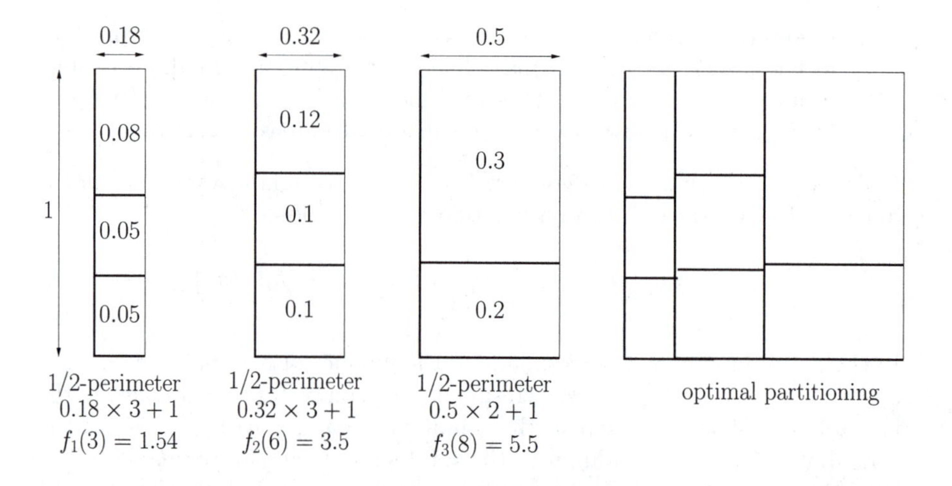

FIGURE 6.10: Optimal column partitioning for our eight-rectangle example. The thicker lines correspond to the rectangle sides that contribute to the sum of the half-perimeters.

Algorithm 6.3 makes it possible to compute the optimal partition with, in the worst case, a complexity of $O(p^3)$. The final partition corresponds to function $f_{C_{opt}}(p) = \min_{1 \leqslant C \leqslant p} f_C(p)$. This partition is found by Algorithm 6.4. We saw in an example how this algorithm operates, namely, by reading Table 6.1 from right to left and by following the values in boldface. The unit square is partitioned in C_{opt} columns. The i-th column contains rectangles $s_{d_i}, s_{d_i+1}, \ldots, s_{d_i+k_i}$ with $d_i = k_1 + k_2 + \ldots + k_{i-1}$.

```
1  PARTITION(s_1, ..., s_p)
2    S = 0
3    for q = 1 to p do
4      S = S + s_q
5      f_1(q) = 1 + S × q
6      f_1^cut(q) = 0
7    for C = 2 to p do
8      for q = C to p do
9        f_C(q) = min_{a∈[C−1,q−1]} ( 1 + (q − a) Σ_{q−a<i≤q} s_i + f_{C−1}(a) )
10       f_C^cut(q) = q − a_0   { where a_0 is the value of a that leads to the
                                   minimum in the previous expression. }
11   return (f_*^cut)
```

ALGORITHM 6.3: Algorithm for ColPeriSum(s): construction of the functions $f_{\mathcal{C}}$. $f_{\mathcal{C}}^{\text{cut}}(q)$ corresponds to the number of rectangles used in the $\mathcal{C}-1$ first columns, which leaves $q - f_{\mathcal{C}}^{\text{cut}}(q)$ rectangles in column $\mathcal{C}$.

Algorithms 6.3 and 6.4 together compute the optimal solution of problem ColPeriSum(s). The only remaining difficulty is to show that one can indeed consider only ordered $s_1 \leqslant s_2 \ldots \leqslant s_p$ sequences:

DEFINITION 6.5. A partitioning is said to be *well-ordered* if for all pair of columns $\mathcal{C}_i$ and $\mathcal{C}_j$, either all elements of $\mathcal{C}_i$ are lower than or equal to all elements of $\mathcal{C}_j$, or the other way around. Figure 6.11 illustrates this definition.

Without loss of generality, consider a partitioning in $\mathcal{C}$ columns of widths $k_1 \geqslant k_2 \geqslant \ldots \geqslant k_{\mathcal{C}}$ (swapping columns around does not change the quality of the partitioning). Assume that the s_i values are indexed so that $s_1 \leqslant s_2 \leqslant \ldots \leqslant s_p$; let τ be a permutation of $\{1, 2, \ldots, p\}$ such that the i-th

```
1  RE-BUILD(f_*^cut, C_opt)
2    q = p
3    for C = C_opt down to 1 do
4      k_C = q − f_C^cut(q)
5      q = f_C^cut(q)
6    return (k_1, ..., k_{C_opt})
```

ALGORITHM 6.4: Reconstructing the optimal solution from functions $f_{\mathcal{C}}^{\text{cut}}$.

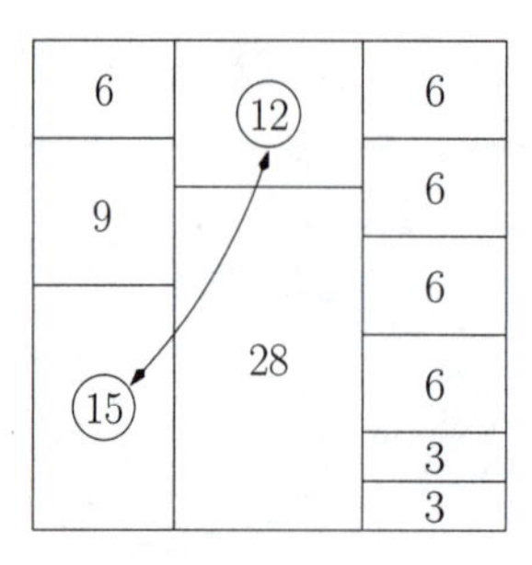

well-ordered

FIGURE 6.11: Two example column partitions. The one on the left is not well-ordered, while the one on the right is.

column of the partitioning contains rectangles $s_{\tau(d_i+1)}, \ldots, s_{\tau(d_i+k_i)}$, where $d_i = k_1 + k_2 + \ldots + k_{i-1}$. Recall that the contribution of column C_i to the total sum of the half-perimeters is $1 + k_i \sum_{j=d_i+1}^{d_i+k_i} s_{\tau(j)}$. The total sum is computed as follows:

$$\begin{aligned}
&\mathcal{C} + k_1 s_{\tau(1)} + k_1 s_{\tau(2)} + \ldots + k_1 s_{\tau(k_1)} \\
&\quad + k_2 s_{\tau(k_1+1)} + k_2 s_{\tau(k_1+2)} + \ldots + k_2 s_{\tau(k_1+k_2)} \\
&\quad + \ldots \\
&\quad + k_{\mathcal{C}} s_{\tau(k_1+\ldots+k_{\mathcal{C}-1}+1)} + k_{\mathcal{C}} s_{\tau(k_1+\ldots+k_{\mathcal{C}-1}+2)} + \ldots + k_{\mathcal{C}} s_{\tau(k_1+\ldots+k_{\mathcal{C}})} .
\end{aligned}$$

Since $k_1 \geqslant k_2 \geqslant \ldots \geqslant k_{\mathcal{C}}$, the above expression is minimized for $\tau =$ Identity, which characterizes a well-ordered partition! The proof is completed via an induction on the number of element swaps in permutation τ. We conclude that, for all partition there is a corresponding well-ordered partition that leads to a total sum of half-perimeters that is lower or equal to that of the original partition. This completes the proof that Algorithms 6.3 and 6.4 compute the optimal solution to ColPeriSum(s).

Performance Guarantee

In this section, we show that column partitioning leads to a good approximation of the optimal (free) partition. This is especially true when the ratio between the largest rectangle area, $\max s_i$, and the smallest area, $\min s_i$, is low.

THEOREM 6.3. *Let $r = \frac{\max s_i}{\min s_i}$. Let $\widehat{C}$ denote the sum of the half-perimeters of the rectangles in the optimal column partition, and let $LB = 2\sum_{i=1}^{p} \sqrt{s_i}$. Then we have*

$$\frac{\widehat{C}}{LB} \leqslant \sqrt{r}\left(1 + \frac{1}{\sqrt{p}}\right) = \sqrt{\frac{\max s_i}{\min s_i}}\left(1 + \frac{1}{\sqrt{p}}\right).$$

If $r = 1$, i.e, if all processors have the same speed, column partitioning is asymptotically optimal. By contrast, if r is large, then the above upper bound is very pessimistic.

Proof. We define $\mathcal{C} = \lceil \sqrt{r} \sum_i \sqrt{s_i} \rceil$. Let us consider the partition that consists of $\mathcal{C}$ columns in which the rectangles are distributed evenly across the columns. Therefore, the number of rectangles per column is either $\lfloor \frac{p}{\mathcal{C}} \rfloor$ or $\lceil \frac{p}{\mathcal{C}} \rceil$. Let $\widehat{C}^*$ denote the sum of the half-perimeters of this partitioning. We then have

$$\begin{aligned}\widehat{C}^* &\leqslant \left\lceil \sqrt{r} \sum_i \sqrt{s_i} \right\rceil + \left\lceil \frac{p}{\lceil \sqrt{r} \sum_i \sqrt{s_i} \rceil} \right\rceil \\ &\leqslant 2 + \sqrt{r} \sum_i \sqrt{s_i} + \frac{p}{\sqrt{r} \sum_i \sqrt{s_i}} \ .\end{aligned}$$

And therefore,

$$\frac{\widehat{C}^*}{2 \sum_i \sqrt{s_i}} \leqslant \frac{1}{\sum_i \sqrt{s_i}} + \frac{\sqrt{r}}{2} + \frac{p}{2\sqrt{r}(\sum_i \sqrt{s_i})^2} \ .$$

Furthermore,

$$\begin{aligned}\sum_i s_i = 1 &\Longrightarrow p \max s_i \geqslant 1 \\ &\Longrightarrow \min s_i \geqslant \frac{1}{pr},\end{aligned}$$

which leads to

$$\sum_i \sqrt{s_i} \geqslant p \sqrt{\min s_i} \geqslant \sqrt{\frac{p}{r}} \ .$$

Finally, we obtain

$$\begin{aligned}\frac{\widehat{C}^*}{2 \sum_i \sqrt{s_i}} &\leqslant \sqrt{\frac{r}{p}} + \frac{\sqrt{r}}{2} + \frac{\sqrt{r}}{2} \\ &\leqslant \sqrt{r} \left(1 + \frac{1}{\sqrt{p}} \right) .\end{aligned}$$

Since $\widehat{C}$ corresponds to the best solution among all the possible column partitionings, we have $\widehat{C} \leqslant \widehat{C}^*$, which completes the proof. □

Bibliographical Notes

The load balancing results for 1-D data distributions presented in the first part of this chapter are well known, and we refer the reader to the referenced work therein for further details. The section on load balancing for 2-D data distributions comes for the most part from the Ph.D. thesis by Rastello [98] and related articles, which contain many results. Generally speaking, the literature is rife with works that study load balancing problems and we explore some of them in the exercises accompanying this chapter.

6.4 Exercises

The first exercise is a straightforward demonstration that rather than facing the difficulty of load balancing for 2-D data distributions, an easier option is to transform a "grid algorithm" into a "ring algorithm." The second exercise studies load balancing for the LU factorization on a heterogeneous grid of processors. Finally, the third exercise discusses optimal load balancing for a stencil application on a heterogeneous ring of processors (inspired by the work in [78], to which we refer the reader for more details).

⋄ Exercise 6.1 : Matrix Product on a Heterogeneous Ring

1. Consider the outer-product matrix multiplication algorithm that we described on a grid of processors in Section 5.3.1. We have seen that heterogeneous grids of processors lead to difficult load balancing problems. So, instead, let us see how this algorithm can be executed on a ring of processors. For starters, explain how this algorithm can be executed on a *homogeneous* ring of p processors for $n \times n$ matrices.

2. Adapt the algorithm in the previous question to the case of a *heterogeneous* ring of p processors with cycle times t_i, $i = 1, \ldots, p$, so that the execution is load balanced. Consider that the execution of the algorithm is blocked using an $r \times r$ block size (as opposed to proceeding column-by-column), where r divides n and $r \leqslant n/p$. Explain the execution of the algorithm in an example.

⋄ Exercise 6.2 : LU Factorization on a Heterogeneous Grid

Consider the execution of the LU factorization of an $n \times n$ matrix A (see Section 4.4) but using a 2-D data distribution on a $p \times q$ heterogeneous grid of processors, with processor $P_{i,j}$'s cycle time denoted by $t_{i,j}$, $i = 1, \ldots, p$, $j = 1, \ldots, q$. Recall that a challenge with the LU factorization is that it updates the lower right sub-matrix of matrix A, the size of which decreases at each step of the algorithm. In Section 6.1.4 we have seen that allocating columns to processors "in reverse" of the allocation computed by the static load balancing algorithm in Section 6.1.1 leads to a data distribution that is optimal at each step. But this result was for a ring of processor and thus a 1-D data distribution. We now know that load balancing is more challenging for 2-D data distributions. But what of the LU factorization with a 2-D distribution?

1. Consider the case in which the matrix of processor cycle times is of rank 1. In this case, we know that perfect load balancing is possible for the whole matrix. In other words, we can find c_i and r_j, $i = 1, \ldots, p$, $j = 1, \ldots, q$, such that $r_i t_{i,j} c_j = 1$ for all i, j, and such that $\sum_i r_i = 1$ and $\sum_j c_j = 1$. Processor $P_{i,j}$ is thus allocated an $r_i \times c_j$ block of the matrix (which is really normalized and must be multiplied by n and rounded to an integer value). See Section 6.2.1 for details. Show that if the matrix of processor cycle times is of rank 1, then it is possible to distribute columns and rows to processors such that the distribution is optimal at each step step of the LU factorization.

2. Show an example in which the matrix of processor cycle times is not of rank 1 and in which it is not possible to achieve a data distribution that is optimal at each step.

Exercise 6.3 : Stencil Application on a Heterogeneous Ring

Consider a stencil application like the one described in Section 4.5 to be executed on a ring of p processors. Each processor P_i, $i = 1, \ldots, p$, can be allocated a block column of the 2-D domain. Let W be the total number of columns; α_i contiguous columns are allocated to processor P_i. At each iteration of the stencil application a processor and its neighboring processors must exchange two columns of the domain, say a total amount D both ways. We assume a 2-port, bidirectional ring with full-duplex links, with non-blocking sends and blocking receives.

1. Let us first consider the case of a ring with heterogeneous processors and homogeneous network links: The time for processor P_i to process a column is w_i; the time for a processor to exchange a column with both its successor and predecessors is $c \times D$ (where c is the time to exchange one unit of data with both its neighbors). Show that the optimal allocation uses either one or all processors. Give a condition sufficient for the optimal allocation to use all processors.

2. We now consider that the underlying platform is fully connected with heterogeneous network links. Therefore, processors P_i and P_j ($i \neq j$) exchange D data items in time $c_{i,j}D$. Assume that *all* processors participate in the computation. It turns out that the load balancing problem in this case is difficult. Indeed, show that finding the best allocation is equivalent to finding the shortest Hamiltonian cycle in the platform (with processors as vertices and network links as edges) for some edge weights that you must define. (Which is to say that solving the load balancing problem is equivalent to solving the well-known NP-hard traveling salesman problem.)

6.5 Answers

Exercise 6.1 (Matrix Product on a Heterogeneous Ring)

▷ **Question 1.** Just consider the ring as a rectangular $1 \times p$ grid and simply run the outer-product algorithm on this grid. For a $C = A \times B$ product, the three matrices are distributed across the processors by column blocks of $r = n/p$ columns. At step $k = 1, \ldots, p$, the processor that holds the k-th column block of A broadcasts it to all other processors. This is the only communication since vertical broadcasts of B are now replaced by local memory accesses within block rows.

▷ **Question 2.** Let us consider an example with $n = 2,528$ and three processors with cycle times $t_1 = 3$, $t_2 = 5$, and $t_3 = 8$. Let us take a block size $r = 32$, so that the matrices consist of $B = 79$ block columns, which must be distributed across the processors.

One can simply use the static load balancing algorithm in Section 6.1.1 to determine how many block columns are allocated to each processor. The load balancing algorithm produces the following allocations of block columns to processors: $c_1 = 40$, $c_2 = 24$, $c_3 = 15$. This allocation is perfectly load balanced. Processor P_1 holds the first c_1 column blocks of all matrices, P_2 the next c_2 column blocks, and P_3 the last c_3 column blocks. The blocked algorithm proceeds in B steps. At step k, the processor that holds the k-th column block broadcasts it to all the other processors. Each processor also has the pieces of the k-th row blocks of B. Therefore, each processor can update the corresponding $r \times r$ block of C. Two steps are shown in Figure 6.12. The time for each step is proportional to $c_1 \times t_1 = c_2 \times t_2 = c_3 \times t_3$.

Exercise 6.2 (LU Factorization on a Heterogeneous Grid)

▷ **Question 1.** It turns out that using the static allocation algorithm in reverse on *both* the rows and the columns does the job. Consider a distribution σ_1 of the matrix rows to the processors, i.e., a function from $\{1, \ldots, n\}$ to $\{1, \ldots, p\}$. Similarly, consider a distribution σ_2 of the matrix columns to the processors, i.e., a function from $\{1, \ldots, n\}$ to $\{1, \ldots, q\}$. Let $k \in \{1, \ldots, n\}$ denote the step of the LU factorization algorithm. We consider the bottom right square sub-matrix of A, $A(k)$, which starts at row k and at column k. Let $r'_i(k)$ denote the number of rows in $A(k)$ that are allocated to processors on processor row i, that is, the number of indices l such that $\sigma_1(l) = i$. Similarly, let $c'_j(k)$ denote the number of columns in $A(k)$ that are allocated to processors on processor column j, that is, the number of indices l such that $\sigma_2(l) = j$. The time needed to complete step k of the LU factorization algorithm is thus $T(k) = \max_{i,j} r'_i(k) t_{i,j} c'_j(k)$, since processor $P_{i,j}$ is responsible for updating

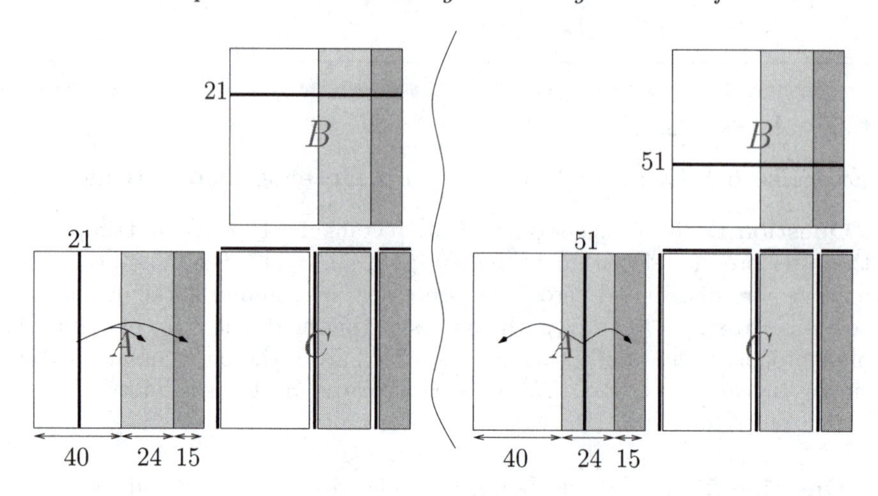

FIGURE 6.12: Steps 21 and 51 of the outer-product algorithm on a heterogeneous ring for our example matrix multiplication.

$r'_i(k) \times c'_j(k)$ elements of sub-matrix $A(k)$. We easily see that

$$T(k) = \max_i \max_j r'_i(k) t_{i,j} c'_j(k) = \max_i \max_j \frac{r'_i(k) c'_j(k)}{r_i c_j},$$

since $r_i t_{i,j} c_j = 1$. One can then separate the maxima, which are independent, to obtain

$$T(k) = \max_i \frac{r'_i(k)}{r_i} \times \max_j \frac{c'_j(k)}{c_j}.$$

The objective is to minimize the above quantity for all k. It turns out that the static load balancing algorithm from Section 6.1.1, whose allocation is used in reverse, can be used to minimize both these maxima because it leads to 1-D data distributions that are optimal at each step of the LU factorization. (Minimizing the maximum of the, for instance, $r'_i(k)/r_i$ ratio for all i is equivalent to ensuring that each processor has an amount of work that is as commensurate as possible to its cycle time.) We conclude that using the allocation produced by this algorithm in reverse and along both dimensions produces a data distribution optimal at each step!

▷ **Question 2.** Consider a 3×3 processor grid with the cycle times depicted in Figure 6.13. There is one fast processor and eight slow processors. We must show that the optimal allocation of matrix elements for a 3×3 matrix cannot be constructed based on the optimal allocation of matrix elements for a 2×2 matrix.

For a 2×2 matrix there are only three possible kinds of allocations of matrix elements to the processors: (i) the fast processor is assigned all matrix elements; (ii) the fast processor is assigned only one matrix element; and

$t_{1,1} = 5$	$t_{1,2} = 5$	$t_{1,3} = 5$
$t_{2,1} = 5$	$t_{2,2} = 5$	$t_{2,3} = 5$
$t_{3,1} = 5$	$t_{3,2} = 5$	$t_{3,3} = 1$

FIGURE 6.13: 3×3 processor grid and cycle times.

(iii) the fast processor is assigned no matrix element. In case (i), the time to update the matrix is 4 (the fast processor updates four elements sequentially). In the other cases, since a slow processor is involved, the time to update the matrix is at least 5. So the best distribution is to allocate all matrix elements to the fast processor.

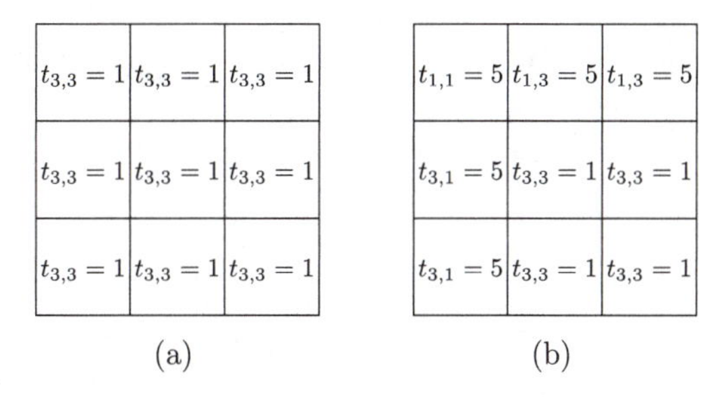

FIGURE 6.14: 3×3 distributions based on the best 2×2 distribution.

Let us now consider the case of a 3×3 distribution. Our objective is to exhibit an allocation that is better than an allocation that uses the above 2×2 allocation for the lower right 2×2 sub-matrix. Let us consider allocations that use the 2×2 allocation. We show the two possible allocations in Figure 6.14. Figure 6.13(a) shows the distribution that involves only the fast processor, whose time to update the matrix is 9. (The figure simply shows which processors hold which matrix element, and also indicates the processor cycle times.) Figure 6.13(b) shows the distribution that also involves processors $P_{1,3}$ and processor $P_{3,1}$, whose time to update the matrix is $\max(5, 2 \times 5, 4 \times 1) = 10$. Therefore, the best 3×3 distribution based on the best 2×2 distribution leads to a time to update the matrix of 9. However, consider the natural distribution in which processor $P_{i,j}$ holds matrix element $A_{i,j}$. Its update time is $\max(5, 1) = 5$! Therefore, the best 3×3 distribution is not based on the best 2×2 distribution. Therefore, with a matrix of cycle times of rank > 1, it is not possible to obtain a distribution that is optimal at all steps of the LU factorization.

Part III

Scheduling

Chapter 7

Scheduling

7.1 Introduction

This chapter presents basic (but important) results on task graph scheduling. We start with a motivating example before providing a quick overview of models and complexity results.

7.1.1 Where Do Task Graphs Come From?

Consider the following algorithm to solve the linear system $Ax = b$, where A is an $n \times n$ nonsingular lower triangular matrix and b is a vector with n components:

```
for i = 1 to n do
    Task T_{i,i}: x_i ← b_i / a_{i,i}
    for j = i + 1 to N do
        Task T_{i,j}: b_j ← b_j − a_{j,i} × x_i
```

For a given value of i, $1 \leqslant i \leqslant n$, each task $T_{i,*}$ represents some computations executed during the i-th iteration of the external loop. The computation of x_i is performed first (task $T_{i,i}$). Then, each component b_j of vector b such that $j > i$ is updated (task $T_{i,j}$).

In the original program, there is a total precedence order between tasks. Let us write $T <_{seq} T'$ if task T is executed before task T' in the original sequential code. We have

$$T_{1,1} <_{seq} T_{1,2} <_{seq} T_{1,3} <_{seq} \ldots <_{seq} T_{1,n} <_{seq} T_{2,2} <_{seq} T_{2,3} <_{seq} \ldots <_{seq} T_{n,n} .$$

However, there are independent tasks that can be executed in parallel. Intuitively, independent tasks are tasks whose execution orders can be interchanged without modifying the result of the program execution. A necessary condition for tasks to be independent is that they do not update the same variable. They can read the same value, but they cannot write into the same memory location (otherwise there would be a race condition and the result would be non-deterministic). For instance, tasks $T_{1,2}$ and $T_{1,3}$ both read x_1 but modify distinct components of b, hence they are independent.

We can express this notion of independence more formally. Each task T has an input set $\text{In}(T)$ (read values) and an output set $\text{Out}(T)$ (written values). In our example, $\text{In}(T_{i,i}) = \{b_i, a_{i,i}\}$ and $\text{Out}(T_{i,i}) = \{x_i\}$. For $j > i$, $\text{In}(T_{i,j}) = \{b_j, a_{j,i}, x_i\}$ and $\text{Out}(T_{i,j}) = \{b_j\}$. Two tasks T and T' are not independent (we write $T \perp T'$) if they share some written variable:

$$T \perp T' \Leftrightarrow \begin{cases} \quad \text{In}(T) \cap \text{Out}(T') & \neq \emptyset \\ \text{or } \text{Out}(T) \cap \text{In}(T') & \neq \emptyset \\ \text{or } \text{Out}(T) \cap \text{Out}(T') & \neq \emptyset \end{cases}$$

These conditions are known as Bernstein's conditions [27]. We come back to these conditions in Section 8.4 and discuss how to determine whether they hold or not.

Let us simply try some examples here. Tasks $T_{1,1}$ and $T_{1,2}$ are not independent because $\text{Out}(T_{1,1}) \cap \text{In}(T_{1,2}) = \{x_1\}$; therefore $T_{1,1} \perp T_{1,2}$. Similarly, $\text{Out}(T_{1,3}) \cap \text{Out}(T_{2,3}) = \{b_3\}$, and hence $T_{1,3}$ and $T_{2,3}$ are not independent; we write $T_{1,3} \perp T_{2,3}$.

Given the dependence relation $\perp$, we can extract a partial order from the total order $<_{seq}$ induced by the sequential execution of the program. If two tasks T and T' are dependent, i.e., $T \perp T'$, we order them according to the sequential execution; we write $T \prec T'$ if both $T \perp T'$ and $T <_{seq} T'$. The precedence relation $\prec$ represents the dependences that must be satisfied to preserve the semantics of the original program; if $T \prec T'$, then T was executed before T' in the sequential code, and it has to be executed before T' even if we have infinitely many processors, because T and T' share a written variable.

To define $\prec$ more accurately, in terms of order relations, we can write

$$\prec \text{ equals } (<_{seq} \cap \perp)^+,$$

where $^+$ denotes the transitive closure. In other words, we take the transitive closure of the intersection of $\perp$ and $<_{seq}$ to derive the set of all constraints that need to be satisfied to preserve the semantics of the original program. In a sense, $\prec$ captures the intrinsic sequentiality of the original program. The original total ordering $<_{seq}$ was unduly restrictive, so only the partial ordering $\prec$ needs to be respected. Why do we need to take the transitive closure of $<_{seq} \cap \perp$ to get a correct definition of $\prec$? In our example, we have $T_{2,4} \perp T_{4,4}$ (which is not a predecessor relationship, as there is $T_{3,4}$ in between) and $T_{4,4} \perp T_{4,5}$, hence a path of dependences from $T_{2,4}$ to $T_{4,5}$, while we do not have $T_{2,4} \perp T_{4,5}$. We need to track dependence chains to define $\prec$ correctly.

We can draw a directed graph to represent the dependence constraints that need to be enforced. The vertices of the graph denote the tasks, while the edges express the dependence constraints. An edge $e : T \rightarrow T'$ in the graph means that the execution of T' must begin only after the end of the execution of T, whatever the number of available processors. We do not usually draw transitivity edges on the graph, as they represent redundant information; if $T \prec T'$ and $T' \prec T''$, and if there exists a dependence $T \perp T''$, then it will be automatically satisfied. We say that T is a predecessor of T' if $T \prec T'$ and if there is no task T'' in between, i.e., such that $T \prec T''$ and $T'' \prec T'$. We will give formal definitions in the next section. In our example, the predecessor relationships are as follows:

- $T_{i,i} \prec T_{i,j}$ for $1 \leqslant i < j \leqslant n$
  (the computation of x_i must be done before updating b_j at step i of the outer loop).

- $T_{i,j} \prec T_{i+1,j}$ for $1 \leqslant i < j \leqslant n$
  (updating b_j at step i of the outer loop must[1] be done before reading it at step $i + 1$).

We end up with the graph shown in Figure 7.1. We will use this graph several times in this chapter for illustrative purposes.

7.1.2 Overview

This chapter presents classic theorems and algorithms from scheduling theory. The communication model used by this theory is rather unrealistic, but it makes it possible to obtain fundamental complexity results.

We start with the most simple (one might say crude) model where all communication delays between processors are neglected. We introduce basic definitions in Section 7.2. When there is no restriction on the number of available processors, optimal schedules can be found in polynomial time, as shown in Section 7.3. Section 7.4 deals with a limited number of processors; the scheduling problem becomes NP-complete, and so-called list scheduling heuristics are the typical approach. An elegant and powerful theorem shows that any list scheduling algorithm generates a schedule that is no longer than twice the optimal schedule (Section 7.4.2). We continue on the theoretical side: In Section 7.4.5, we discuss the scheduling of independent tasks, and we derive arbitrarily good approximation algorithms, i.e., polynomial algorithms whose performance can be guaranteed within a $(1+\varepsilon)$ of the optimal, for any arbitrary $\varepsilon > 0$.

Next we move to the classical scheduling model in which communication costs are taken into account each time two dependent tasks are not assigned to the same processor. We detail this model in Section 7.5. In this case, even the problem with unlimited processors is NP-complete, as explained in Section 7.6. We present heuristics for p identical processors in Section 7.7 and briefly discuss how to extend these heuristics to handle heterogeneous processors in Section 7.8.

[1] Well, "must" is slightly exaggerated, because the addition is an associative and commutative operation. This is true the way the program is written, reusing the same memory location b_j for all the updates. Technically, we have an output dependence that can be removed using standard techniques [92, 119].

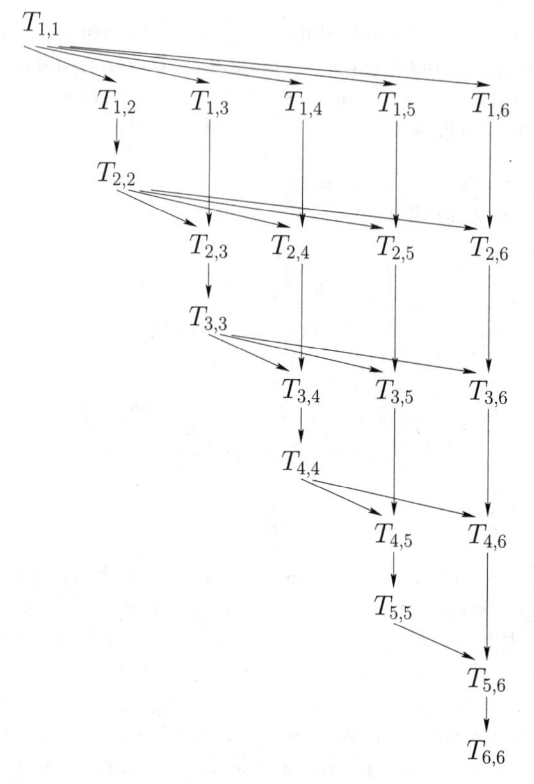

FIGURE 7.1: Task graph for the triangular system ($n = 6$).

7.2 Scheduling Task Graphs

DEFINITION 7.1. A task system, or task graph, is a directed vertex-weighted graph $G = (V, E, w)$, where:

- the set V of vertices represents the tasks (note that V is finite).
- the set E of edges represents precedence constraints between tasks: $e = (u, v) \in E$ if and only if $u \prec v$.
- the weight function $w : V \longrightarrow \mathbb{N}^*$ gives the weight (or duration) of each task. Task weights are assumed to be positive integers.[2]

For the triangular system (Figure 7.1), we can assume that all tasks have equal weight; let $w(T_{i,j}) = 1$ for $1 \leqslant i \leqslant j \leqslant n$. We could also consider

[2]This is not a restriction; tasks weights can be rational numbers. However, because there is a finite number of tasks, the weights can always be scaled up to integers.

that a division is more costly than a multiply-add and give extra weight to the diagonal tasks $T_{i,i}$.

A schedule σ of a task system is a function that assigns a start time to each task:

DEFINITION 7.2. A schedule of a task system $G = (V, E, w)$ is a function $\sigma : V \longrightarrow \mathbb{N}^*$ such that $\sigma(u) + w(u) \leqslant \sigma(v)$ whenever $e = (u, v) \in E$.

In other words, a schedule must preserve the *dependence constraints* induced by the precedence relation $\prec$ and embodied by the edges of the dependence graph; if $u \prec v$, then the execution of u begins at time $\sigma(u)$ and requires $w(u)$ units of time, and the execution of v at time $\sigma(v)$ must start after the end of the execution of u.

Often there are other constraints that must be met by schedules, namely, *resource constraints*. When there is an infinite number of processors,[3] we say that we have a problem with unlimited processors, denoted Pb(∞). When there is only a fixed number p of available processors, we speak of a problem with limited processors, Pb(p). In this case we need an allocation function $\mathsf{alloc} : V \longrightarrow \mathcal{P}$, where $\mathcal{P} = \{1, \ldots, p\}$ denotes the set of available processors. This function assigns a target processor to each task. The resource constraints simply specify that no processor can be allocated more than one task at the same time. This translates into the following conditions:

$$\mathsf{alloc}(T) = \mathsf{alloc}(T') \Rightarrow \begin{cases} \quad \sigma(T) + w(T) \leqslant \sigma(T') \\ \text{or } \sigma(T') + w(T') \leqslant \sigma(T). \end{cases}$$

This condition expresses the fact that if two tasks T and T' are allocated to the same processor, then their executions cannot overlap in time.

Given a task system, there is a basic condition for a schedule to exist, regardless of resource constraints:

THEOREM 7.1. *Let $G = (V, E, w)$ be a task system. There exists a schedule if and only if G contains no cycle.*

Proof. Clearly, if there is a cycle $v_1 \to v_2 \ldots \to \ldots \to v_k \to v_1$, then $v_1 \prec v_1$ (v_1 depends on itself), and a schedule σ would satisfy $\sigma(v_1) + w(v_1) \leqslant \sigma(v_1)$, which is impossible because we assumed $w(v_1) > 0$.

Conversely, if G has no cycle, then there exist vertices with no predecessor (V is finite) and we can thus topologically sort the vertices and schedule them one after the other (i.e., assuming a single processor) according to the topological order. Formally, if $v_{\pi(1)}, v_{\pi(2)}, \ldots, v_{\pi(n)}$ is the ordered list of vertices obtained by the topological sort, let $\sigma(v_{\pi(1)}) = 0$ and $\sigma(v_{\pi(i)}) = \sigma(v_{\pi(i-1)}) + w(v_{\pi(i-1)})$ for $2 \leqslant i \leqslant n$. Dependence constraints are respected, because if $v_i \prec v_j$, then the topological sort ensures that $\pi(i) < \pi(j)$. □

[3] In fact we need no more processors than the total number of tasks.

Theorem 7.1 states that scheduling deals with *directed acyclic graphs* (or DAGs).

DEFINITION 7.3. A DAG $G = (V, E, w)$ is a task system (as in Definition 7.2) where G is a directed acyclic graph. We have the following:

1. Let σ be a schedule for G. Assume σ uses at most p processors (let $p = \infty$ if the processors are unlimited). The makespan $\text{MS}(\sigma, p)$ of σ is its total execution time:
$$\text{MS}(\sigma, p) = \max_{v \in V}\{\sigma(v) + w(v)\} - \min_{v \in V}\{\sigma(v)\}\ .$$

2. $\text{Pb}(p)$ is the problem of determining a schedule σ of minimal makespan $\text{MS}(\sigma, p)$ assuming p processors (let $p = \infty$ if the processors are unlimited). Let $\text{MS}_{opt}(p)$ be the value of the makespan of an optimal schedule with p processors:
$$\text{MS}_{opt}(p) = \min_{\sigma} \text{MS}(\sigma, p)\ .$$

If the first task is scheduled at time 0, which is a common assumption, the expression of the makespan can be reduced to $\text{MS}(\sigma, p) = \max_{v \in V}\{\sigma(v) + w(v)\}$. We extend weights to paths in G as usual; if $\Phi = (T_1, T_2, \ldots, T_n)$ denotes a path in G, then $w(\Phi) = \sum_{i=1}^{n} w(T_i)$. Because schedules respect dependences, we have the following easy bound on the makespan:

PROPOSITION 7.1. *Let $G = (V, E, w)$ be a DAG and σ a schedule for G with p processors. Then, $MS(\sigma, p) \geqslant w(\Phi)$ for all paths Φ in G.*

Proof. Consider any path $\Phi = (T_1, T_2, \ldots, T_n)$ in G: $e = (T_i, T_{i+1}) \in E$ for $1 \leqslant i < n$. Then, $\sigma(T_i) + w(T_i) \leqslant \sigma(T_{i+1})$ for $1 \leqslant i < n$, and thus $\text{MS}(\sigma, p) \geqslant w(T_n) + \sigma(T_n) - \sigma(T_1) \geqslant \sum_{i=1}^{n} w(T_i) = w(\Phi)$. □

Our last definition introduces the notions of speedup and efficiency for schedules (see [76] for a detailed discussion of speedup and efficiency):

DEFINITION 7.4. Let $G = (V, E, w)$ be a DAG, and σ a schedule for G with p processors:

1. The speedup is the ratio $s(\sigma, p) = \frac{\text{Seq}}{\text{MS}(\sigma, p)}$, where $\text{Seq} = \sum_{v \in V} w(v)$ is the sum of all task weights.

2. The efficiency is the ratio $e(\sigma, p) = \frac{s(\sigma, p)}{p} = \frac{\text{Seq}}{p \times \text{MS}(\sigma, p)}$.

Seq is the optimal execution time $\text{MS}_{opt}(1)$ of a schedule with a single processor. We have the following well-known result:

THEOREM 7.2. *Let $G = (V, E, w)$ be a task system. For any schedule σ with p processors,*
$$0 \leqslant e(\sigma, p) \leqslant 1\ .$$

Proof. Consider the execution of σ as illustrated in Figure 7.2 (this is a fictitious example, not related to the triangular system example). At any time during execution, some processors are active, and some are idle. At the end, all tasks have been processed. Let Idle denote the cumulated idle time of the p processors during the whole execution. Because Seq is the sum of all task weights, the quantity Seq + Idle is equal to the area of the rectangle in Figure 7.2, i.e., the product of the number of processors by the makespan of the schedule: $\text{Seq}+\text{Idle} = p \times \text{MS}(\sigma, p)$. Hence, $e(\sigma, p) = \frac{\text{Seq}}{p\times\text{MS}(\sigma,p)} \leqslant 1$. □

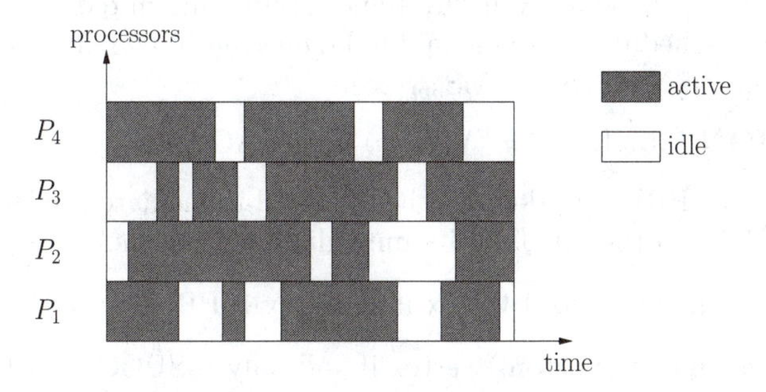

FIGURE 7.2: Active and idle processors during execution.

Another way to state Theorem 7.2 is to say that the speedup with p processors is always bounded by p. No superlinear speedup with our model! Here is an easy result to conclude this section: The more processors, the smaller (or equal) the optimal makespan.

THEOREM 7.3. *Let $G = (V, E, w)$ be a task system. We then have*

$$Seq = MS_{opt}(1) \geqslant \ldots \geqslant MS_{opt}(p) \geqslant MS_{opt}(p+1) \geqslant \ldots \geqslant MS_{opt}(\infty) .$$

Proof. The proof is straightforward. Consider an optimal schedule σ with p processors, and view it as a schedule with $p + 1$ processors where the last processor is kept idle. Then, $\text{MS}(\sigma, p + 1) = \text{MS}(\sigma, p) = \text{MS}_{opt}(p)$; hence, $\text{MS}_{opt}(p + 1) \leqslant \text{MS}_{opt}(p)$. □

Theorem 7.3 can be refined as follows. The number of processors actually used by a schedule σ is $|\mathsf{alloc}(V)|$, i.e., the number of processors that execute at least one task. If we define $\text{MS}'(p)$ as the minimum makespan of all schedules that use exactly p processors, we have $\text{MS}'(p) = \text{MS}_{opt}(p)$ for $1 \leqslant p \leqslant |V|$, so that Theorem 7.3 holds when replacing $\text{MS}_{opt}(p)$ by $\text{MS}'(p)$. Intuitively, it cannot hurt to make use of more processors in a model where communication costs are not taken into account! In Section 7.5, we introduce communication costs, and we give an example where $\text{MS}'(p) < \text{MS}'(p')$ with $p < p'$; the refined version of Theorem 7.3 is no longer true with this new model.

We are now ready to address the search for optimal schedules. Not surprisingly, it turns out that the problem $\mathrm{Pb}(p)$ with limited processors is more difficult than $\mathrm{Pb}(\infty)$, whose solution is explained in the next section.

7.3 Solving $\mathrm{Pb}(\infty)$

Let $G = (V, E, w)$ be a given DAG and assume unlimited processors. Remember that a schedule σ for G is said to be *optimal* if its makespan $\mathrm{MS}(\sigma, \infty)$ is minimal, i.e., if $\mathrm{MS}(\sigma, \infty) = \mathrm{MS}_{opt}(\infty)$.

DEFINITION 7.5. Let $G = (V, E, w)$ be a DAG.

1. For $v \in V$, $\mathrm{PRED}(v)$ denotes the set of all immediate predecessors of v, and $\mathrm{SUCC}(v)$ the set of all its immediate successors.
2. $v \in V$ is an entry (top) vertex if and only if $\mathrm{PRED}(v) = \emptyset$.
3. $v \in V$ is an exit (bottom) vertex if and only if $\mathrm{SUCC}(v) = \emptyset$.
4. For $v \in V$, the top level $tl(v)$ is the largest weight of a path from an entry vertex to v, excluding the weight of v.
5. For $v \in V$, the bottom level $bl(v)$ is the largest weight of a path from v to an output vertex, including the weight of v.

In the example of the triangular system, there is a single entry vertex, $T_{1,1}$, and a single exit vertex, $T_{n,n}$. The top level of $T_{1,1}$ is 0, and $tl(T_{1,2}) = tl(T_{1,1}) + w(T_{1,1}) = 1$. We have

$$tl(T_{2,3}) = \max\{w(T_{1,1}) + w(T_{1,2}) + w(T_{2,2}), w(T_{1,1}) + w(T_{1,3})\} = 3$$

because there are two paths from the entry vertex to $T_{2,3}$.

The top level of a vertex can be computed by a traversal of the DAG; the top level of an entry vertex is 0, while the top level of a non-entry vertex v is

$$tl(v) = \max\{tl(u) + w(u); u \in \mathrm{PRED}(v)\} .$$

Similarly, $bl(v) = \max\{bl(u); u \in \mathrm{SUCC}(v)\} + w(v)$ (and $bl(v) = w(v)$ for an exit vertex v). The top level of a vertex is the earliest possible time at which it can be executed, while its bottom level represents a lower bound of the remaining execution time once starting its execution. This can be stated more formally as follows.

THEOREM 7.4. *Let $G = (V, E, w)$ be a DAG and define σ_{free} as follows:*

$$\forall v \in V,\ \sigma_{free}(v) = tl(v) .$$

Then, σ_{free} is an optimal schedule for G.

Proof. The proof has two parts. First we show that σ_{free} is indeed a schedule, then we derive its optimality. Both are easy:

1. σ_{free} respects all dependence constraints by construction; if $(u, v) \in E$, then $u \in \text{PRED}(v)$ and the constraint $\sigma_{free}(v) \geqslant \sigma_{free}(u) + w(u)$ is taken into account in the computation of $tl(v) = \sigma_{free}(v)$.

2. To prove that σ_{free} is optimal, we use a topological sort order of the vertices of G, and prove by induction that all vertices are scheduled as soon as possible, i.e., as soon as the execution of all their predecessors has been completed.

The free schedule σ_{free} is also known as the *as soon as possible* (ASAP) schedule. □

From Theorem 7.4 we have

$$\text{MS}_{opt}(\infty) = \text{MS}(\sigma_{free}, \infty) = \max_{v \in V}\{tl(v) + w(v)\} .$$

Hence, $\text{MS}_{opt}(\infty)$ is simply the maximal weight of a path in the graph. Note that σ_{free} is not the only optimal schedule. For example, the *as late as possible* (ALAP) schedule σ_{late} is also optimal. We define σ_{late} as follows:

$$\forall v \in V, \sigma_{late}(v) = \text{MS}(\sigma_{free}, \infty) - bl(v) .$$

To understand the definition, note that $bl(v)$ is the maximal weight of a path from v to exit nodes, hence the need to start the execution of v no later than $\text{MS}(\sigma_{free}, \infty) - bl(v)$ if all tasks must have completed within $\text{MS}(\sigma_{free}, \infty)$ time units.

CORROLARY 7.1. *Let $G = (V, E, w)$ be a directed acyclic graph. $Pb(\infty)$ can be solved in time $O(|V| + |E|)$.*

Proof. From Theorem 7.4 we know that the optimal schedule σ_{free} can be computed using top levels and that $\text{MS}_{opt}(\infty)$ is the maximal weight of a path in the graph. Because G is acyclic, these quantities can be computed by a traversal of the graph, hence the complexity $O(|V| + |E|)$. □

Going back to the triangular system (Figure 7.1), because all tasks have weight 1, the weight of a path is equal to its length plus 1. The longest path is

$$T_{1,1} \to T_{1,2} \to T_{2,2} \to \ldots \to T_{n-1,n-1} \to T_{n,-1,n} \to T_{n,n},$$

whose weight is $2n - 1$. Note that we do not need as many processors as tasks to achieve execution within $2n - 1$ time units. For example, we can use only $n - 1$ processors. Let $1 \leqslant i \leqslant n$; at time $2i - 2$, processor P_1 starts the execution of task $T_{i,i}$, while at time $2i - 1$, the first $n - i$ processors P_1, P_2, ..., P_{n-i} execute tasks $T_{i,j}$, $i + 1 \leqslant j \leqslant n$.

7.4 Solving Pb(p)

Let $G = (V, E, w)$ be a DAG. It turns out that the problem with limited processors Pb(p) is NP-complete. Hence, a polynomial algorithm to determine an optimal schedule is unlikely to exist (unless P = NP!). Therefore, in the next sections we will introduce heuristics to compute approximate solutions.

7.4.1 NP-completeness of Pb(p)

In the following proofs, we use two well-know *partition* problems, although their names are somewhat misleading:

2-Partition Given n positive integer numbers $\{a_1, a_2, \ldots, a_n\}$, is there a subset I of indices such that $\sum_{i \in I} a_i = \sum_{i \notin I} a_i$?

3-Partition Given $3n$ positive integer numbers $\{a_1, a_2, \ldots, a_{3n}\}$ and a bound B, assuming that $\frac{B}{4} < a_i < \frac{B}{2}$ for all i and that $\sum_{i=1}^{3n} a_i = nB$, is there a partition of these numbers into n subsets $I_1, I_2, \ldots, I_n$ of sum B? In other words, are there n subsets $I_1, I_2, \ldots, I_n$ such that $I_1 \cup I_2 \ldots \cup I_n = \{1, 2, \ldots, 3n\}$, $I_i \cap I_j = \emptyset$ if $i \neq j$ and $\sum_{j \in I_i} a_j = B$ for all i?

We see that 3-Partition does not amount to partitioning the data into three sets of equal sums, which would have been the natural extension of 2-Partition. The technical condition $\frac{B}{4} < a_i < \frac{B}{2}$ ensures that all subsets will be of cardinal 3, hence the name 3-Partition. The last condition $\sum_{i=1}^{3n} a_i = nB$ is obviously needed for a solution to exist.

Why deal with these two variants then? It turns out that 2-Partition, although NP-complete, is *easier*, if we may say so, than 3-Partition: There exists a pseudo-polynomial algorithm to solve 2-Partition, while 3-Partition is NP-complete in the strong sense [57]. Intuitively, problems involving numbers may turn out to be hard to solve only when these numbers are very large, or they may remain of combinatorial nature even for small numbers. The latter are said to be NP-hard in the strong sense. We refer the reader to [57] for a primer on NP-complete number problems.

Partitioning problems and scheduling independent tasks are closely related topics, and we come back to this important notion in Section 7.4.5. Beforehand, we turn to the complexity of Pb(p) and to list scheduling heuristics:

DEFINITION 7.6. The decision problem Dec(p) associated with Pb(p) is as follows. Given a DAG $G = (V, E, w)$, a number of processors $p \geqslant 1$, and an execution bound $K \in \mathbb{N}^*$, does there exist a schedule σ for G using at most p processors, such that $\mathrm{MS}(\sigma, p) \leqslant K$? The restriction of Dec(p) to independent tasks (no dependence, i.e., when $E = \emptyset$) is denoted Indep-tasks(p). In both problems, p is arbitrary (it is part of the problem instance). When p is fixed a priori, say $p = 2$, we note Dec(2) and Indep-tasks(2).

THEOREM 7.5.

- *Indep-tasks(2) is NP-complete but can be solved by a pseudo-polynomial algorithm.*
- *Indep-tasks(p) is NP-complete in the strong sense.*
- *Dec(2) (and hence Dec(p)) is NP-complete in the strong sense.*

Proof. First, Dec(p) (and hence all the other problems, which are restrictions of it) belongs to NP. If we are given a schedule σ whose makespan is less than or equal to K, we can check in polynomial time that both dependences and resource constraints are satisfied. Indeed, we have to ensure that each dependence constraint (each edge in E) is satisfied, which is straightforward. Also, we need to check that no more than p tasks ever execute simultaneously. We can sort the tasks by their starting times, and check the latter condition by scanning the sorted array. This can easily be done in time polynomial in the size of the problem instance.

For proving the NP-completeness of Indep-tasks(2), we show that 2-Partition can be polynomially reduced to Indep-tasks(2). Consider an arbitrary instance Inst_1 of 2-Partition, with n integers $\{a_1, a_2, \ldots, a_n\}$, and let $S = \sum_{i=1}^{n} a_i$ be even (otherwise we know there is no solution). We build an instance Inst_2 of Indep-tasks(2) as follows. We let $p = 2$ (of course), $G = (V, E, w)$ with $V = \{v_1, v_2, \ldots, v_n\}$, $E = \emptyset$, and $w(v_i) = a_i, 1 \leqslant i \leqslant n$. We also let $K = \frac{S}{2}$. The construction of Inst_2 is polynomial (and even linear) in the size of Inst_1. Moreover, Inst_1 has a solution if and only if there exists a schedule that meets the bound K, hence if and only if Inst_2 has a solution.

The pseudo-polynomial algorithm to solve Indep-tasks(2) is a simple dynamic programming algorithm. For $1 \leqslant i \leqslant n$ and $0 \leqslant T \leqslant S$, let the boolean variable $c(i, T)$ be true if there exists a subset of $\{a_1, a_2, \ldots, a_i\}$ whose sum is T. We need to determine the value of $c(n, \frac{S}{2})$. We use the induction

$$c(i,T) = c(i-1,T) \text{ or } ((T \geqslant a_i) \text{ and } c(i-1, T-a_i)),$$

which basically states that either a_i is involved in the target subset, or not. The initialization is $c(1, a_1) = 1$, $c(i, 0) = 1$ for all i and all other boundary values set to 0. The complexity of the algorithm is $O(nS)$, which is not polynomial in the problem size, whose typical binary encoding would be $O(n + \sum_{i=1}^{n} \log a_i)$. (However, if the a_i's are encoded in unary, we have polynomial complexity, which is the definition of a pseudo-polynomial algorithm.)

The reduction for the strong NP-completeness of Indep-tasks(p) is straightforward. Consider an arbitrary instance Inst_1 of 3-Partition, with $3n$ integers $\{a_1, a_2, \ldots, a_{3n}\}$ and bound B as stated above. The instance Inst_2 of Indep-tasks(p) is built with $3n$ independent tasks of weight a_i, $p = n$ processors and $K = B$. Clearly, Inst_1 has a solution if and only if there exists a schedule that meets the bound K, hence if and only if Inst_2 has a solution.

The reduction for the strong NP-completeness of Dec(2) is more interesting. Consider again an arbitrary instance Inst_1 of 3-Partition, with $3n$ integers $\{a_1, a_2, \ldots, a_{3n}\}$ and bound B. The instance Inst_2 of Dec(2) is built with $p = 2$ processors; $3n$ independent tasks $\{T_1, \ldots, T_{3n}\}$, where the weight of T_i is a_i; and $3n$ other tasks, $\{X_1, Y_1, Z_1, \ldots, X_n, Y_n, Z_n\}$, all of weight B, linked by the following dependences:

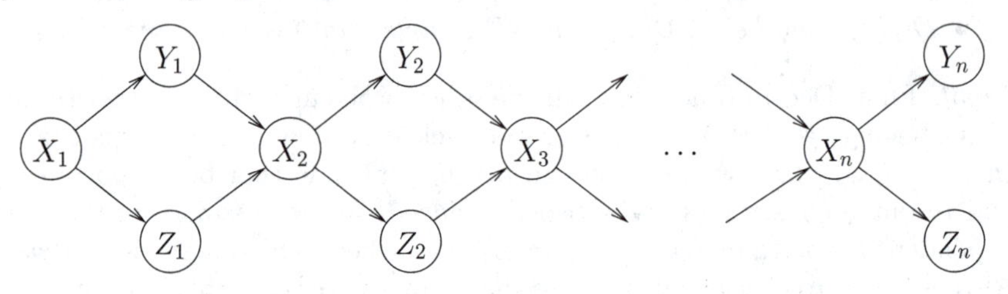

Finally, we let $K = 2nB$.

Assume first that Inst_1 has a solution $I_1 \cup I_2 \ldots \cup I_n$, where each I_i is composed of three numbers whose sum is B. The solution to Inst_2 is the following schedule σ:

- The first processor P_1 executes all $2n$ tasks X_i and Y_i. These tasks are totally ordered along a dependence path of length $K = 2nB$.
- The second processor P_2 executes Z_i while P_1 executes Y_i. While P_1 executes a task X_i, it has a slot of size B to execute the three tasks T_j that belong to I_i.

Altogether, all dependence and resource constraints are satisfied, and σ is a valid schedule of makespan K.

Now, assume that Inst_2 has a solution schedule σ. We see that the schedule σ is quite constrained. Because $X_1 \to Y_1 \to X_2 \to \ldots \to X_n \to Y_n$ is a dependence path of length K, these tasks must be processed as soon as possible, without any idle time in between. But $X_1 \to Z_1 \to X_2 \to \ldots \to X_n \to Z_n$ is another dependence path of the same length, so these tasks must be processed as soon as possible too. This enforces that $\sigma(X_i) = (2i - 2)B$ and that $\sigma(Y_i) = \sigma(Z_i) = (2i - 1)B$. Up to some exchanges, we can assume that P_1 executes all the X_i and Y_i, and that P_2 executes all the Z_i. We see that the $3n$ tasks T_i are executed by P_2 during n intervals of length B, hence the solution to Inst_1. □

7.4.2 List Scheduling Heuristics

Because Pb(p) is NP-complete, we rely on heuristics to schedule DAGs with limited processors. The most natural idea is to use greedy strategies: At each instant, we try to schedule as many tasks as possible onto available processors. Such strategies deciding *not to deliberately keep a processor idle* are called list scheduling algorithms. Of course there are different possible strategies to decide which tasks are given priority in the (frequent) case where there are more free tasks than available processors. But a key result due to Coffman [42] is that any list algorithm can be shown to achieve at most twice the optimal makespan. We express this more formally after giving some definitions.

DEFINITION 7.7. Let $G = (V, E, w)$ be a DAG and let σ be a schedule for G. A task $v \in V$ is free at time t (we note $v \in FREE(\sigma, t)$) if and only if its execution has not yet started ($\sigma(v) \geqslant t$) but all its predecessors have been executed ($\forall\, u \in \text{PRED}(v),\ \sigma(u) + w(u) \leqslant t$).

A list schedule is a schedule such that no processor is deliberately left idle; at each time t, if $|FREE(\sigma, t)| = r \geqslant 1$, and if q processors are available, then $\min(r, q)$ free tasks start executing.

THEOREM 7.6. *Let $G = (V, E, w)$ be a DAG and assume there are p available processors. Let σ be any list schedule of G. Let $MS_{opt}(p)$ be the makespan of an optimal schedule. Then,*

$$MS(\sigma, p) \leqslant \left(2 - \frac{1}{p}\right) MS_{opt}(p) \ .$$

It is important to point out that Theorem 7.6 holds for *any* list schedule, regardless of the strategy to choose among free tasks when there are more free tasks than available processors.

Proof. We first need a lemma:

LEMMA 7.1. *There exists a dependence path Φ in G whose weight $w(\Phi)$ satisfies*

$$Idle \leqslant (p - 1) \times w(\Phi),$$

where Idle is the cumulated idle time of the p processors during the whole execution of the list schedule.

Proof. Let T_{i_1} be a task whose execution terminates at the end of the schedule:

$$\sigma(T_{i_1}) + w(T_{i_1}) = \text{MS}(\sigma, p) \ .$$

Let t_1 be the largest time smaller than $\sigma(T_{i_1})$ and such that there exists an idle processor during the time interval $[t_1, t_1 + 1[$ (let $t_1 = 0$ if such a time does not exist). Why is this processor idle? Because σ is a list schedule, no task is free at t_1, otherwise the idle processor would start executing a free task. Therefore, there must be a task T_{i_2} that is an ancestor[4] of T_{i_1} and that is being executed at time t_1; otherwise T_{i_1} would have been started at time t_1 by the idle processor. Because of the definition of t_1 we know that all processors

[4]The ancestors of a task are its predecessors, the predecessors of its predecessors, and so on.

are active between the end of the execution of T_{i_2} and the beginning of the execution of T_{i_1}.

We start the construction again from T_{i_2} so that we obtain a task T_{i_3} such that all processors are active between the end of T_{i_3} and the beginning of T_{i_2}. Iterating the process, we end up with r tasks $T_{i_r}, T_{i_{r-1}}, \ldots, T_{i_1}$ that belong to a dependence path Φ of G and such that all processors are active except perhaps during their execution. In other words, the idleness of some processors can only occur during the execution of these r tasks, during which at least one processor is active (the one that executes the task). Hence, $\text{Idle} \leqslant (p-1) \times \sum_{j=1}^{r} w(T_{i_j}) \leqslant (p-1) \times w(\Phi)$. □

Going back to the proof of Theorem 7.6, we know that $p \times \text{MS}(\sigma, p) = \text{Idle} + \text{Seq}$, where $\text{Seq} = \sum_{v \in V} w(v)$ is the sequential time, i.e., the sum of all task weights (see Figure 7.2). Now take the dependence path Φ constructed in Lemma 7.1. We have $w(\Phi) \leqslant \text{MS}_{opt}(p)$, because the makespan of any schedule is greater than the weight of all dependence paths in G (simply because dependence constraints are met). Furthermore, $\text{Seq} \leqslant p \times \text{MS}_{opt}(p)$ (with equality only if all p processors are active all the time). Putting this together, we get

$$\begin{aligned} p \times \text{MS}(\sigma, p) &= \text{Idle} + \text{Seq} \\ &\leqslant (p-1)w(\Phi) + \text{Seq} \leqslant (p-1)\text{MS}_{opt}(p) + p\text{MS}_{opt}(p) \\ &= (2p-1)\text{MS}_{opt}(p), \end{aligned}$$

which proves the theorem. □

Fundamentally, Theorem 7.6 says that any list schedule is within 50% of the optimum. Therefore, list scheduling is guaranteed to achieve half the best possible performance, regardless of the strategy to choose among free tasks. Before presenting the most widely used strategy to perform this choice (in order to obtain a practical scheduling algorithm), we make a short digression to show that the bound $\frac{2p-1}{p}$ cannot be improved.

PROPOSITION 7.2. *Let $MS_{list}(p)$ be the shortest possible makespan produced by a list scheduling algorithm. The bound*

$$MS_{list}(p) \leqslant \frac{2p-1}{p} MS_{opt}(p)$$

is tight.

Proof. Let K be an arbitrarily large integer. We build a DAG $G = (V, E, w)$, for which any list schedule σ has a makespan $\text{MS}(\sigma, p) \approx \frac{2p-1}{p}\text{MS}_{opt}(p)$ (see Figure 7.3). There are $2p+1$ vertices, whose weights are as follows: $w(T_i) = K(p-1)$ for $1 \leqslant i \leqslant p-1$; $w(T_p) = 1$; $w(T_i) = K$ for $p+1 \leqslant i \leqslant 2p$; and $w(T_{2p+1}) = K(p-1)$. Precedence edges are indicated in the figure. There are exactly p entry vertices, hence $\sigma(T_i) = 0, 1 \leqslant i \leqslant p$ for any list schedule

σ. At time 1, the execution of T_p is complete and the free processor (the one that executed T_p) will be successively assigned $p-1$ of the p free tasks $T_{p+1}, T_{p+2}, \ldots, T_{2p}$. Note that this processor starts the execution of the last of its $p-1$ tasks at time $1+K(p-2)$ and terminates it at time $1+K(p-1)$. Therefore, the remaining p-th task will be executed at time $K(p-1)$ by another processor. Only at time $K(p-1)+K = Kp$ will task T_{2p+1} be free, which leads to $\text{MS}(\sigma, p) = Kp + K(p-1) = K(2p-1)$.

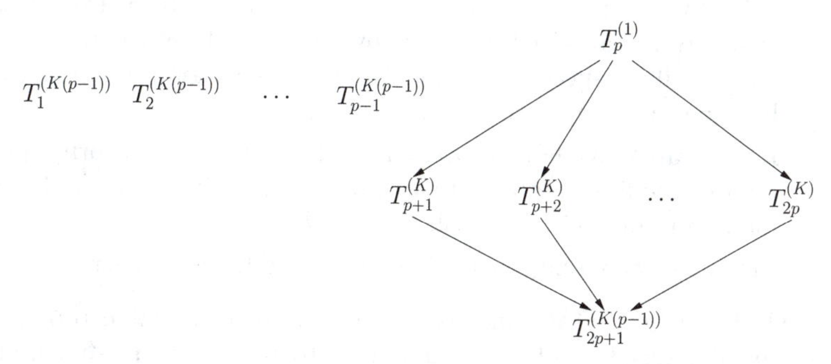

FIGURE 7.3: The DAG used to bound list scheduling performance; task weights are indicated as exponents inside parentheses.

However, the DAG can be scheduled in only $Kp+1$ time units. The key is to deliberately keep $p-1$ processors idle while executing task T_p at time 0 (which is forbidden in a list schedule). Then, at time 1, each processor executes one of the p tasks $T_{p+1}, T_{p+2}, \ldots, T_{2p}$. At time $1+K$, one processor starts executing T_{2p+1} while the other $p-1$ processors execute tasks $T_1, T_2, \ldots, T_{p-1}$. This defines a schedule with a makespan equal to $1+K+K(p-1) = Kp+1$, which is optimal because it is equal to the weight of the path $T_p \rightarrow T_{p+1} \rightarrow T_{2p+1}$. Hence, we obtain the ratio

$$\frac{\text{MS}(\sigma, p)}{\text{MS}_{opt}(p)} \geqslant \frac{K(2p-1)}{Kp+1} = \frac{2p-1}{p} - \frac{2p-1}{p(Kp+1)} = \frac{2p-1}{p} - \varepsilon(K),$$

where $\lim_{K \to +\infty} \varepsilon(K) = 0$. □

7.4.3 Implementing a List Schedule

In this section, we show how to implement a "generic" list scheduling algorithm, meaning that we do not specify how to choose free tasks when there are more free tasks than available processors. We simply assume that free tasks are placed in a *priority queue*, according to some to-be-defined priority. We give an example instantiation of this generic algorithm in Section 7.4.4.

The implementation is not difficult but is somewhat lengthy to describe. The overall scheme can be outlined as follows:

1. *Initialization*:
   (a) Compute the priority of all tasks, for some definition of priority.
   (b) Place the free tasks in a priority queue, sorted by non-increasing priority.
   (c) Let t be the current time: $t = 0$.
2. *While there remain tasks to schedule*:
   (a) Add new free tasks, if any, to the priority queue. If the execution of a task terminates at time t, remove this task from the predecessor lists of all its successors. Add those tasks whose predecessor lists have become empty to the priority queue.
   (b) If there are q available processors and r tasks in the priority queue, remove the first $\min(q, r)$ tasks from the priority queue and schedule them; for each such task T set $\sigma(T) = t$.
   (c) Increment t by one (recall that task weights are integers).

Let $G = (V, E, w)$ be a DAG and assume there are p available processors. Let σ be any list schedule of G. Our aim is to derive an implementation whose complexity is $O(|V| \log |V| + |E|)$ for computing the schedule. Clearly, the above scheme must be modified because time varies from $t = 0$ up to $t = \text{MS}(\sigma, p)$, implying that the complexity depends on task weights. Indeed, $\text{MS}(\sigma, p)$ may be of the order of Seq, the sum of all task weights, and we would have a pseudo-polynomial algorithm instead of a true polynomial algorithm; a binary encoding of the problem instance is of size log(Seq), not of size Seq. We outline a possible solution written in pseudo-code in Algorithm 7.1. Rather than using time we use *events*, which correspond to times when tasks become free or processors become idle.

A few words of explanation are in order for Algorithm 7.1. We use a heap $\mathcal{Q}$ (see [44]) to store free tasks for two reasons. We can access the task with the highest priority in constant time, and we can insert a task in the heap, according to its priority level, in time proportional to the logarithm of the heap size, which is bounded by $|V|$. We use another heap $\mathcal{P}$ to handle active processors; a processor executing a task $v \in V$ is valued by the time at which the execution of v terminates. Thereby we can compute the next event in constant time, and we can insert a new active processor in the heap in $O(\log |\mathcal{P}|) \leqslant O(\log |V|)$ time. When we extract a processor from the processor heap, meaning that a task v has terminated, we need to update the in-degree of each successor of v in array A. On the fly, if the in-degree of a given successor v' becomes zero, we insert v' in the priority heap $\mathcal{Q}$. This way, we process each edge of G only once, for a global cost $O(|E|)$. Overall, each task causes two insertions: the first is the insertion of the task itself in heap $\mathcal{Q}$; the second is the insertion of the processor that executes it in heap $\mathcal{P}$. Because each operation costs at most $O(\log |V|)$, we obtain the desired complexity $O(|V| \log |V| + |E|)$ for computing the schedule.

```
1  LIST_SCHEDULE(G = (V, E, w), p)
       { p processors, with 1 ⩽ p ⩽ |V| }
       { In addition to the data structure representing G, store in array A the
         number of predecessors of each task (its in-degree) }
2      Insert all the tasks without predecessors in priority heap Q
3      Initialize processor heap P to Empty
4      t ← −1
5      while Q ≠ Empty do
6          t' ← NEXT_EVENT(P, t)
7          UPDATE(t', A, Q)
8          ALLOCATE_TASKS(t', P, Q)
9          t ← t'
```

ALGORITHM 7.1: Outline of a list scheduling algorithm.

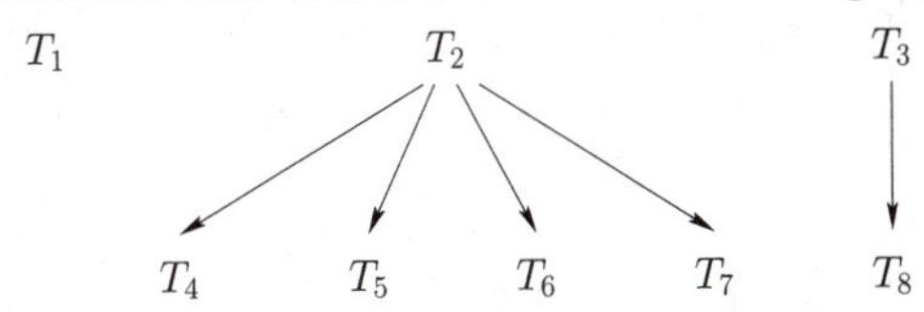

FIGURE 7.4: A small example DAG.

7.4.4 Critical Path Scheduling

In this section, we detail a widely used list scheduling technique, known as *critical path scheduling.* We have seen the basic principle of the list scheduling technique and assessed its performance. We now have to explain how to choose among free tasks to get a practical list scheduling algorithm. The most popular selection criterion is based on the value of the bottom level of the tasks. Intuitively, the larger the bottom level, the more "urgent" the task. The *critical path* of a task is defined as its bottom level and is used to assign priority levels to tasks. Critical path scheduling is list scheduling where the priority level of a task is given by the value of its critical path. Ties are broken arbitrarily.

Let us work out a small example. Consider the DAG shown in Figure 7.4. There are eight tasks, whose weights and critical paths are listed in Table 7.1. Assume there are $p = 3$ available processors and let $\mathcal{Q}$ be the priority queue of free tasks. At $t = 0$, $\mathcal{Q}$ is initialized as $\mathcal{Q} = (T_3, T_2, T_1)$. Because $q = r = 3$,

TABLE 7.1: Weights and critical paths for DAG in Figure 7.4.

Tasks	T_1	T_2	T_3	T_4	T_5	T_6	T_7	T_8
Weights	3	2	1	3	4	4	3	6
Critical Paths	3	6	7	3	4	4	3	6

we execute these three tasks. At $t = 1$, we add T_8 to the queue: $\mathcal{Q} = (T_8)$. There is one processor available, which starts the execution of T_8. At $t = 2$, we add the four successors of T_2 to the queue: $\mathcal{Q} = (T_5, T_6, T_4, T_7)$. Note that we have broken ties arbitrarily (using task indices in this case). The available processor picks the first task T_5 in $\mathcal{Q}$. Following this scheme, the execution goes on up to $t = 10$, as summarized in Figure 7.5.

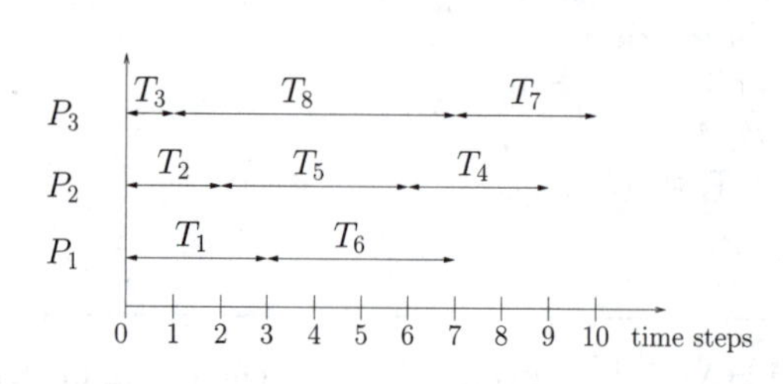

FIGURE 7.5: Critical path schedule for the example DAG in Figure 7.4.

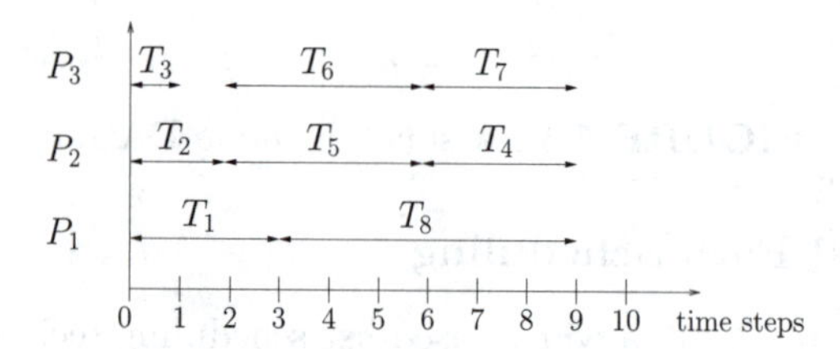

FIGURE 7.6: Optimal schedule for the example DAG in Figure 7.4.

Note that it is possible to schedule the DAG in only 9 time units, as shown in Figure 7.6. The trick is to leave a processor idle at time $t = 1$ deliberately; although it has the highest critical path, T_8 can be delayed by two time units. T_5 and T_6 are given preference to achieve a better load balance between processors. How do we know that the schedule shown in Figure 7.6 is optimal? Because Seq $= 26$, so that three processors require at least $\lceil \frac{26}{3} \rceil = 9$ time units. This small example illustrates the difficulty of scheduling with a limited number of processors.

7.4.5 Scheduling Independent Tasks

We come back to the problem Indep-tasks(p) introduced in Section 7.4.1. Recall that the objective is to schedule a set of independent tasks (no dependence relation). In addition to being a scheduling problem, this problem can be viewed as a load balancing problem where we aim at partitioning the tasks so as to assign (almost) equal load to each processor.

For fixed p, we have shown that Indep-tasks(p) is NP-complete but solvable in pseudo-polynomial time. In this section, we aim at deriving λ-approximation algorithms, i.e., polynomial-time schedules whose makespans are at most λ times the optimal makespan for any problem instance.

Because all tasks are independent, any greedy (or list scheduling) algorithm will keep some processors idle only at the end of the execution. Can we improve the performance bound $\lambda = 2 - \frac{1}{p}$ of Theorem 7.6 for this particular case? For the sake of simplicity we only consider the case $p = 2$ in this section. Although simple, this case enables us to introduce approximation schemes that play a major role in scheduling theory.

Given n independent tasks $T_1, T_2, \ldots, T_n$, where the weight of T_i is a_i, how do we schedule these tasks on two identical processors? Let MS_{opt} be the optimal schedule and $P_{sum} = \sum_{i=1}^{n} a_i$. Of course $\mathrm{MS}_{opt} \geqslant \frac{P_{sum}}{2}$. There are two natural greedy algorithms:

- GREEDY: Consider the tasks in some arbitrary order; at each step, schedule the current task on the least-loaded processor.

- SORTED-GREEDY: Same as above, but considering the tasks sorted by non-increasing weight.

The rationale to sort the tasks is that the arrival of a big task in the end may unbalance the whole execution. However, we need to know all the tasks (for sorting) before starting the execution of the algorithm. For this reason SORTED-GREEDY is called an *off-line* algorithm. By contrast, GREEDY can be applied to an *on-line* problem in which new tasks dynamically arrive (e.g., to a dual-processor computer).

THEOREM 7.7. *GREEDY is a $\frac{3}{2}$-approximation and SORTED-GREEDY is a $\frac{7}{6}$-approximation for Indep-tasks(2), and these approximation factors cannot be improved.*

Proof. We first show that the bounds are tight (note that the list scheduling bound is $2 - \frac{1}{2} = \frac{3}{2}$). For GREEDY, take an instance with three tasks of weights 1, 1, and 2: GREEDY has a makespan of 3 while the optimal is 2. For SORTED-GREEDY, take five tasks, three of weight 2 and two of weight 3. SORTED-GREEDY has a makespan of 7, while the optimal is 6.

For the approximations, recall that $\mathrm{MS}_{opt} \geqslant \frac{P_{sum}}{2}$ and that $\mathrm{MS}_{opt} \geqslant a_i$ for all i. Let us start with GREEDY. Let P_1 and P_2 be the two processors. Assume that P_1 finishes execution last. Let M_1 be the execution time on P_1 (the sum of all task weights assigned to it) and M_2 the execution time on P_2. Because P_1 terminates last, $M_1 \geqslant M_2$. Of course $M_1 + M_2 = P_{sum}$.

Let T_j be the last task that executes on P_1. Let $M_0 = M_1 - a_j$ be the load of P_1 before T_j (of weight a_j) is assigned to it. Why does the GREEDY algorithm choose to assign T_j to P_1? It can only be because at that time

P_2 has more load than (or the same load as) P_1: M_0 is not larger than the current load of P_2 at that time, which itself is not larger than its final load M_2 (note that P_2 may have been assigned more tasks after T_j was scheduled on P_1). Therefore, $M_0 \leqslant M_2$. To summarize, the makespan of the schedule computed by GREEDY is M_1, and

$$\begin{aligned} M_1 = M_0 + a_j &= \frac{((M_0 + M_0 + a_j) + a_j}{2} \\ &\leqslant \frac{((M_0 + M_2 + a_j) + a_j}{2} = \frac{P_{sum} + a_j}{2} \\ &\leqslant \mathrm{MS}_{opt} + \frac{a_j}{2} \leqslant \mathrm{MS}_{opt} + \frac{\mathrm{MS}_{opt}}{2}, \end{aligned}$$

hence proving the $\frac{3}{2}$-approximation result for GREEDY.

For SORTED-GREEDY the same line of reasoning is used, but with a tighter bounding of a_j than by MS_{opt}. First, if $a_j \leqslant \frac{1}{3}\mathrm{MS}_{opt}$, we obtain what we need, i.e., $M_1 \leqslant \frac{7}{6}\mathrm{MS}_{opt}$. But if $a_j > \frac{1}{3}\mathrm{MS}_{opt}$, then necessarily $j \leqslant 4$. Indeed, if T_j was the fifth task or higher, because task weights are sorted, there would be at least five tasks of weight greater than $\frac{1}{3}\mathrm{MS}_{opt}$; in any schedule, including the optimal schedule, one processor would receive three of these tasks, a contradiction. Next we observe that the makespan achieved by SORTED-GREEDY is the same when scheduling all tasks as when scheduling only the first four tasks. But for any problem instance with $n \leqslant 4$, SORTED-GREEDY is optimal, and $M_1 = \mathrm{MS}_{opt}$. □

We now show that Indep-tasks(2) has a $(1+\varepsilon)$-approximation for any value of $\varepsilon > 0$. We say that Indep-tasks(2) has a *polynomial time approximation scheme*, or PTAS. For any value of $\varepsilon > 0$, we build an algorithm whose makespan is guaranteed up to a factor $1+\varepsilon$ from the optimal, and whose complexity is polynomial. A word of caution here: "polynomial" means polynomial in the problem size only. Even if $\frac{1}{\varepsilon}$ can tend to infinity, ε is a constant for the algorithm. Later we present a *fully polynomial time approximation scheme*, or FPTAS, where the complexity of the $1 + \varepsilon$ approximation is polynomial both in the problem size and in $\frac{1}{\varepsilon}$.

THEOREM 7.8. *$\forall \varepsilon > 0$, Indep-tasks(2) has a $(1 + \varepsilon)$-approximation.*

Proof. Let $P_{max} = \max_i a_i$ and $L = \max(\frac{P_{sum}}{2}, P_{max})$. We already know that $L \leqslant \mathrm{MS}_{opt}$. We call *big jobs* those tasks T_i whose weights are such that $a_i > \varepsilon L$ and *small jobs* those such that $a_i \leqslant \varepsilon L$. The number of big jobs is at most

$$\left\lceil \frac{P_{sum}}{\varepsilon L} \right\rceil \leqslant \left\lceil \frac{2L}{\varepsilon L} \right\rceil = \left\lceil \frac{2}{\varepsilon} \right\rceil = B\,.$$

Because ε is fixed, B is a constant, so there is a (possibly large but) constant number of big jobs. We temporarily forget about small jobs, consider only big jobs and search for the best schedule. There are 2^B possible schedules (each big job can be assigned to either processor), which is a constant number again, so we try them all and keep the best one, say σ_{opt}^{big}. The resulting makespan MS_{opt}^{big} satisfies $\mathrm{MS}_{opt}^{big} \leqslant \mathrm{MS}_{opt}$ because there are fewer jobs than in the original problem.

Now we extend σ_{opt}^{big} and schedule the small jobs after the big jobs, using GREEDY, and we obtain a schedule σ for the original problem. We claim that σ solves the problem, i.e., that $\mathrm{MS}(\sigma) \leqslant (1+\varepsilon)\mathrm{MS}_{opt}$. If the makespans of σ_{opt}^{big} and σ are equal, then σ is optimal. Otherwise, σ terminates with a small job T_j, say on the first processor P_1. But the load of P_1 before this last assignment could not exceed $\frac{P_{sum}}{2}$, otherwise GREEDY would have assigned T_j on P_2 (same proof as in Theorem 7.7). Hence,

$$\mathrm{MS}(\sigma) \leqslant \frac{P_{sum}}{2} + a_j \leqslant L + \varepsilon L \leqslant (1+\varepsilon)\mathrm{MS}_{opt},$$

which proves the theorem. □

Theorem 7.8 is interesting, but the combinatorial search for the best assignment of big jobs can be prohibitively expensive when ε tends to 0. This motivates our last result, which states that Indep-tasks(2) has an FPTAS. (This will conclude our initiation to the fascinating world of approximation schemes.)

THEOREM 7.9. $\forall \varepsilon > 0$, *Indep-tasks*(2) *has a* $(1+\varepsilon)$*-approximation whose complexity is polynomial in* $\frac{1}{\varepsilon}$.

Proof. We encode schedules as *vector sets (VS)*. The first component of each vector represents the load of the first processor P_1, and the second component the load of P_2. Here is the construction:

Initialization Set $VS_1 = \{[a_1, 0], [0, a_1]\}$.

Phase k **for** $2 \leqslant k \leqslant n$ For every vector $[x, y] \in VS_{k-1}$, put $[x + a_k, y]$ and $[x, y + a_k]$ in VS_k.

Output Vector $[x, y] \in VS_n$ that minimizes $\max(x, y)$.

Instead of building all possible vectors (whose number is exponential), we will prune some of them during the construction. The difficulty is to retain a vector close to the optimal when pruning. Let $\Delta = 1 + \frac{\varepsilon}{2n}$. All considered vectors lie in the rectangle $[0, P_{sum}] \times [0, P_{sum}]$. We subdivide this rectangle into many boxes. Horizontal and vertical cuts are made at the coordinates Δ^i for $1 \leqslant i \leqslant M$, where

$$M = \lceil \log_\Delta P_{sum} \rceil = \left\lceil \frac{\ln P_{sum}}{\ln \Delta} \right\rceil \leqslant \left\lceil \left(1 + \frac{2n}{\varepsilon}\right) \ln(P_{sum}) \right\rceil .$$

The last inequality comes from the fact that $\ln(z) \geqslant 1 - \frac{1}{z}$ for all $z \geqslant 1$.

The idea is to add new vectors during each phase of the construction only if they fall into empty boxes. In other words, boxes are small enough so that we do not need several vectors per box. Two vectors, $[x_1, y_1]$ and $[x_2, y_2]$, fall into the same box if and only if

$$\begin{cases} \dfrac{x_1}{\Delta} \leqslant x_2 \leqslant \Delta x_1 \\ \\ \dfrac{y_1}{\Delta} \leqslant y_2 \leqslant \Delta y_1. \end{cases}$$

The pruned construction is as follows:

Initialization Set $VS_1^{\#} = VS_1$.

Phase k for $2 \leqslant k \leqslant n$ For every vector $[x, y] \in VS_{k-1}^{\#}$, put $[x + a_k, y]$ in $VS_k^{\#}$ if and only if the box that it falls into does not already contain another element of $VS_k^{\#}$. Apply the same procedure for $[x, y + a_k]$.

Output Vector $[x, y] \in VS_n^{\#}$ that minimizes $\max(x, y)$.

We generate at most one vector per box. As a result, the total number of vectors is bounded by nM^2. Because M has polynomial size in $\frac{1}{\varepsilon}$ and in $\log P_{sum}$, we have the required complexity. Let us prove that the pruned construction leads to a $(1+\varepsilon)$-approximation of the optimal schedule.

LEMMA 7.2. *$\forall [x, y] \in VS_k$, $\exists [x^{\#}, y^{\#}] \in VS_k^{\#}$ such that $x^{\#} \leqslant \Delta^k x$ and $y^{\#} \leqslant \Delta^k y$.*

Proof. We proceed by induction. The case $k = 1$ is clear. Assume the property holds for $k - 1$, and consider $[x, y] \in VS_k$. Assume that $x = u + a_k$ and $y = v$ for some $[u, v] \in VS_{k-1}$ (the other case, $x = u$ and $y = v + a_k$, is symmetric). By induction, there exists $[u^{\#}, v^{\#}] \in VS_{k-1}^{\#}$ such that $u^{\#} \leqslant \Delta^{k-1} u$ and $v^{\#} \leqslant \Delta^{k-1} v$. We do not know whether $[u^{\#} + a_k, v]$ belongs to $VS_k^{\#}$, but we do know that there is at least one element in the corresponding box:

$$\exists [x^{\#}, y^{\#}] \in VS_k^{\#} \text{ such that } x^{\#} \leqslant \Delta \left(u^{\#} + a_k\right) \text{ and } y^{\#} \leqslant \Delta v^{\#}$$

$$\Rightarrow \begin{cases} x^{\#} \leqslant \Delta^k u + \Delta a_k \leqslant \Delta^k (u + a_k) = \Delta^k x \\ y^{\#} \leqslant \Delta v^{\#} \leqslant \Delta^k y. \end{cases}$$

□

As a consequence of Lemma 7.2, the vector $[x^{\#}, y^{\#}] \in VS^{\#}$ output by the pruned construction is such that $\max(x^{\#}, y^{\#}) \leqslant \Delta^n \mathrm{MS}_{opt}$. There remains to show that $\Delta^n \leqslant (1 + \varepsilon)$. We have $\Delta^n = \left(1 + \dfrac{\varepsilon}{2n}\right)^n$.

LEMMA 7.3. *For $0 \leqslant z \leqslant 1$, $f(z) = \left(1 + \dfrac{z}{n}\right)^n - 1 - 2z \leqslant 0$.*

Proof. $f'(z) = \frac{1}{n} n \left(1 + \frac{z}{n}\right)^{n-1} - 2$ and $f''(z) \geqslant 0$, so f is convex and its maximum is either $f(0) = 0$ or $f(1) = g(n) = \left(1 + \frac{1}{n}\right)^n - 3$. But $g(n) \leqslant e - 3 < 0$ (where $e = 2.718$ is the base of natural logarithms). □

We use the lemma with $z = \frac{\varepsilon}{2}$ to complete the proof of Theorem 7.9. □

7.5 Taking Communication Costs into Account

Distributed-memory parallel computing platforms pose many challenges to the algorithm designer and the programmer. An obvious factor contributing to this complexity is the need for network communication, whose performance is difficult to model in a way that is both precise and conducive to understanding the performance of algorithms (see Chapter 3). Older parallel computers used a store-and-forward approach to communicate messages, which was not efficient but simple to understand and to model. Essentially, the time for sending a message from a processor p to a processor p' is $c(p, p') = \text{dist}(p, p') \times (L + \mathsf{b}b)$, where b is the length of the message, $\text{dist}(p, p')$ is the distance between p and p' in number of hops, L is the communication start-up cost, and b is the inverse of the steady-state bandwidth. In modern computers, messages are split into packets that are dynamically routed between processors, possibly using different paths. Messages can be routed efficiently if there are no contentions on the communication links (or "hot spots"). The distance between communicating processors is no longer the single most important factor for communication performance. In fact, if several processors are to exchange data simultaneously, then the more structured the communication patterns, the more efficient they are, making the role of locality on performance at best indirect.

In light of the complexity of performance modeling for network communications, the vast majority of scheduling works and results are for a very simple model, which is as follows. If a task T communicates data to a successor task T', the cost is modeled as

$$\text{cost}(T, T') = \begin{cases} 0 & \text{if } \mathsf{alloc}(T) = \mathsf{alloc}(T') \\ c(T, T') & \text{otherwise,} \end{cases}$$

where $\mathsf{alloc}(T)$ denotes the processor that executes task T (see Section 7.2), and $c(T, T')$ is defined by the application specification. The above model states that the time for communication between two tasks running on the same processor is negligible. The model also assumes that the processors are part of a fully connected clique. This so-called *macro-dataflow* model makes two main assumptions: (i) communication can occur as soon as data are available; and (ii) there is no contention for network links. Assumption (i) is reasonable as communication can overlap with (independent) computations in most modern computers. Assumption (ii) is much more questionable. Indeed, there is no physical device capable of sending, say, 1,000 messages to 1,000 distinct processors, at the same speed as if there were a single message. In the worst case, it would take 1,000 times longer (serializing all messages). In the best case, the output bandwidth of the network card of the sender would be a limiting factor. In other words, assumption (ii) amounts to assuming infinite network resources! Nevertheless, this assumption is omnipresent in the traditional scheduling literature. Perhaps it is the price to pay to derive tractable mathematical results? We use this mode for now and turn to more realistic communication models in Chapter 8.

We conclude this discussion by stating the model more formally.

DEFINITION 7.8. A communication DAG (or cDAG) is a direct acyclic graph $G = (V, E, w, c)$, where vertices represent tasks and edges represent precedence constraints. The computation weight function is $w : V \longrightarrow \mathbb{N}^*$ and the communication cost function is $c : E \longrightarrow \mathbb{N}^*$. A schedule σ must preserve dependences, which is written as

$$\forall e = (T, T') \in E, \begin{cases} \sigma(T) + w(T) \leqslant \sigma(T') & \text{if } \mathsf{alloc}(T) = \mathsf{alloc}(T') \\ \sigma(T) + w(T) + c(T, T') \leqslant \sigma(T') & \text{otherwise.} \end{cases}$$

The expression of resource constraints is the same as in the no-communication case.

7.6 Pb(∞) with communications

Including communication costs in the model makes everything difficult, including solving Pb(∞). The intuitive reason is that we hesitate between allocating tasks to either many processors (hence balancing the load but communicating intensively) or few processors (leading to less communication but less parallelism as well). We illustrate this with a small example, borrowed from [60].

Consider the cDAG in Figure 7.7. Task weights are indicated close to the tasks within parentheses, and communication costs are shown along the edges, underlined. For the sake of this example, we use two non-integer communication costs: $c(T_4, T_6) = c(T_5, T_6) = 1.5$. Of course, we could scale every weight w and cost c to have only integer values. We can check the following:

- On the one hand, if we allocate all tasks to the same processor, the makespan will be equal to the sum of all task weights, i.e., 13.

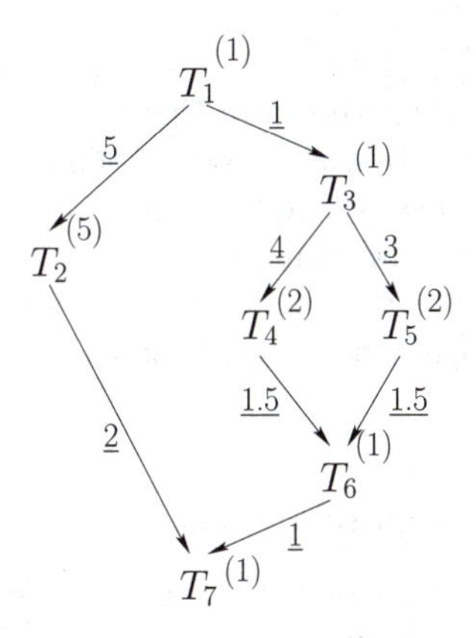

FIGURE 7.7: An example cDAG.

- On the other hand, if we have as many processors as we want (we need no more than seven processors because there are seven tasks), we can allocate one task per processor. Then, we can check that the makespan of the ASAP schedule is equal to 14. To see this, it is important to point out that once the allocation of tasks to processors is given, we can compute the makespan easily: For each edge $e : T \rightarrow T'$, add a virtual node of weight $c(T, T')$ if the edge links two different processors ($\mathsf{alloc}(T) \neq \mathsf{alloc}(T')$), and do nothing otherwise. Then, consider the new graph as a DAG (without communications) and traverse it to compute the length of the longest path, as explained in Section 7.3. In our case, because all tasks are allocated to different processors, we add a virtual node on each edge. The longest path is $T_1 \rightarrow T_2 \rightarrow T_7$, whose length is $w(T_1) + c(T_1, T_2) + w(T_2) + c(T_2, T_7) + w(T_7) = 14$.

There is a difficult trade-off between executing tasks in parallel (hence with several distinct processors) and minimizing communication costs. In our example, it turns out that the best solution is to use two processors, according to the schedule in Figure 7.8, whose makespan is equal to 9.

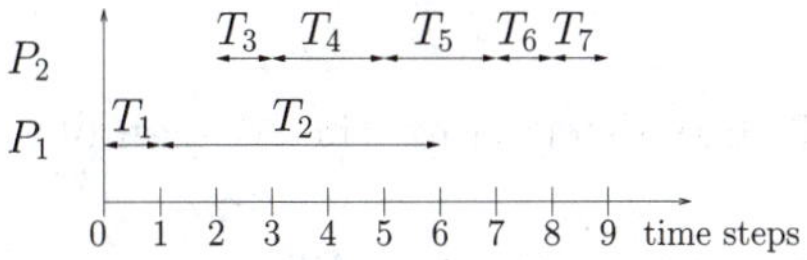

P_2: T_3 T_4 T_5 T_6 T_7
P_1: T_1 T_2
0 1 2 3 4 5 6 7 8 9 time steps

FIGURE 7.8: An optimal schedule for the example.

Note that dependence constraints are satisfied in Figure 7.8. For example, T_2 can start at time 1 on processor P_1 because this processor executes T_1, hence there is no need to pay the communication cost $c(T_1, T_2)$. By contrast, T_3 is executed on processor P_2, hence we need to wait until time 2 to start it even though P_2 is idle: $\sigma(T_1) + w(T_1) + c(T_1, T_3) = 0 + 1 + 1 = 2$.

How did we find the schedule shown in Figure 7.8? And how do we know it is optimal? By a tedious case-by-case analysis! This example shows that using more processors does not always lead to a shorter execution time. Using the notations of Section 7.2, the minimum makespan of a schedule making actual use of seven processors is $\text{MS}'(7) = 14$ while $\text{MS}'(1) = 13$ (or $\text{MS}'(2) = 9$). In other words, Theorem 7.3 is no longer true when communication costs are taken into account.

7.6.1 NP-Completeness of Pb(∞)

Communication costs make solving Pb(∞), the scheduling problem with unlimited processors, very difficult.

THEOREM 7.10. *Pb(∞) is NP-complete.*

Proof. The decision problem Comm(∞) associated with Pb(∞) is the following. Given a cDAG $G = (V, E, w, c)$ and an execution bound $K \in \mathbb{N}^*$, does there exist a schedule σ for G such that $\text{MS}(\sigma, \infty) \leqslant K$? We want to show that Comm(∞) is NP-complete. First, Comm(∞) belongs to NP. If we are given a schedule σ whose makespan is less than or equal to K, we can check in polynomial time that dependence constraints are satisfied. For each task we know the beginning $\sigma(T)$ of its execution and the processor $\mathsf{alloc}(T)$ that executes it, hence we just have to check for constraints.

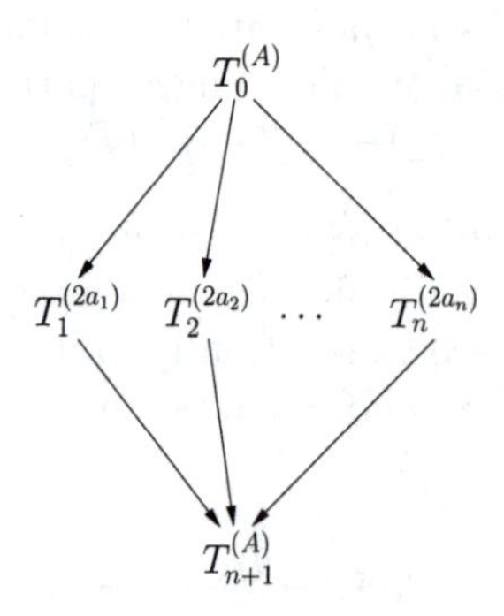

FIGURE 7.9: Reduction for the NP-completeness proof.

To prove NP-completeness, we use 2-Partition as in the proof of Theorem 7.5, but the reduction is more involved. Consider any instance Inst_1 of 2-Partition. Given n positive integer numbers $\{a_1, \ldots, a_n\}$, is there a subset I of indices such that $\sum_{i \in I} a_i = \sum_{i \notin I} a_i$? We build an instance Inst_2 of Comm(∞) as follows. We let $G = (V, E, w, c)$ be a fork-join graph (see Figure 7.9). There are $n + 2$ tasks: $V = \{T_0, T_1, T_2, \ldots, T_n, T_{n+1}\}$. The task weights are defined as follows: $w(T_i) = 2a_i$ for $1 \leqslant i \leqslant n$, and $w(T_0) = w(T_{n+1}) = A$, where A

is a positive integer. There are $2n$ edges, and the communication costs are all equal: $c(T_0, T_i) = c(T_i, T_{n+1}) = C$ for $1 \leqslant i \leqslant n$. Here, C is any integer in the interval $]\alpha - \min_{1\leqslant i\leqslant n} 2a_i, \alpha[$, where $\alpha = \sum_{i=1}^{n} a_i$. Note that this interval does contain an integer, as $\min_{1\leqslant i\leqslant n} 2a_i \geqslant 2$. Finally, let $K = 2A + C + \alpha$. The size of Inst_2 is clearly polynomial in the size of Inst_1. The difficult part is to show that Inst_1 has a solution if and only if there exists a schedule that meets the bound K, i.e., if and only if Inst_2 has a solution.

First, assume that Inst_1 has a solution. Let I be a subset of indices such that $\sum_{i\in I} a_i = \sum_{i\notin I} a_i = \frac{\alpha}{2}$. Let $\mathcal{T}_1 = \{T_i, i \in I\}$ and $\mathcal{T}_2 = \{T_i, i \notin I\}$. By hypothesis, $w(\mathcal{T}_1) = w(\mathcal{T}_2) = \alpha$. Consider the schedule in Figure 7.10.

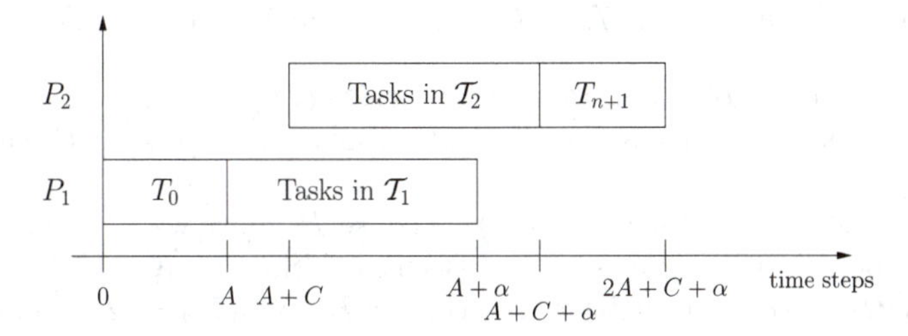

FIGURE 7.10: A schedule with makespan $K = 2A + C + \alpha$.

The makespan of this schedule is equal to $K = 2A+C+\alpha$. All dependence constraints are satisfied. Indeed:

- Processor P_2 starts executing the tasks in $\mathcal{T}_2$ at time $A+C = w(T_0)+C$.
- Processor P_1 terminates executing the tasks in $\mathcal{T}_1$ at time $A+\alpha$. Hence, at time $A + \alpha + C$, task T_{n+1} is ready to be executed by processor P_2. Its execution can start right then, as P_2 terminates the tasks in $\mathcal{T}_2$ at time $A + C + \alpha$.

Conversely, assume that Inst_2 has a solution. Let σ be a schedule whose makespan $\text{MS}(\sigma)$ is less than or equal to K. We need the two following lemmas.

LEMMA 7.4. *Tasks T_0 and T_{n+1} are not executed by the same processor in schedule σ.*

Proof. Assume that the same processor P executes both T_0 and T_{n+1}. Then, P executes all n other tasks T_i, $1 \leqslant i \leqslant n$. Otherwise, let T_{i_0} be a task executed by another processor. The makespan of σ is greater than or equal to the length of the path $T_0 \to T_{i_0} \to T_{n+1}$:

$$\text{MS}(\sigma) \geqslant A+C+2a_{i_0}+C+A = (2A+C)+(2a_{i_0}+C) > (2A+C)+\alpha = K,$$

hence a contradiction. Therefore, P does execute the $n+2$ tasks, which is a contradiction, as the sum of all task weights is $2A+2\alpha > 2A+\alpha+C = K$. □

Let P_1 be the processor that executes T_0 and P_2 be the processor that executes T_{n+1}.

LEMMA 7.5. *Each task T_i, $1 \leqslant i \leqslant n$, is executed by either P_1 or P_2.*

Proof. Assume that there exists a task T_{i_0}, $1 \leqslant i_0 \leqslant n$, executed by a processor other than P_1 and P_2. Then, the makespan of σ is greater than or equal to the length of the path $T_0 \to T_{i_0} \to T_{n+1}$: $\mathrm{MS}(\sigma) \geqslant A+C+2a_{i_0}+C+A > K$ (as in Lemma 7.4), hence a contradiction. □

Let $\mathcal{T}_1$ be the set of tasks T_i, $1 \leqslant i \leqslant n$, executed by P_1. Define similarly $\mathcal{T}_2$ for P_2. The makespan of σ satisfies:

$$\mathrm{MS}(\sigma) \geqslant w(T_0) + w(\mathcal{T}_1) + C + w(T_{n+1}) = 2A + C + w(\mathcal{T}_1) .$$

To understand this note that P_1 takes at least $w(T_0) + w(\mathcal{T}_1)$ time units to execute its tasks, and that a communication must occur before P_2 can start T_{n+1}.

Similarly, $\mathrm{MS}(\sigma) \geqslant 2A + C + w(\mathcal{T}_2)$, because P_2 must wait at least $A + C$ time units before starting execution. Because $\mathrm{MS}(\sigma) \leqslant K = 2A + C + \alpha$, we have $w(\mathcal{T}_1) \leqslant \alpha$ and $w(\mathcal{T}_2) \leqslant \alpha$. But $w(\mathcal{T}_1) + w(\mathcal{T}_2) = 2\alpha$. Therefore, $w(\mathcal{T}_1) = w(\mathcal{T}_2) = \alpha$. Let I denote the set of indices of the tasks in $\mathcal{T}_1$; I is a solution to Inst_1, our instance of 2-Partition. □

Theorem 7.10 only shows that $PB(\infty)$ is NP-complete in the weak sense. In fact, $PB(\infty)$ is NP-complete in the strong sense. Even the problem in which all task weights and communication costs have the same (unit) value, the so-called UET-UCT problem (unit execution time-unit communication time), is NP-hard [93, 94].

7.6.2 A Guaranteed Heuristic for Pb(∞)

In this section, we present a guaranteed heuristic, due to Hanen and Munier [66], to solve Pb(∞). The heuristic is guaranteed within a factor at most $\frac{4}{3}$ of the optimal under the assumption that all communication costs are smaller than all computation costs. Such a task graph is said to be coarse-grain, as stated in the following definition:

DEFINITION 7.9. Let $G = (V, E, w, c)$ be a cDAG. The granularity of G is the computation to communication ratio

$$g(G) = \frac{\min_{T \in V} w(T)}{\max_{T,T' \in V} c(T, T')} .$$

G is coarse-grain if $g(G) \geqslant 1$.

Before stating Hanen and Munier's heuristic formally, we explain the main idea, that of "favorite successors."

TABLE 7.2: Favorite successors for the schedule in Figure 7.8.

Task	T_1	T_2	T_3	T_4	T_5	T_6	T_7
Favorite Successor	T_2	—	T_4	—	T_6	T_7	—

Favorite successors – Let $G = (V, E, w, c)$ be a coarse-grain cDAG and σ be any schedule for G. Let $T \in V$ be any task. The favorite successor of T, if it exists, is the unique immediate successor T' of T such that

$$\sigma(T') < \sigma(T) + w(T) + c(T, T'). \qquad \text{(FS)}$$

If it exists, the favorite successor of T is executed by the same processor as T, otherwise a communication cost would be paid and condition (FS) would not hold. Now, to see why the favorite successor is unique (if it exists), assume that two successors T' and T'' of T satisfy condition (FS). T' and T'' are executed by the same processor. Without loss of generality, assume that T' is executed before T''. This implies that $\sigma(T) + w(T) \leqslant \sigma(T')$, and that $\sigma(T') + w(T') \leqslant \sigma(T'')$. But, by hypothesis, $\sigma(T'') < \sigma(T) + w(T) + c(T, T'')$; hence $w(T') < c(T, T'')$, a result that contradicts the fact that G is coarse-grain. Table 7.2 gives all the favorite successors for the optimal schedule shown in Figure 7.8.

For each edge $e = (T, T') \in E$, we introduce a boolean variable $x_{T,T'}$: $x_{T,T'} = 0$ if T' is the favorite successor of T, and $x_{T,T'} = 1$ otherwise. The inequality

$$\sigma(T) + w(T) + x_{T,T'}\ c(T, T') \leqslant \sigma(T')$$

holds for any edge $(T, T') \in E$, whether T' is the favorite successor of T or not. Casting all such inequalities into a linear program is the main idea underlying the heuristic.

Hanen and Munier's heuristic – Given a coarse-grain cDAG, we define the following integer linear program.

DEFINITION 7.10. Let $G = (V, E, w, c)$ be a cDAG. We define the integer linear program ILP(G) as follows:

Minimize M_∞ subject to

$$\begin{cases} \forall (T, T') \in E & x_{T,T'} \in \{0, 1\} & (A) \\ \forall T \in V & s(T) \geqslant 0 & (B) \\ \forall (T, T') \in E & s(T) + w(T) + x_{T,T'} c(T, T') \leq s(T') & (1) \\ \forall T \in V s.t.\ \text{SUCC}(T) \neq \emptyset & \sum_{T' \in \text{SUCC}(T)} x_{T,T'} \geqslant |\text{SUCC}(T)| - 1 & (2) \\ \forall T \in V s.t.\ \text{PRED}(T) \neq \emptyset & \sum_{T' \in \text{PRED}(T)} x_{T',T} \geqslant |\text{PRED}(T)| - 1 & (3) \\ \forall T \in V & s(T) + w(T) \leqslant M_\infty & (4) \end{cases}$$

We refer to [106] for a primer on (integer) linear programming. Intuitively, the makespan M_∞ is the maximum of the completion times of all tasks, which is expressed by constraint (4).

LEMMA 7.6. *Let $G = (V, E, w, c)$ be a cDAG. The solution M_∞ of the integer linear program ILP(G) is equal to the optimal makespan with unlimited processors $MS_{opt}(\infty)$.*

Proof. We show that there is a one-to-one correspondence between valid schedules for G (with unlimited processors) and solutions to the integer linear program ILP(G).

Let σ be a valid schedule for G. Let $s(T) = \sigma(T)$ for each task T, and let $x_{T,T'} = 0$ if T' is the favorite successor of T, and $x_{T,T'} = 1$ otherwise. Then, all constraints of ILP(G) are met:

- Constraints (A) and (B) are met by construction.
- Constraint (1) was derived above.
- Constraint (2) expresses the fact that each task has at most one favorite successor.
- Constraint (3) expresses the fact that each task is the favorite successor of at most one task, which can be proved quite similarly to constraint (2).
- The definition of the makespan is $\text{MS}(\sigma, \infty) = \max_{T \in V}(\sigma(T) + w(T))$, hence constraint (4) is met.

Let $M_\infty(\sigma)$ be the value returned by ILP(G) when all variables $s(T)$ and $x_{T,T'}$ are defined as above. Because constraint (4) is the only constraint on the objective function, we have $M_\infty(\sigma) = \text{MS}(\sigma, \infty)$.

Reciprocally, consider a solution of the optimization problem ILP(G). To define the induced schedule σ, we need to determine for each task both a starting time and the processor that executes it. For starting times, we simply let $\sigma(T) = s(T)$ for any task $T \in V$. We define the allocation function as follows:

$$\forall e = (T, T') \in E,\ x_{T,T'} = 0 \Leftrightarrow \mathsf{alloc}(T) = \mathsf{alloc}(T') \ .$$

To be more precise, we allocate entry tasks to different processors, and we traverse the graph to compute the allocation function as follows: If T' is an immediate successor of T and $x_{T,T'} = 1$, we allocate T' to a new processor, otherwise we allocate T' to the same processor as T. We have no conflict during this traversal. Indeed, due to condition (2), for each task $T \in V$, there is at most one immediate successor of T allocated to the same processor as T. This successor, if it exists, is the unique task T' such that $x_{T,T'} = 0$ (T' is then the favorite successor of T). Similarly, due to condition (3), for each task $T' \in V$, there is at most one predecessor of T' allocated to the same processor as T'. Furthermore, constraint (1) together with the choice of the allocation function ensure that all dependence constraints are met: σ is a valid schedule for G. Finally, condition (4) shows that M_∞ is equal to the makespan of σ. □

Given a solution to ILP(G), we can interpret $s(T)$ as the top level of T, where the bottom and top levels are computed according to the allocation function induced by the variables $x_{T,T'}$. We add the communication cost $c(T,T')$ into the weight of a path going from T to T' if and only if $\mathsf{alloc}(T) \neq \mathsf{alloc}(T')$, i.e., if and only if $x_{T,T'} = 1$. Condition (4) shows that the solution to ILP(G) is indeed equal to the value of the maximal weight of a path in the dependence graph computed using the previous rules. The difficulty lies in determining the allocation function, i.e., in determining the $x_{T,T'}$ values. Because these values are integers (and even restricted to 0 or 1), the ILP problem is an integer linear program, which is NP-hard [106]. However, if we relax the condition that the $x_{T,T'}$'s are integers, we obtain a linear program with rational unknowns, whose complexity is known to be polynomial [106].

DEFINITION 7.11. Let $G = (V, E, w, c)$ be a cDAG:

- We define the relaxed linear program RLP(G) as the program obtained by replacing equation (A) in the definition of ILP(G) by the equation

$$\forall (T,T') \in E,\ 0 \leqslant x_{T,T'} \leqslant 1 \ .$$

  Now the variables $x_{T,T'}$ are rational numbers instead of integers.

- We let $(x^{rel}_{T,T'}, s^{rel}(T), M^{rel}_{\infty})$ denote the solution of the relaxed problem RLP(G) over the rational numbers.

Hanen and Munier define their schedule σ^{hm} directly from the solution of the relaxed linear program RLP(G). Let $T \in V$ be any task. Constraint (2) ensures that there is at most one successor T' of T such that $x^{rel}_{T,T'} < \frac{1}{2}$; and constraint (3) ensures that there is at most one predecessor T'' of T such that $x^{rel}_{T'',T} < \frac{1}{2}$. Therefore, let $x^{hm}_{T,T'} = 0$ for any edge $e = (T, T') \in E$ such that $x^{rel}_{T,T'} < \frac{1}{2}$, and $x^{hm}_{T,T'} = 1$ otherwise. For any task $T \in V$, define σ^{hm}_T to be the top level of T, where the bottom and top levels are computed according to the allocation function induced by the $x^{hm}_{T,T'}$. We add the communication cost $c(T,T')$ in the weight of a path going from T to T' if and only if $\mathsf{alloc}(T) \neq \mathsf{alloc}(T')$, i.e., if and only if $x^{hm}_{T,T'} = 1$. As explained earlier, this defines a valid schedule for G.

THEOREM 7.11. *Let $G = (V, E, w, c)$ be a coarse-grain cDAG, with granularity $g(G) \geqslant 1$. Let σ^{hm} be the schedule defined by Hanen and Munier. Then,*

$$MS(\sigma^{hm}, \infty) \leqslant MS_{opt}(\infty)\ \frac{2g(G)+2}{2g(G)+1} \ .$$

Proof. For any path in the graph going from a task T to one of its successors T', we have the communication cost $x^{hm}_{T,T'}c(T,T')$ for Hanen and Munier's schedule, instead of $x^{rel}_{T,T'}c(T,T')$ for the solution of RLP(G). Two cases occur:

- $x^{hm}_{T,T'} = 0$: then $w(T) + x^{hm}_{T,T'}c(T,T') \leqslant w(T) + x^{rel}_{T,T'}c(T,T')$.
- $x^{hm}_{T,T'} = 1$: then $x^{rel}_{T,T'} \geqslant \frac{1}{2}$. We have

$$\frac{w(T) + x^{hm}_{T,T'}c(T,T')}{w(T) + x^{rel}_{T,T'}c(T,T')} \leqslant \frac{w(T) + c(T,T')}{w(T) + c(T,T')/2} = \frac{1 + \frac{c(T,T')}{w(T)}}{1 + \frac{c(T,T')}{2w(T)}} \quad \text{and}$$

$$\frac{1 + \frac{c(T,T')}{w(T)}}{1 + \frac{c(T,T')}{2w(T)}} \leqslant \frac{1 + \frac{1}{g(G)}}{1 + \frac{1}{2g(G)}} = \frac{2g(G) + 2}{2g(G) + 1}.$$

In all cases, $w(T) + x^{hm}_{T,T'}c(T,T') \leqslant \frac{2g(G)+2}{2g(G)+1}(w(T) + x^{rel}_{T,T'}c(T,T'))$, and this inequality extends to all paths in the graph, which proves the theorem. □

An immediate consequence of Theorem 7.11 is that Hanen and Munier's heuristic is guaranteed with a factor at most $\frac{4}{3}$ for coarse-grain graphs.

7.7 List Heuristics for Pb(p) with Communications

As expected, the limited processors scheduling problem Pb(p) remains NP-complete when introducing communication costs. Pb(p) does remain in the NP class; once the allocation is known, dependence and resource constraints can be checked by traversing the graph. The true problem is to determine a good allocation. The most natural idea is to extend the critical path list algorithm introduced in Section 7.4.4. We explain how to modify it in a straightforward fashion and then how to design a much improved version.

7.7.1 Naïve Critical Path

The idea of critical path scheduling remains the same: list scheduling with task priorities equal to the task bottom levels. The problem is that we do not know how to compute bottom levels. Without knowing the allocation, it is not possible to decide which communication costs should be taken into account. A conservative approach is to include all communication costs when computing bottom levels (which amounts to assuming one distinct processor per task).

Consider again the example DAG in Figure 7.7, with corresponding task bottom levels shown in Table 7.3. We check that the bottom level of task T_1 is 14, the length of the ASAP schedule with one processor per task.

Let us build a list schedule based on the values of these bottom levels. The algorithm proceeds as explained in Section 7.4.4, but with an important

TABLE 7.3: Task bottom levels for the DAG in Figure 7.7.

Task	T_1	T_2	T_3	T_4	T_5	T_6	T_7
Critical Path	14	8	11.5	6.5	6.5	3	1

difference. A task is free when all its predecessors have been executed. But a free task cannot start execution as soon as it is free, even if there are available processors; depending on the allocation decision, we may or may not need to wait for some communication delay.

Assume $p = 3$ available processors P_1, P_2, and P_3. When there are more available processors than free tasks, assign the tasks to the processors with (say) the lowest indices. Let $\mathcal{Q}$ be the priority queue of free tasks. In our example, there is $r = 1$ free task at time $t = 0$: $\mathcal{Q} = (T_1)$. Processor P_1 executes T_1 at $t = 0$. At $t = 1$, we update $\mathcal{Q}$ as $\mathcal{Q} = (T_3, T_2)$ (T_3 is given priority over T_2 because its bottom level is larger). All processors are available, so according to our rule we allocate T_3 to P_1 and T_2 to P_2. Note that we are lucky to assign T_3 to P_1: Because P_1 has executed T_1, there is no communication delay to pay, hence it can start T_3 at time $t = 1$. On the other hand, P_2 must wait until time $t = 6$ to start T_2. At time $t = 2$, we have $\mathcal{Q} = (T_4, T_5)$ (breaking ties arbitrarily, using task numbers), and two available processors P_1 and P_3. Indeed, although still idle, P_2 has been marked busy until it returns from the execution of T_2. Hence, we allocate T_4 to P_1 (execution can start immediately) and T_5 to P_3 (execution cannot start before $t = 5$). In the end, we obtain the schedule shown in Figure 7.11.

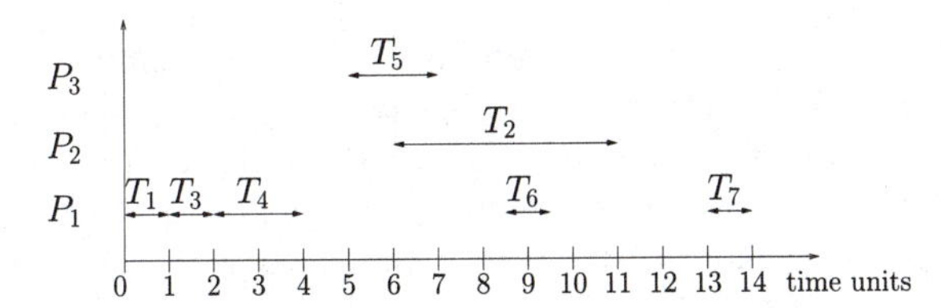

FIGURE 7.11: Naïve critical path scheduling for the example DAG in Figure 7.7.

We obtain a makespan equal to 14 time units. Note that the naïve critical path (naïve CP) scheduling with two processors leads to the same result: T_5 would have been executed by P_1 at time $t = 4$ rather than by P_3 at time $t = 5$: this is the only difference. In both cases, we obtain the same makespan, even worse than the execution on a single processor! There must be room for improvement.

7.7.2 Modified Critical Path

If we analyze the execution of naïve CP scheduling on our small example, we see that we made a wrong decision when assigning T_2 to P_2. Indeed, at time $t = 1$ we had $\mathcal{Q} = (T_3, T_2)$. The first allocation, that of T_3 to P_1, is fine. But the second, that of T_2 to P_2, is not. We should allocate it to P_1 again, even though it is not available. The reason is that P_1 can start T_2 earlier than P_2. The rule of modified critical path (MCP) is: *Allocate a free task to the processor that allows its earliest execution, given previous task allocation decisions.* It is important to explain further what "previous task allocation decisions" means. Free tasks from the queue are processed one after the other. At any moment, we know which processors are available and which ones are busy. Moreover, for the busy processors, we know when they will finigh computing their currently allocated tasks. Hence, we can always select the processor that can start the execution of the task under consideration the earliest. It may well be the case that we select a processor that is currently busy, as discussed earlier for allocating task T_2.

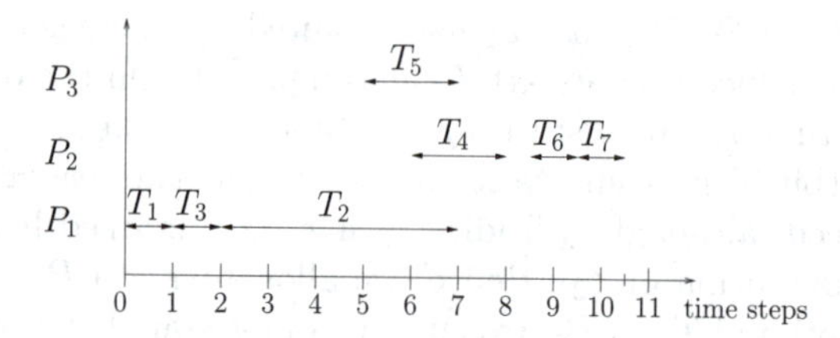

FIGURE 7.12: Modified critical path scheduling for the example DAG in Figure 7.7.

Rather than writing the details of MCP in algorithmic form, we apply it in our example with three processors. We obtain the schedule shown in Figure 7.12. As already discussed, T_2 is allocated to P_1. Similarly, T_6 is allocated to P_2, because it allows execution at time $t = 8.5$, against 9.5 on P_1 or P_3. The new makespan is 10.5, to be compared with the makespan 14 of naïve CP.

7.7.3 Hints for Comparison

The comparison of naïve CP and MCP for our little example should not lead to drastic conclusions. We are comparing two heuristics; neither of them is always superior to the other. There are examples where naïve CP is better than MCP: For the example shown in Figure 7.13 with two processors, one can check that the makespan of MCP is 15, while that of naïve CP is 14.

Our intuition, however, shows that MCP is likely to outperform naïve CP in most cases. How do we quantify this assertion? To be less specific than with particular examples, let us compare CP and MCP on simple graph types. We analyze two very simple cases, a fork with two nodes and a join with two nodes. In Figures 7.14(a) and 7.14(b), we have three tasks of the same weight w. The communication costs are all equal to c. Assume two processors are

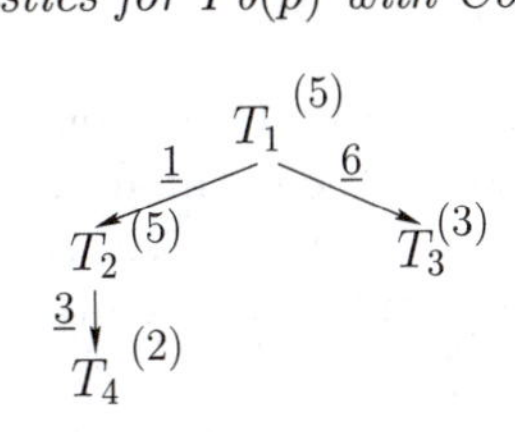

FIGURE 7.13: An example where naïve CP is better than MCP.

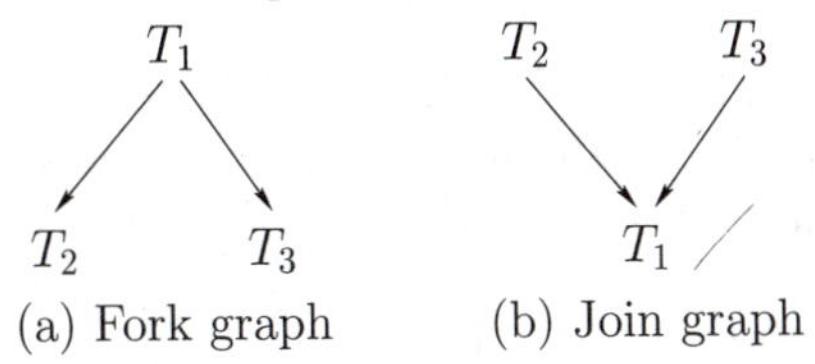

(a) Fork graph (b) Join graph

FIGURE 7.14: Elementary graphs for comparing naïve CP and MCP.

available, P_1 and P_2:

1. **Fork Graph** (Figure 7.14(a)). Naïve CP schedules T_1 on P_1 at time $t = 0$. Then, it schedules T_2 on P_1 at time $t = w$, and T_3 on P_2 at time $w + c$, hence a makespan equal to $2w + c$. MCP does the same as naïve CP if $w > c$. But if $w < c$, the earliest execution time for T_3 is $t = 2w$ on P_1, hence MCP schedules all tasks on P_1, and its makespan is equal to $3w < 2w + c$. Conclusion: MCP outperforms naïve CP if $w < c$ and ties it otherwise.

2. **Join Graph** (Figure 7.14(b)). Naïve CP and MCP perform identically. At time $t = 0$, they schedule T_2 on P_1 and T_3 on P_2. At time $t = w + c$, T_1 is scheduled on either P_1 or P_2, and the makespan is $2w + c$. Note that is not optimal if $w < c$; it is better to schedule the three tasks on the same processor! Of course this is forbidden in list scheduling: No processor can be deliberately kept idle.

This short discussion shows how difficult it is to draw conclusions. The scheduling problem is NP-complete, and we only compare heuristics. Generally speaking, there are three possible approaches:

1. **Theoretical:** Prove that the heuristic is guaranteed, i.e., that it always leads to a makespan within some fraction of the optimal (in other words, make it an approximation algorithm).

2. **Experimental:** Use random graphs and "standard benchmark" graphs to compare heuristics. The difficulty is that there is little consensus on which benchmarks are representative of large and relevant application classes.

3. **Tricky:** Prove that the heuristic is optimal for certain classes of graphs: forks, joins, fork-joins, trees, etc.

The first approach is the strongest: One is ensured that the heuristic will perform within a certain factor of the optimal in the worst case. The second approach is quite useful (and used) in practice. And the third approach helps to tune the heuristics so as to be optimal for certain graph classes (and maybe to publish nice research papers!).

A small step in the first direction is the following (rather weak) counterpart of Theorem 7.6 :

THEOREM 7.12. *Let $G = (V, E, w, c)$ be a cDAG of granularity $g(G)$ (see Section 7.6.2), and let $MS_{opt}(p)$ be the makespan of an optimal schedule. Then, we can derive a schedule σ with p processors whose makespan verifies*

$$MS(\sigma, p) \leqslant \left(2 - \frac{1}{p}\right)(1 + g(G))MS_{opt}(p) .$$

Proof. The proof is straightforward. Neglect all communication costs and construct a list schedule σ: its makespan is such that

$$\mathrm{MS}(\sigma, p) \leqslant \left(2 - \frac{1}{p}\right)\mathrm{MS}^*_{opt}(p) \leqslant \left(2 - \frac{1}{p}\right)\mathrm{MS}_{opt}(p),$$

where $\mathrm{MS}^*_{opt}(p)$ is the optimal makespan without communication costs. Then, we stretch the schedule by a factor $1 + g(G)$, which allows us to pay for the communication cost incurred from the predecessors of each task T_i. We have an interval of length $(1 + g(G))w_i$ to communicate data from the predecessors of T_i and execute it. Therefore, we have derived a valid schedule whose makespan meets the desired bound. Note that this schedule is not necessarily a list schedule because we may have waited longer than needed to execute some tasks. □

7.8 Extension to Heterogeneous Platforms

This section explains how to extend list scheduling techniques to heterogeneous platforms, i.e., to platforms that consist of processors with different speeds and interconnection links with different bandwidths. We have discussed the issue of static load balancing on platforms with heterogeneous processors (but homogeneous network links) in Chapter 6. Rather than defining all notations precisely, we proceed rather informally and focus on the key differences with the homogeneous case.

We start with a cDAG with n tasks $T_1, \ldots, T_n$. The goal is to schedule this cDAG on a platform with p heterogeneous processors $P_1, \ldots, P_p$. There are many parameters to instantiate:

Computation costs – The execution cost of T_i on P_q is modeled as w_{iq}. Therefore, an $n \times p$ matrix of values is needed to specify all computation costs. This matrix comes directly for the specific scheduling problem at hand. However, when attempting to evaluate competing scheduling heuristics over a large number of synthetic scenarios, one must generate this matrix. One can distinguish two approaches. In the first approach on generates a *consistent* (or *uniform*) matrix with $w_{iq} = w_i \times \gamma_q$, where w_i represents the number of operations required by T_i and γ_q is the inverse of the speed of P_q (in operations per second). With this definition the relative speed of the processors does not depend on the particular task they execute. If instead some processors are faster for some tasks than some other processors, but slower for other tasks, one speaks of an *inconsistent* (or *non-uniform*) matrix. This corresponds to the case in which some processors are specialized for some tasks (e.g., specialized hardware or software).

Communication costs – Just as processors have different speeds, communication links may have different bandwidths. However, while the speed of a processor may depend upon the nature of the computation it performs, the bandwidth of a link does not depend on the nature of the bytes it transmits. It is therefore natural to assume *consistent* (or *uniform*) links. If there is a dependence $e_{ij} : T_i \rightarrow T_j$, if T_i is executed on P_q and T_j executed on P_r, then the communication time is modeled as

$$\text{comm}(i, j, q, r) = \text{data}(i, j) \times v_{qr} \ ,$$

where $\text{data}(i, j)$ is the data volume associated to e_{ij} and v_{qr} is the communication time for a unit-size message from P_q to P_r (i.e., the inverse of the bandwidth). Like in the homogeneous case, we let $v_{qr} = 0$ if $q = r$, i.e., if both tasks are assigned the same processor. If one wishes to generate synthetic scenarios to evaluate competing scheduling heuristics, one then must generate two matrices: one of size $n \times n$ for data and one of size $p \times p$ for v_{qr}.

The main list scheduling principle is unchanged. As before, we need to compute the priority of each task so as to decide which one to execute first when there are more free tasks than available processors. In other words, we need to find the equivalent of bottom levels for MCP. The most natural idea is to compute averages of computation and communication times, and use these to compute priority levels exactly as in the homogeneous case. We define:

- $\overline{w_i} = \frac{\sum_{q=1}^{p} w_{iq}}{p}$, the *average* execution time of T_i;
- $\overline{\text{comm}_{ij}} = \text{data}(i, j) \times \frac{\sum_{1 \leqslant q, r \leqslant p, q \neq r} v_{qr}}{p(p-1)}$, the *average* communication cost for edge $e_{ij} : T_i \rightarrow T_j$.

The last (but important) modification concerns the way in which tasks are assigned to processors: Instead of assigning the current task to the processor that will *start* its execution first (given all already taken decisions), we should assign it to the processor that will *complete* its execution first (given all already taken decisions). Both choices are equivalent with homogeneous processors, but intuitively the latter is likely to be more efficient in the heterogeneous case.

Altogether, we have re-discovered the list heuristic called HEFT, for *heterogeneous earliest finish time* [115]. The complexity of the algorithm as we have outlined it here is the same as that of MCP. More sophisticated versions attempt to insert tasks in intervals of time during which processors are idle. This technique is called *insertion scheduling*: Instead of scheduling a new task after those already assigned to a given processor, a good idea may be to try and schedule it at some earlier time, provided that there exists an interval long enough to accommodate the task and during which the processor was idle (most likely waiting for some communication to complete).

Bibliographical Notes

All the material covered in this chapter is rather basic. Without communication costs, we mention that pioneering work includes the book by Coffman [42]. Chapter 9 of [82]; the book by El-Rewini, Lewis, and Ali [53]; and the IEEE compilation of papers [107] provide additional material. On the theoretical side, Appendix A5 of Garey and Johnson [57] provides a list of NP-complete scheduling problems. Also, the book by Brucker [37] offers a comprehensive overview of many complexity results.

The literature with communication costs is more recent. Theorem 7.10 is due to Chrétienne [40]. Picouleau [93, 94] proves that $Pb(\infty)$ remains NP-complete even when we assume all task weights and communication costs to have the same (unit) value — this is the so-called UET-UCT problem (unit execution time-unit communication time) — or even if communication costs are arbitrarily small (but non-zero). Several extensions to Theorem 7.10 are discussed in the survey paper by Chrétienne and Picouleau [41]. Hanen and Munier's heuristic can be extended to cope with limited processors; see [66]. See also the book by Darte, Robert and Vivien [49], where many clustering heuristics are surveyed. Finally, a recent book by Sinnen [108] provides a thorough discussion on communication models. In particular, it describes several extensions for modeling and accounting for communication contention.

7.9 Exercises

We start with two exercises on scheduling without any communication costs. Exercise 7.1 studies some properties of the free schedule and its variants, and Exercise 7.2, borrowed from [64], shows surprising anomalies of the list scheduling approach. Communication costs are involved in Exercise 7.3, which studies the complexity of scheduling a simple fork graph using various communication models. We conclude with a more difficult problem in Exercise 7.4, namely, scheduling in-trees using Hu's algorithm. This exercise is borrowed from [49]. The original reference is [89], which gives a simpler proof of Hu's algorithm [69].

◇ Exercise 7.1 : Free Schedule

1. Consider a DAG $G = (V, E, w)$ and assume unlimited processors. Show that any optimal schedule σ satisfies

$$\forall v \in V, \quad \sigma_{free}(v) \leqslant \sigma(v) \leqslant \sigma_{late}(v),$$

where σ_{free} and σ_{late} are the ASAP and ALAP schedules defined in Section 7.3.

2. Give an example of a DAG $G = (V, E, w)$ that has at least three different optimal schedules with unlimited processors.

3. Consider the DAG in Figure 7.15. Assume that all tasks have unit weight. What is the optimal execution time $\mathrm{MS}_{opt}(\infty)$? How many processors are needed for the ASAP scheduling? For the ALAP scheduling? Determine the minimum number p_{opt} of processors needed to achieve execution in optimal time $\mathrm{MS}_{opt}(\infty)$.

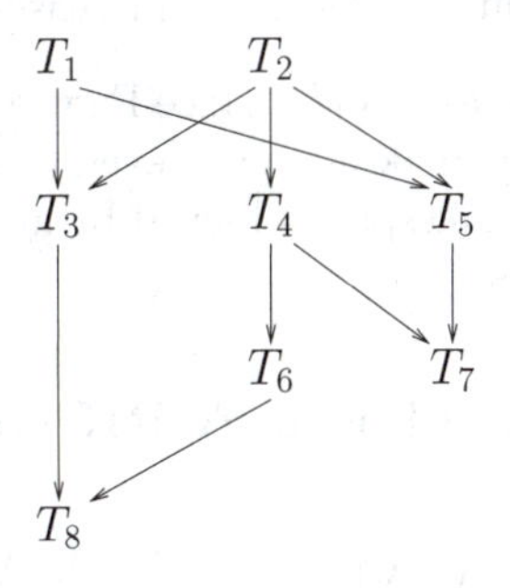

FIGURE 7.15: The DAG to schedule in Exercise 7.1.

4. Consider a DAG $G = (V, E, w)$ and let p_{opt} be the minimum number of processors required to achieve execution in optimal time. Formally,

$$p_{opt} = \min\{p \mid \mathrm{MS}_{opt}(p) = \mathrm{MS}_{opt}(\infty)\} .$$

Show that the problem of determining p_{opt} is NP-complete.

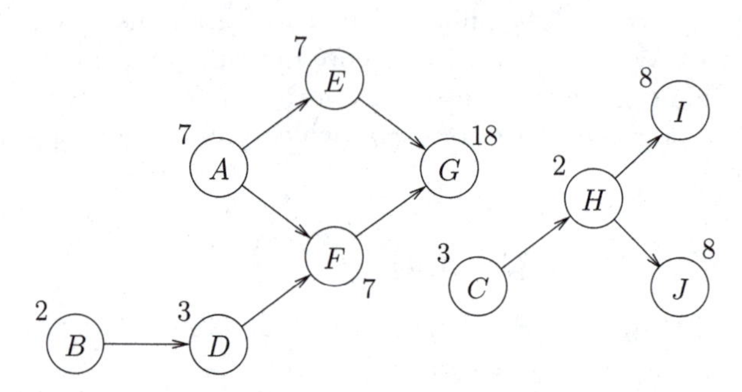

FIGURE 7.16: A DAG to reveal anomalies of list scheduling.

◇ Exercise 7.2 : List Scheduling Anomalies

Consider the DAG in Figure 7.16 where a pair X/w means that task X has weight w. For instance A has weight 8.

1. What is the makespan achieved by critical path list scheduling with 2 processors? Is it optimal?

2. Assume that each task weight is decreased by one unit (now A has weight 7, B has weight 1, and so on). Show that the makespan achieved by critical path list scheduling increases. Show that, somewhat shockingly, the makespan achieved by any list scheduling algorithm increases.

3. Going back to original task weights (see Figure 7.16), assume that we have 3 processors. Show that the makespan achieved by critical path list scheduling increases. Show that the makespan achieved by any list scheduling algorithm, shockingly again, increases.

◇ Exercise 7.3 : Scheduling a FORK Graph with Communications

DEFINITION 7.12 (FORK with n children)**.** A FORK graph with n children is a cDAG with $n+1$ tasks $T_0, T_1, \ldots, T_n$ as shown in Figure 7.17. There is an edge from T_0 to each child T_i, $1 \leqslant i \leqslant n$. The weight of task T_i is w_i. The weight of edge (T_0, T_i) is d_i.

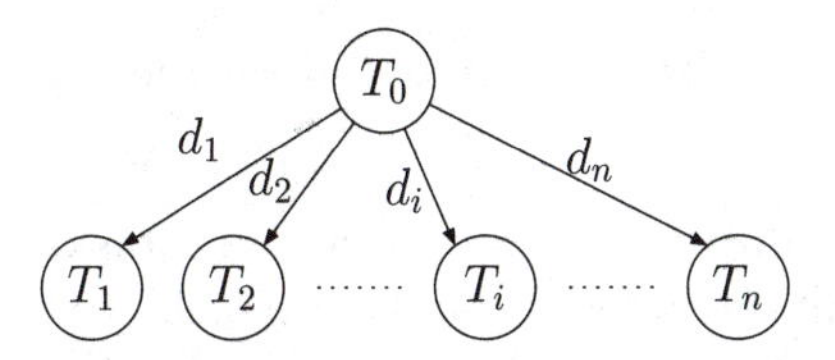

FIGURE 7.17: FORK graph with n children.

First we assume an unlimited number of (identical) processors. We define the following optimization problem:

DEFINITION 7.13 (ForkSched-$\infty(G)$)**.** Given a FORK graph G with n children and an unlimited number of processors, what is the optimal makespan?

1. Give a polynomial algorithm to solve ForkSched-$\infty(G)$.

Now we target the same problem, but with a bounded number of processors:

DEFINITION 7.14 (ForkSchedBounded(G,p))**.** Given a FORK graph G with n children and a set of p processors, what is the optimal makespan?

2. Show that the decision problem associated to ForkSchedBounded(G,p) is NP-complete.

We come back to the problem with infinite processors, but we introduce a new model for communications: We assume that a processor can send or receive a single message at a time. Communications are serialized in this *1-port* model (see Section 3.2.3 for more details).

DEFINITION 7.15 (ForkSched-1-Port-$\infty(G)$)**.** Given a FORK graph G with n children and an unlimited number of 1-port processors, what is the optimal makespan?

3. Show that the decision problem associated to ForkSched-1-Port-$\infty(G)$ is NP-complete.

Exercise 7.4 : Hu's Algorithm

1. In this exercise, we study the problem of scheduling an in-tree $G = (V, E, w)$ (a DAG where each vertex has at most one successor). We assume that all vertices have unit execution time: $w(v) = 1$ for all $v \in V$. We denote by level(v) the maximal length of a path starting from v, and we let level(v) $= 0$ if v has no successor. We let $h = \max_{v \in V}$ level(v) be the maximal level in G. We assume there are p identical resources and we denote by $\mathrm{MS}_{opt}(p)$ the minimal makespan of a schedule for p resources and by $\mathrm{MS}(\sigma, p)$ the makespan of a schedule σ for p resources.

Generalizing Proposition 7.1 and Theorem 7.2, show that

$$\forall i,\ 0 \leqslant i \leqslant h,\ \mathrm{MS}_{opt}(p) \geqslant \frac{|V_i|}{p} + i,$$

where $V_i = \{v \in V \mid \text{level}(v) \geqslant i\}$ is the set of the vertices whose level is at least i.

2. Let σ be a list schedule with a priority queue of tasks ordered by decreasing level. This means that if two tasks u and v are ready to be scheduled at a given time, then $\text{level}(u) \geqslant \text{level}(v)$ implies $\sigma(u) \leqslant \sigma(v)$. For $0 \leqslant t < \text{MS}(\sigma, p)$, we denote by S_t the set of tasks executed at time t by σ: $S_t = \{v \in V \mid \sigma(v) = t\}$.

We first assume that there exists an integer t, $0 \leqslant t < \text{MS}(\sigma, p)$, such that S_t has exactly p tasks with the same level. We denote by k the largest such integer and by L the (common) level of the tasks in S_k. We also assume that $k < \text{MS}(\sigma, p) - 1$ and we denote by L' the maximal level of a task v not yet scheduled at time k, i.e., such that $\sigma(v) > k$.

(a) Give the maximum level of a task v in S_t for $k < t < \text{MS}(\sigma, p)$ and show that $\text{MS}(\sigma, p) = k + L' + 2$.

(b) Show that, for all integers t, $0 \leqslant t \leqslant k$, $|S_t| = p$. (Use the fact that G is an in-tree, i.e., each vertex has at most one successor.)

(c) Infer from the previous two questions that $\text{MS}(\sigma, p) = \text{MS}_{opt}(p)$. Show that this optimality result still holds even if $k = \text{MS}(\sigma, p) - 1$ (in which case L' is not defined) or if k does not exist.

7.10 Answers

Exercise 7.1 (Free Schedule)

▷ **Question 1.** Consider any task $v \in V$. By definition, $\sigma_{free}(v)$ is the length of the longest path from an entry node up to v, so $\sigma_{free}(v) \leqslant \sigma(v)$ for any schedule σ, be it optimal or not. Now read the definition of σ_{late} carefully: There is a path of length $\sigma_{late}(v) = \mathrm{MS}_{opt}(\infty) - bl(v)$ from v to an exit node. Hence, if a schedule σ starts executing v later than time $\sigma_{late}(v)$, its makespan will be greater than $\mathrm{MS}_{opt}(\infty)$, implying that σ is not optimal.

▷ **Question 2.** Consider a DAG with four tasks T_1, T_2, T_3, T_4 of unit weights. The only dependences are

$$T_1 \rightarrow T_2 \rightarrow T_3.$$

We have $\mathrm{MS}_{opt}(\infty) = 3$, $\sigma_{free}(T_1) = \sigma_{late}(T_1) = 0$, $\sigma_{free}(T_2) = \sigma_{late}(T_2) = 1$, and $\sigma_{free}(T_3) = \sigma_{late}(T_3) = 2$. However, T_4 is independent from the other tasks. We have $\sigma_{free}(T_4) = 0$ and $\sigma_{late}(T_4) = 2$. There is room for a third optimal schedule σ that coincides with the other two on T_1, T_2, and T_3, and such that $\sigma(T_4) = 1$.

▷ **Question 3.** The longest path is $T_2 \rightarrow T_4 \rightarrow T_6 \rightarrow T_8$, thus $\mathrm{MS}_{opt}(\infty) = 4$. The following table shows the starting times for σ_{free} and σ_{late}:

Tasks	T_1	T_2	T_3	T_4	T_5	T_6	T_7	T_8
σ_{free}	0	0	1	1	1	2	2	3
σ_{late}	1	0	2	1	2	2	3	3

We see from the table that we need three processors for σ_{free} (at time 1) and also three processors for σ_{late} (at time 2). Any schedule whose makespan is $\mathrm{MS}_{opt}(\infty) = 4$ requires at least two processors, since there are eight tasks. The schedule σ below is valid (check that all dependences are satisfied) and achieves a makespan of 4 with only two processors:

Tasks	T_1	T_2	T_3	T_4	T_5	T_6	T_7	T_8
σ	0	0	1	1	2	2	3	3

We conclude that $p_{opt} = 2$.

▷ **Question 4.** The problem is obviously in NP. We use a reduction from 2-Partition. Consider an arbitrary instance Inst_1 with n integers $\{a_1, a_2, \ldots, a_n\}$. Let $S = \sum_{i=1}^{n} a_i$. We assume that S is even and that $a_i \leqslant \frac{S}{2}$ for all i (otherwise we know there is no solution). We build an instance Inst_2 of our problem as follows: We have a DAG of $n+1$ independent tasks T_1 to T_{n+1}. We let $w(T_i) = a_i$ for $1 \leqslant i \leqslant n$, and $w(T_{n+1}) = \frac{S}{2}$. The size of Inst_2 is linear in the size of Inst_1. We have $\mathrm{MS}_{opt}(\infty) = w(T_{n+1}) = \frac{S}{2}$ and we ask whether $p_{opt} \leqslant K = 3$. Obviously, there is a solution to Inst_1 if and only if there is one to Inst_2.

Exercise 7.2 (List Scheduling Anomalies)

▷ **Question 1.** It is not difficult to compute the bottom levels and execution times. We obtain:

Task	A	B	C	D	E	F	G	H	I	J
Bottom level	33	30	13	28	25	25	18	10	8	8
σ	0	0	5	2	8	8	15	15	17	25
Processor	P_1	P_2	P_2	P_2	P_1	P_2	P_1	P_2	P_2	P_2

The makespan is 33, while the sequential time is 66. There is no idle time in this schedule obtained with critical path list scheduling, and thus it is optimal.

▷ **Question 2.** Here again, we can compute the bottom levels and execution times. We obtain:

Task	A	B	C	D	E	F	G	H	I	J
Bottom level	30	26	10	25	23	23	17	8	7	7
σ	0	0	3	1	7	13	19	5	6	13
Processor	P_1	P_2	P_2	P_2	P_1	P_1	P_1	P_2	P_2	P_2

The makespan is 36, three time units more than with larger task weights. It turns out that any list scheduling algorithm leads to a makespan of at least 36. This is painful to prove: There is no other way than trying all possibilities. At time 0, either we schedule A and B, or we schedule A and C, or we schedule B and C. In this way we explore a tree of possibilities and eventually prove the result.

▷ **Question 3.** With three processors there is a single list schedule: We have no freedom at all. We obtain:

Task	A	B	C	D	E	H	I	J	F	G
σ	0	0	0	2	8	3	5	5	13	20
Processor	P_1	P_2	P_3	P_2	P_1	P_3	P_2	P_3	P_2	P_1

The makespan is 38, five time units more than with two processors.

Exercise 7.3 (Scheduling a FORK Graph with Communications)

▷ **Question 1.** Let G be a FORK graph with n children. Let P_0 be the processor executing T_0. Which other tasks will be executed by P_0? We first make the following simplifying assumption: Any other processor P_i executes at most one task. Indeed, if two tasks are executed on $P_i \neq P_0$, the makespan is not increased if we reassign the tasks to two distinct processors (remember that we have infinite resources). Therefore, there exists an optimal schedule such that a subset $\mathcal{I} = \{T_{i_k}, \ldots, T_{i_n}\}$ of the tasks is executed by P_0 and all the other tasks are executed on distinct processors. The makespan of such a schedule is

$$T = \max\left(\sum_{i\in\mathcal{I}} w_i, \{w_0 + (d_j + w_j) \mid j \notin \mathcal{I}\}\right) .$$

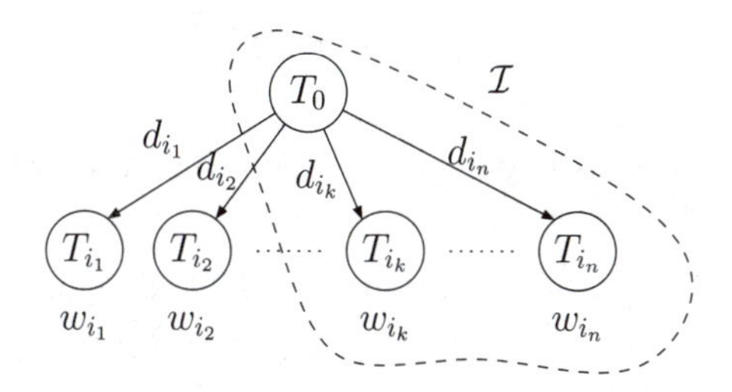

Assume that there are two tasks $T_j \notin \mathcal{I}$ and $T_k \in \mathcal{I}$ such that $w_k + d_k < w_j + d_j$. Then, the makespan does not increase if we remove task T_k from $\mathcal{I}$ and assign it to a new processor. Hence, there exists an optimal schedule where all tasks in $\mathcal{I}$ have a value $w_i + d_i$ larger than that of tasks not in $\mathcal{I}$.

To derive an optimal schedule, we need to sort tasks T_i, $1 \leqslant i \leqslant n$, by non-decreasing value of $w_i + d_i$, and to find k minimizing $\max(\sum_{i=k}^{n} w_i, w_{k-1} + d_{k-1})$. Then, the $n - k$ remaining tasks will be assigned to P_0.

▷ **Question 2.** This is not difficult. The problem is in NP, and we use a reduction from 2-Partition, just as in Section 7.4.1. Given an arbitrary instance Inst_1 of 2-Partition, with n integers $\{a_1, a_2, \ldots, a_n\}$, we construct the following instance Inst_2 of ForkSchedBounded(G,p) with $p = 2$ processors:

- G is a FORK graph $\{T_0, \ldots, T_n\}$ with n children.
- The parent node T_0 has zero weight $w_0 = 0$.
- For $1 \leqslant i \leqslant n$, node T_i has weight $w_i = a_i$ and zero communication cost: $d_i = 0$.

Obviously, the size of Inst_2 is linear in the size of Inst_1. It is easy to check that Inst_1 has a solution if and only if we can schedule G in time $K = \frac{1}{2}\sum_{i=1}^{n} a_i$.

▷ **Question 3.** The problem is in NP, and we use a reduction from 2-Partition. Given an arbitrary instance Inst_1 of 2-Partition, with n integers $\{a_1, a_2, \ldots, a_n\}$, let $S = \frac{1}{2}\sum_{i=1}^{n} a_i$ (if S is not an integer, Inst_1 has no solution), $M = \max a_i$, and $m = \min a_i$. We construct the following instance Inst_2 of ForkSched-1-Port-∞(G):

- G is a FORK graph $\{T_0, \ldots, T_{n+3}\}$ with $n + 3$ children.
- The father node T_0 has zero weight $w_0 = 0$.
- For $1 \leqslant i \leqslant n$, node T_i has weight $w_i = 10(M + a_i + 1)$.

- The last three children T_{n+1}, T_{n+2} and T_{n+3} have weights

$$w_{n+1} = w_{n+2} = w_{n+3} = 10(M + m) + 1.$$

- Communication costs are equal to the weights: $d_i = w_i$ for $1 \leqslant i \leqslant n+3$.
- The bound on the makespan is $K = \frac{1}{2}\sum_{i=1}^{n} w_i + 2w_{n+1} = 5n(M+1) + 10S + 20(M+m) + 2$.

Clearly, the size of Inst_2 is linear in the size of Inst_1. We show that Inst_1 has a solution if and only if Inst_2 has a solution:

$\boxed{\Leftarrow}$ Assume that Inst_1 has a solution: Let $\mathcal{I}_1$ and $\mathcal{I}_2$ be two subsets partitioning $\{1, \ldots, n\}$ and such that $\sum_{\mathcal{I}_1} a_i = \sum_{\mathcal{I}_2} a_i = S$. We build the schedule as follows:

- Processor P_0 executes task T_0, tasks T_i for $i \in \mathcal{I}_1$, and tasks T_{n+1} and T_{n+2}. P_0 needs exactly K time units to process all these tasks since $K = \frac{1}{2}\sum_{i=1}^{n} w_i + 2w_{n+1}$ and $\frac{1}{2}\sum_{i=1}^{n} w_i = \sum_{i \in \mathcal{I}_1} w_i$.
- Each remaining task is assigned a distinct processor. Thus, we use $|\mathcal{I}_2| + 1$ processors in addition to P_0.
- Communications are performed according to non-decreasing task indices. Thus, the last message sent by P_0 is for task T_{n+3}.
- The processor in charge of T_{n+3} is ready to start its execution at time $\sum_{\mathcal{I}_2} d_i + d_{n+3}$. It terminates the execution of T_{n+3} at time

$$\sum_{\mathcal{I}_2} d_i + d_{n+3} + w_{n+3} = K\ .$$

- All the other processors terminate their executions no later than K. Indeed, they receive their messages no later than at time $\sum_{\mathcal{I}_2} d_i$, and their weight w_i is no larger than $2w_{n+3}$.

We have built a valid schedule for Inst_2.

$\boxed{\Rightarrow}$ Reciprocally, assume that Inst_2 has a solution, i.e., a schedule σ whose makespan is no greater than K. Let P_0 be the processor that executes T_0. Define $\mathcal{I} = \{i \mid 1 \leqslant i \leqslant n+3 \text{ and } T_i \text{ is executed by } P_0\}$ as the index set of the tasks assigned to P_0. The execution time of P_0 is at least $A = \sum_{i \in \mathcal{I}} w_i$. The processor receiving the last message from P_0 to execute task T_{last} (whose index is not in $\mathcal{I}$) cannot terminate its execution before time $B = \sum_{i \notin \mathcal{I}} d_i + w_{\text{last}}$. Since σ is a solution to Inst_2, we have $\max(A, B) \leqslant K$. But $A + B = \sum_i w_i + w_{\text{last}} = 2K + w_{\text{last}} - w_{n+1} \geqslant 2K$. Therefore, $A = B = K$ and $w_{n+1} = w_{\text{last}}$. Because $A = B$, we have $A = B \bmod 10$. Hence, $\mathcal{I}$ contains two indices $\{n_1, n_2\}$ taken from $\{n+1, n+2, n+3\}$. Letting $\mathcal{I}_1 = \mathcal{I} \backslash \{n_1, n_2\}$ and $\mathcal{I}_2 = \{1, \ldots, n\} \backslash \mathcal{I}_1$, we obtain a solution to Inst_1.

Chapter 8

Advanced Scheduling

In this chapter, we discuss several scheduling topics that are more advanced than those studied in Chapter 7. We strongly believe that the macro-dataflow task graph scheduling model should be modified to better account for network resources consumption. It would be unrealistic to expect that a single model realistically models all kinds of architecture/software combinations. But bandwidths of network cards and of communication links are always limiting factors, exactly as CPU speeds limit computing resource consumption.

As stated in Section 3.2.3, a common approach is to use 1-port models (either uni- or bidirectional) for single-threaded programs using single-threaded communication libraries, and to use multi-port models (with bandwidth bounds and/or overheads) for asynchronous multi-threaded programs. But recall that the work in [105] casts doubts on the ability to achieve true asynchronous communications. Let us point out that serialized communications in the 1-port model have a dramatic impact on application execution time (makespan). For example, in the traditional macro-dataflow model, scheduling a fork graph with an unlimited number of homogeneous processors has polynomial complexity, while in the 1-port model this problem becomes NP-hard (see Exercise 7.3).

In this chapter, we address four important (but largely independent) topics:

- Scheduling of *divisible load* applications, that is, master-worker applications in which task granularity can be chosen arbitrarily (Section 8.1)/
- *Throughput optimization* for master-worker applications in *steady-state* (Section 8.2).
- Scheduling of *workflow applications* that consist of a large number of identical DAGs, where each DAG is executed on a different data set. The goal is to optimize steady-state throughput and/or response time.
- *Loop nest* scheduling, which is a classical topic at the intersection of scheduling and compiler techniques (Section 8.4). At first glance, this last section stands somewhat apart from the rest of the chapter, in particular because it does not account for communications between processors. Instead, it focuses on extracting parallelism out of a sequential program, and on doing so automatically in a compiler. The discovery of parallelism and of dependences in a program is a fundamental basis

for developing parallel algorithms and for constructing schedules. The automation of this discovery turns out to be a fascinating topic that provides several insights into the fundamentals of parallel computing.

We mostly use 1-port models, either bidirectional (Sections 8.1 and 8.2) or unidirectional (Section 8.3). For the sake of completeness, we also use the bounded multi-port model in Section 8.2. All application models considered in this chapter are *structured*, in that applications exhibit some intrinsic regularity. This is the main reason why we succeed in establishing deeper results than for the scheduling of a single DAG. We refer the adventurous reader to the bibliographical notes at the end of this chapter for more information and for some pointers to recent papers representative of the ongoing research in the field.

8.1 Divisible Load Scheduling

8.1.1 Motivation

In this section, we study the problem of scheduling independent tasks on master-worker platforms. We consider two models. The first model was already introduced in Section 7.4.1 and denoted by Indep-tasks(p). Recall that in that model the number of tasks and the task computational costs are set in advance. In the second model the number of tasks and the task sizes can be chosen arbitrarily; this scenario is relevant when the application consists of an amount of computation, or *load*, that can be divided arbitrarily into sub-tasks. In other words, the application is a perfectly parallel job: any sub-task can itself be processed in parallel, and on any number of workers. In practice, this model is an approximation of an application that consists of a large number of identical, low-granularity computations.

The second model is called the *divisible load* model and has been widely studied after being popularized by the landmark book written in 1996 by Bharadwaj, Ghose, Mani, and Robertazzi [32]. The divisible load model is really a *relaxation* of the first, more traditional model. As a result of this relaxation it is possible to give optimal solutions to several makespan minimization problems, such as the master-worker scheduling problem that we discuss hereafter.

Many applications are approximated reasonably well by the divisible load model. One example among many is an application for studying Earth seismic tomography. The goal of the application is to validate a model for the internal structure of the Earth. This can be done by comparing the propagation time of seismic waves estimated by the model with the actual time measured by physical instruments. This is done for a set of recorded seismic events and for a three-dimensional domain. Below is a simplified pseudo-code for this application implemented in parallel in a master-worker fashion. This pseudo-code is written using the same template as in Chapter 3 and using the SCATTER function (see Section 3.3.2) to distribute pieces of an array among the processors:

```
p ←NUM_PROCS()
myrank ←MY_NUM()
   { The master reads input data into an array }
if myrank = MASTER then
 └ data ←n seismic events of size L read from an input file
   { Each processor receives one piece of the array }
SCATTER(myrank, data, rbuff, n × L/p)
   { Every processor computes its piece of the array }
COMPUTE(rbuff)
```

The master processor reads n data items and scatters them among the p processors (in this case, the master operates as a worker for the computation). Each processor then computes the results independently. This application can be modeled sufficiently accurately as a divisible load only if the number of tasks, n, is large when compared to the number of processors. For this particular application n is expected to be large. For instance, during 1999, as many as $n = 817,101$ seismic events were recorded, each of which can be used to validate the seismic model.

This example application is representative of a large class of applications that consist of very large, some may say enormous, numbers of fine-grain computations. A common and often reasonable assumption is that the execution time of each computation is proportional to the size of the data to be processed in that computation. Since the computations are independent, there is no need for either synchronizations or communications among the processors; only the input messages from the master to the workers need to be taken into account when scheduling the application.

8.1.2 Classical Approach

We target a master-worker platform as illustrated in Figure 8.1. We assume that the processors have different computation speeds. All processors are interconnected via a bus network, meaning that communications between different pairs of processors are serialized and that the latencies and bandwidths of the network between the master and all the workers are identical. We introduce some notations:

- The platform is composed of a master processor M, which holds all application data initially, and of a set P_1, ..., P_p of workers. In a practical application the master also participates in the computation rather than merely doling out work to the workers. As a result, it is common and

FIGURE 8.1: A master-worker computing platform structured as a bus.

convenient to also denote the master by P_0, essentially considering that it is a $(p+1)$-st worker.

- The time for worker P_i to execute a unit-size work is w_i, $i = 0, \ldots, p$. The time needed to send a unit message from M to P_i, $i = 1, \ldots, p$, is c (recall that all workers communicate at the same speed with the master). These notations are depicted in Figure 8.2.

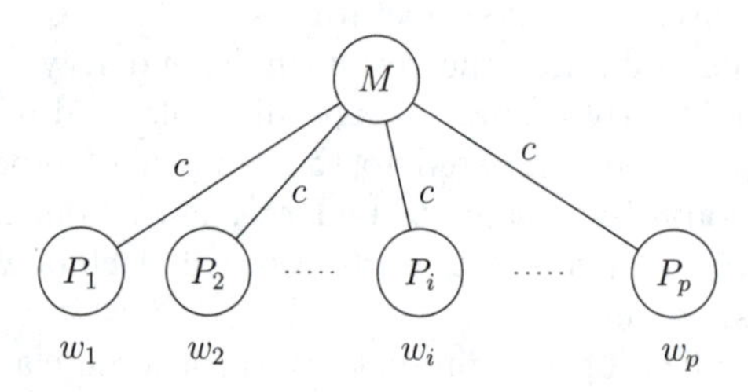

FIGURE 8.2: Abstract model of a bus-structured master-worker platform.

- The total number of tasks is W_{total}. Processor P_i receives n_i tasks (to be determined), where $n_i \in \mathbb{N}$ and $\sum_{i=0}^{p} n_i = W_{\text{total}}$. Note that this sum includes index 0 because the master $M = P_0$ will execute n_0 tasks itself. The computation time on P_i is $n_i.w_i$.

- The master, M, sends a *single* message to each participating worker. Furthermore, we use the 1-port model, which means that these messages are transferred on the bus sequentially. As mentioned in Section 3.2.3, this 1-port model corresponds to synchronous, non-multi-threaded message passing.

- The master, M, can simultaneously compute tasks and send data to a single worker. A worker cannot start processing before receiving all of the data from M.

- To simplify indexing, and without loss of generality, we assume that the master serves the workers in the order P_1, ..., P_p. Hence, it sends a message containing input data for n_1 tasks to P_1, then for n_2 tasks to P_2, and so on. In the meantime, M executes n_0 tasks itself.

Our goal is to determine the n_i values so that the overall execution time, i.e., the makespan, is minimized. Figure 8.3 shows an example execution for $p = 3$ workers. Let T_i denote the execution time of processor P_i (recall that $M = P_0$). Accounting for the serialization of communications on the bus and the order in which the master "serves" the workers, we obtain the following expression for T_i:

- P_0: $T_0 = n_0.w_0$
- P_1: $T_1 = n_1.c + n_1.w_1$
- P_2: $T_2 = (n_1.c + n_2.c) + n_2.w_3$
- P_i: $T_i = \sum_{j=1}^{i} n_j.c + n_i.w_i$ for $i \geqslant 1$

To make the above formulas homogeneous, we define $c_0 = 0$ and $c_i = c$ for $i \geqslant 1$, so that

$$T_i = \sum_{j=0}^{i} n_j.c_j + n_i.w_i \text{ for } i = 0, 1, \ldots, p\,.$$

The total execution time is

$$T = \max_{0 \leqslant i \leqslant p} \left(\sum_{j=0}^{i} n_j.c_j + n_i.w_i \right) ,$$

and we look for a distribution of work to workers, n_0, ..., n_p, which minimizes T. We rewrite T as

$$T = \max \left(n_0.c_0 + n_0.w_0, \max_{1 \leqslant i \leqslant p} \left(\sum_{j=0}^{i} n_j.c_j + n_i.w_i \right) \right) ,$$

and finally as

$$T = n_0.c_0 + \max \left(n_0.w_0, \max_{1 \leqslant i \leqslant p} \left(\sum_{j=1}^{i} n_j.c_j + n_i.w_i \right) \right) .$$

The above equation shows that an optimal solution for the distribution of W_{total} tasks over $p + 1$ processors is obtained by distributing n_0 tasks to processor P_0, and then optimally distributing $W_{\text{total}} - n_0$ tasks to processors $P_1, \ldots, P_p$. This observation would easily lead to a dynamic programming algorithm for computing the optimal n_i values. However, this solution would only be partially satisfactory. First, it is not a closed-form solution. Second, and more importantly, it does not solve the question of the ordering of the workers. Indeed, we have arbitrarily decided that the master M would send messages in the order P_1, P_2, ..., P_p. But the workers have different computing powers, so one should expect the ordering to matter. Unfortunately,

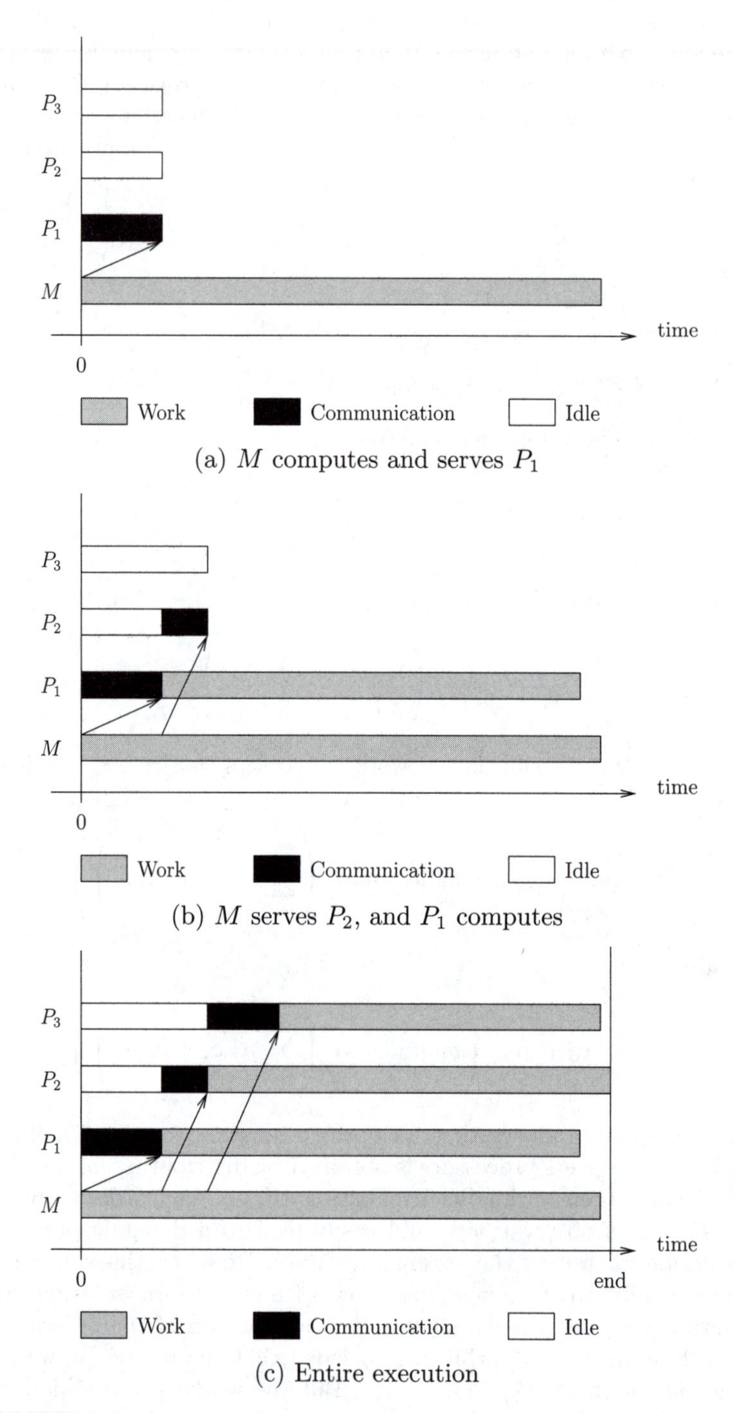

(a) M computes and serves P_1

(b) M serves P_2, and P_1 computes

(c) Entire execution

FIGURE 8.3: An example execution with a master and $p = 3$ workers.

there are $p!$ possible orderings, way too many for resorting to an exhaustive search for the best ordering with reasonable complexity!

One way to address this challenge is to realize that we do not actually need a precise solution where the n_i are integers. Let us think of our seismic model validation application, with 817, 101 tasks and, say, 10 processors. With such a large number of tasks relative to the number of processors, one can simply search for rational n_i values, at the price of some rounding in the end to derive a feasible work allocation. This simple relaxation of the problem is the quintessence of the divisible load approach and turns out to be surprisingly successful, as seen in the next section.

8.1.3 Divisible Load Approach

We keep exactly the same setting and notations as before, except that processor P_i, $0 \leqslant i \leqslant p$, now receives an amount of work $n_i = \alpha_i W_{\text{total}}$ with $\alpha_i \in \mathbb{Q}$ and $\sum_{i=0}^{p} \alpha_i = 1$. We allow for rational values of α_i, the fraction of work assigned to P_i. Note that n_i is rational, too, so we cannot really think in terms of atomic tasks any longer.

The total execution time is

$$T = \max_{0 \leqslant i \leqslant p} \left(\sum_{j=0}^{i} \alpha_j . c_j + \alpha_i . w_i \right) W_{\text{total}}$$

(recall that $c_1 = 0$ and $c_i = c$ for $i \geqslant 1$). We look for a data distribution α_0, ..., α_p that minimizes T.

LEMMA 8.1. *In an optimal solution, all processors stop computing at the same time.*

Proof. We prove this lemma by contradiction. Consider an optimal solution and assume that there are two workers, P_i and P_{i+1} ($i \geqslant 1$), that do not terminate at the same time. Suppose first that $T_i < T_{i+1}$. Figure 8.4 shows an example with $i = 1$, i.e., P_1 terminates earlier than P_2. Let us decrease α_{i+1} by ε, and increase α_i by ε. Furthermore, let us choose ε so that P_i and P_{i+1} terminate simultaneously in the new solution:

$$(\alpha_i + \varepsilon)(c + w_i)W_{\text{total}} = ((\alpha_i + \varepsilon)c + (\alpha_{i+1} - \varepsilon)(c + w_{i+1}))\, W_{\text{total}} \,.$$

Because $c_{i+1} = c_i = c$, the communication time for the following processors is unchanged, and in the new solution, P_i and P_{i+1} terminate strictly earlier than P_{i+1} in the previous schedule (that is, at a time earlier than the former value of T_{i+1}).

Now if we had $T_i > T_{i+1}$, we would proceed similarly and decrease the load of P_i by ε so that P_i and P_{i+1} terminate simultaneously in the new solution:

$$(\alpha_i - \varepsilon)(c + w_i)W_{\text{total}} = ((\alpha_i - \varepsilon)c + (\alpha_{i+1} + \varepsilon)(c + w_{i+1}))\, W_{\text{total}} \,.$$

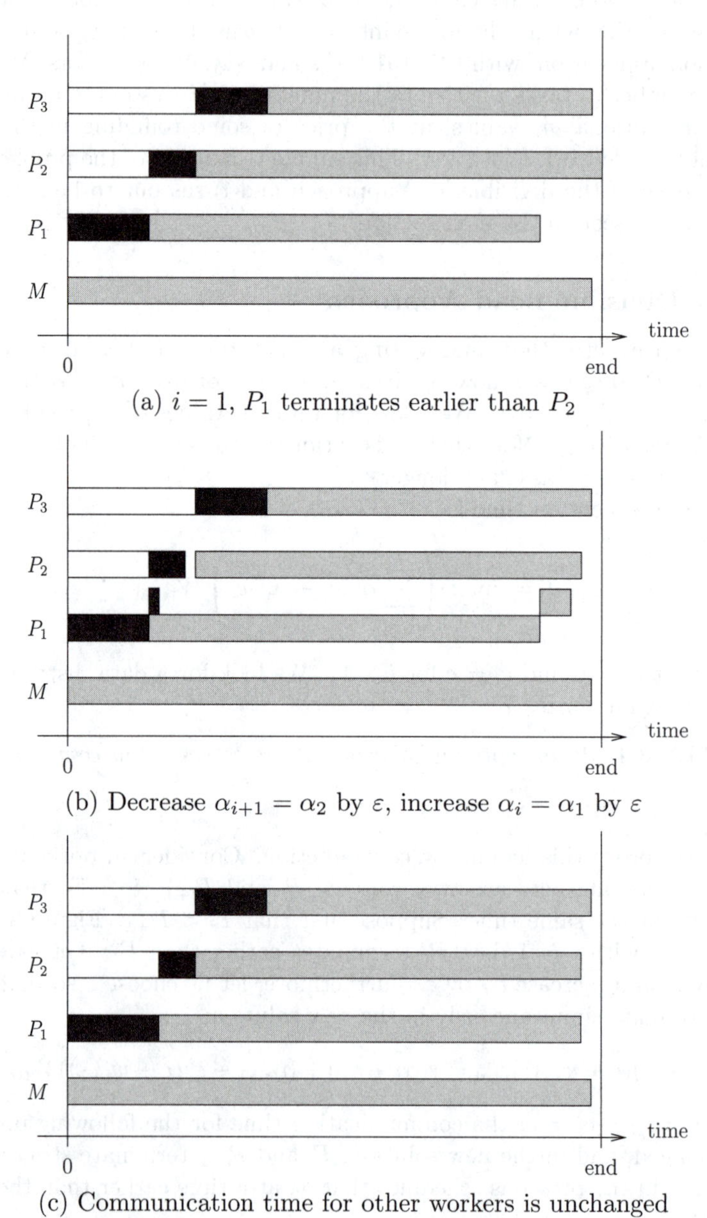

(a) $i = 1$, P_1 terminates earlier than P_2

(b) Decrease $\alpha_{i+1} = \alpha_2$ by ε, increase $\alpha_i = \alpha_1$ by ε

(c) Communication time for other workers is unchanged

FIGURE 8.4: Illustration of the proof of Lemma 8.1.

Note that the reasoning also works if $i = 0$, i.e., with the master P_0. If it finishes before P_1, suppress ε from the load of P_1, and if it finishes after P_1, suppress ε from the load of P_0. In both cases, they finish simultaneously, and strictly earlier than $\max(T_0, T_1)$.

We are almost done with the proof. We start from an optimal solution whose execution time is $T = T_{\text{opt}}$. There exists at least one processor whose end time is T. We apply the exchange procedure to any processor pair P_i and P_{i+1} such that $\min(T_i, T_{i+1}) < \max(T_i, T_{i+1}) = T$. Both termination times of P_i and P_{i+1} are smaller than T after the exchange. We continue applying this procedure until there remains no such pair. In the end, we have found a solution whose total execution time is smaller than T, a contradiction. We conclude that all processors do have the same end time in any optimal solution. □

LEMMA 8.2. *In an optimal solution all processors participate in the computation.*

Proof. This is a direct consequence of Lemma 8.1. □

Equipped with Lemmas 8.1 and 8.2, we can characterize the best way of assigning loads to the master P_0 and to workers $P_1, \ldots, P_p$:

- $T = \alpha_0 w_0 W_{\text{total}}$;
- $T = \alpha_1(c + w_1)W_{\text{total}}$. Therefore, $\alpha_1 = \frac{w_0}{c+w_1}\alpha_0$.
- $T = (\alpha_1 c + \alpha_2(c + w_2))W_{\text{total}}$. Therefore, $\alpha_2 = \frac{w_1}{c+w_2}\alpha_1$.
- For all $i \geqslant 1$ we derive $\alpha_i = \frac{w_{i-1}}{c+w_i}\alpha_{i-1}$.
- We use the normalization equation $\sum_{i=0}^{p} \alpha_i = 1$ to derive

$$\alpha_0 \left(1 + \frac{w_0}{c + w_1} + \ldots + \prod_{k=1}^{i} \frac{w_{k-1}}{c + w_k} + \ldots\right) = 1 \,.$$

We still do not know in which order the master should communicate with the workers. The intuition is that faster workers should be served first, so that they can work longer. Is it correct? Let us assess the impact of the communication order analytically. To do so, consider the load processed by workers P_i and P_{i+1} during a time t:

Processor P_i – We have $\alpha_i(c + w_i)W_{\text{total}} = t$. Therefore, $\alpha_i = \frac{1}{c+w_i}\frac{t}{W_{\text{total}}}$.

Processor P_{i+1} – We have $\alpha_i c W_{\text{total}} + \alpha_{i+1}(c + w_{i+1})W_{\text{total}} = t$. Hence, $\alpha_{i+1} = \frac{1}{c+w_{i+1}}\left(\frac{t}{W_{\text{total}}} - \alpha_i c\right) = \frac{w_i}{(c+w_i)(c+w_{i+1})}\frac{t}{W_{\text{total}}}$.

Sum of the loads – We compute

$$\alpha_i + \alpha_{i+1} = \frac{c + w_i + w_{i+1}}{(c + w_i)(c + w_{i+1})} \frac{t}{W_{\text{total}}} .$$

We see that the formula is symmetric, and we conclude that the communication order has absolutely no impact on the solution, a surprising conclusion indeed! We can perform a similar analysis for the master $M = P_0$ and the first worker P_1:

Processor P_0 – We have $\alpha_0 w_0 W_{\text{total}} = t$. Then, $\alpha_0 = \frac{1}{w_0} \frac{t}{W_{\text{total}}}$.

Processor P_1 – We have $\alpha_1 (c + w_1) W_{\text{total}} = t$. Hence, $\alpha_1 = \frac{1}{c+w_1} \frac{t}{W_{\text{total}}}$.

Sum of the loads – We compute

$$\alpha_0 + \alpha_1 = \frac{c + w_0 + w_1}{w_0(c + w_1)} \frac{t}{W_{\text{total}}} = \frac{c + w_0 + w_1}{cw_0 + w_0 w_1} \frac{t}{W_{\text{total}}} .$$

We see that the sum of loads is larger when $w_0 < w_1$. Perhaps, in some situations, the master is fixed a priori. But for some other applications it may be chosen among all available processors. In the latter case we should select the most powerful (or fastest) processor as the master. We conclude this section by summarizing our results:

THEOREM 8.1. *For divisible load applications on bus-structured networks, the master should be selected (if possible) as the fastest processor. In an optimal solution, all workers participate and terminate simultaneously. The communication order from the master to the workers has no impact. Closed-form formulas can be established for $\alpha_0, \alpha_1, \ldots, \alpha_p$.*

8.1.4 Extension to Star Networks

In this section, we aim at extending previous results to "star-shaped" platforms, an example of which is depicted in Figure 8.5. The key difference with the bus network model is that the links between the master and the workers have *different* characteristics. Of course the workers still have different computational powers. Our star-shaped platform is fully heterogeneous.

We keep the same notations as in Section 8.1.3: We have a master M, also denoted as P_0, and p workers $P_1, \ldots, P_p$. The total load is W_{total}. Worker P_i needs w_i time units to execute a single load unit and c_i time units to receive it from the master. We assign α_i load units to P_i for $1 \leqslant i \leqslant p$, where $\alpha_i \in \mathbb{Q}$ and $\sum_{i=1}^{p} \alpha_i = 1$. Note that for convenience we do not assign any load to the master here. Because links have different communication times, we can always create a fictitious worker P_{p+1} with $w_{p+1} = w_0$ and $c_{p+1} = 0$ to account for the master's computing capacity. This convention makes the approach more symmetric with respect to the master and the workers. We keep the same

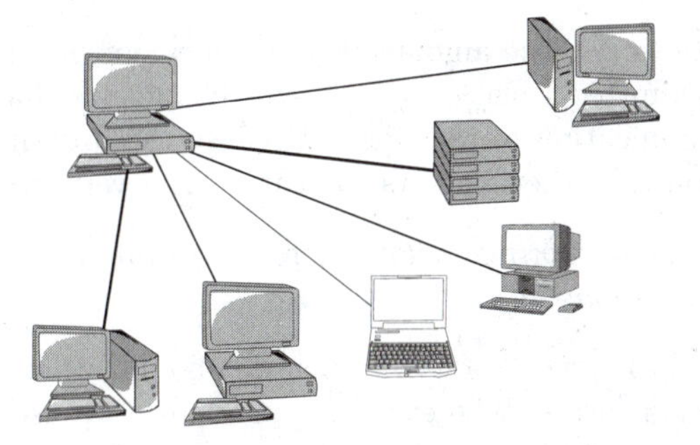

FIGURE 8.5: A star-shaped master-worker platform.

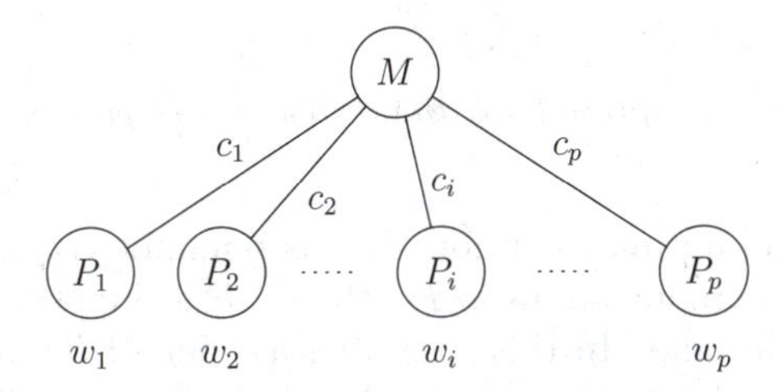

FIGURE 8.6: Abstract model of the star-shaped master-worker platform.

communication model too: M sends a single message to each participating worker, and under the 1-port model these messages are sequentialized.

We start by searching for the best communication order. We have learned to be cautious with our intuition! Furthermore, deciding whether to favor fast-communicating or fast-computing workers is not clear. Let us use some analytical formulas just as for bus-shaped platforms. Consider the load processed by workers P_i and P_{i+1} during a time t:

Processor P_i – We have $\alpha_i(c_i + w_i)W_{\text{total}} = t$. Therefore, $\alpha_i = \frac{1}{c_i+w_i}\frac{t}{W_{\text{total}}}$.

Processor P_{i+1} – We have $\alpha_i c_i W_{\text{total}} + \alpha_{i+1}(c_{i+1} + w_{i+1})W_{\text{total}} = t$. Hence, $\alpha_{i+1} = \frac{1}{c_{i+1}+w_{i+1}}(\frac{t}{W_{\text{total}}} - \alpha_i c_i) = \frac{w_i}{(c_i+w_i)(c_{i+1}+w_{i+1})}\frac{t}{W_{\text{total}}}$.

Sum of the loads – We compute

$$\alpha_i + \alpha_{i+1} = \frac{c_{i+1} + w_i + w_{i+1}}{(c_i + w_i)(c_{i+1} + w_{i+1})}\frac{t}{W_{\text{total}}} .$$

Sum of the communications – We derive

$$\alpha_i c_i + \alpha_{i+1} c_{i+1} = \frac{c_i c_{i+1} + c_i w_{i+1} + c_{i+1} w_i}{(c_i + w_i)(c_{i+1} + w_{i+1})}\frac{t}{W_{\text{total}}} .$$

We see that the formula is symmetric for the total communication time: The network is occupied the same amount of time whether P_i comes before or after P_{i+1}. However, the amount of work does depend upon the ordering: $\alpha_i + \alpha_{i+1}$ is maximized when $c_i \leqslant c_{i+1}$, which suggests that we should serve the faster communicating worker first. Because the ordering of P_i and P_{i+1} has no impact on the other workers, we can infer a very important result:

LEMMA 8.3. *In an optimal solution, participating workers must be served by non-decreasing values of c_i.*

However, we do not yet know whether we should utilize all workers as in the case of bus platforms. The exchange procedure we used in Section 8.1.3 does not work so easily here. This is because communication times change when shifting fractions of load from one worker to another. Nevertheless, the result still holds:

LEMMA 8.4. *In an optimal solution, all workers participate in the computation.*

Proof. Consider an optimal solution. Let us renumber the processors so that the ordering of communications is $P_1, P_2, \ldots, P_p$. Suppose that at least one worker is kept fully idle. In this case, at least one of the α_i's, $1 \leqslant i \leqslant p$, is zero. Let us denote by k the largest index such that $\alpha_k = 0$.

Case $k < p$ – We add P_k at the end of the initial solution, thus using the ordering $P_1, \ldots, P_{k-1}, P_{k+1}, \ldots, P_p, P_k$. By construction, $\alpha_p \neq 0$. Therefore, the network is not used during at least the last $\alpha_p w_p$ time units. It would thus be possible to process at least $\frac{\alpha_p w_p}{c_k + w_k} > 0$ additional units of load with worker P_k, which contradicts the assumption that the original solution was optimal: P_k does more work than in the original solution in which it was kept idle, and all the other processors do the same amount of work.

Case $k = p$ – We modify the initial solution by giving some work to P_p without increasing the execution time. Let k' be the largest index such that $\alpha_{k'} \neq 0$. By construction, the communication medium is not used during at least the last $\alpha_{k'} w_{k'} > 0$ time units. Thus, as previously, it would be possible to process at least $\frac{\alpha_{k'} w_{k'}}{c_p + w_p} > 0$ additional units of load with worker P_p, which leads to a similar contradiction.

Therefore, in an optimal solution all workers participate in the computation. □

It is worth pointing out that the above property does not hold true if we consider solutions in which the communication ordering is fixed a priori. For instance, consider a platform comprising two workers: P_1 (with $c_1 = 4$ and $w_1 = 1$) and P_2 (with $c_2 = 1$ and $w_2 = 1$). If the first chunk has to be sent to P_1 and the second chunk to P_2, the optimal number of units of load that can be processed within 10 time units is 5, and P_1 is kept fully idle in this solution. On the other hand, if the communication ordering is not fixed, then

6 units of load can be performed within 10 time units (5 units of load are sent to P_2, and then 1 to P_1). In the optimal solution, both workers perform some computation, and both workers finish computing at the same time. This is a general result:

LEMMA 8.5. *In an optimal solution, all workers finish computing simultaneously.*

Proof. The reader may want to skip this proof as it requires more involved mathematical arguments. Consider an optimal solution. All α_i's have strictly positive values (Lemma 8.4). Consider the following linear program:

$$\begin{array}{l}\textsc{Maximize } \sum \beta_i,\\ \text{SUBJECT TO}\\ \begin{cases}\mathrm{LB}(i)\ \forall i, & \beta_i \geqslant 0\\ \mathrm{UB}(i)\ \forall i, & \sum_{k=1}^{i} \beta_k c_k + \beta_i w_i \leqslant T\end{cases}\end{array}$$

The α_i's satisfy the set of constraints above, and from any set of β_i's satisfying the set of inequalities, we can build a valid schedule that processes exactly $\sum \beta_i$ units of load. Therefore, if we denote by $(\beta_1, \ldots, \beta_p)$ an optimal solution of the linear program, then $\sum \beta_i = \sum \alpha_i$.

It is known that one of the extremal solutions $\mathcal{S}_1$ of the linear program is one of the convex polyhedron $\mathcal{P}$ induced by the inequalities [106, Chapter 11]: This means that in the solution $\mathcal{S}_1$, there are at least p inequalities among the $2p$ equalities. Since we know that for any optimal solution all the β_i's are strictly positive (Lemma 8.4), then this vertex is the solution of the following (full rank) linear system:

$$\forall i, \quad \sum_{k=1}^{i} \beta_k c_k + \beta_i w_i = T.$$

Thus, we conclude that there exists an optimal solution in which all workers finish their work at the same time.

Let us denote by $\mathcal{S}_2 = (\alpha_1, \ldots, \alpha_p)$ another optimal solution, with $\mathcal{S}_1 \neq \mathcal{S}_2$. As already pointed out, $\mathcal{S}_2$ belongs to the polyhedron $\mathcal{P}$. Now, consider the following function f:

$$f : \begin{cases}\mathbb{R} \to \mathbb{R}^n\\ x \mapsto \mathcal{S}_1 + x(\mathcal{S}_2 - \mathcal{S}_1)\ .\end{cases}$$

By construction, we know that $\sum \beta_i = \sum \alpha_i$. Thus, with the notation $f(x) = (\gamma_1(x), \ldots, \gamma_p(x))$:

$$\forall i, \gamma_i(x) = \beta_i + x(\alpha_i - \beta_i),$$

and therefore

$$\forall x, \ \sum \gamma_i(x) = \sum \beta_i = \sum \alpha_i.$$

Therefore, all the points $f(x)$ that belong to $\mathcal{P}$ are extremal solutions of the linear program.

Since $\mathcal{P}$ is a convex polyhedron and both $\mathcal{S}_1$ and $\mathcal{S}_2$ belong to $\mathcal{P}$, then $\forall\, 0 \leqslant x \leqslant 1,\ \ f(x) \in \mathcal{P}$. Let us denote by x_0 the largest value of $x \geqslant 1$ such that $f(x)$ still belongs to $\mathcal{P}$: At least one constraint of the linear program is an equality in $f(x_0)$, and this constraint is not satisfied for $x > x_0$. Could this constraint be one of the UB(i) constraints? The answer is no, because otherwise this constraint would be an equality along the whole line $(\mathcal{S}_2 f(x_0))$, and would remain an equality for $x > x_0$. Hence, the constraint of interest is one of the LB(i)'s. In other terms, there exists an index i such that $\gamma_i(x_0) = 0$. This is a contradiction since we have proved that the γ_i's correspond to an optimal solution of the problem. Therefore, $\mathcal{S}_1 = \mathcal{S}_2$, the optimal solution is unique, and in this solution, all workers finish computing simultaneously. □

According to Lemma 8.3, we re-order the workers so that $c_1 \leqslant c_2 \leqslant \ldots \leqslant c_p$. The following linear program makes it possible to compute the optimal distribution of the load:

$$
\begin{array}{l}
\text{MINIMIZE } T_f, \\
\text{SUBJECT TO} \\
\left\{\begin{array}{lll}
(1) & \alpha_i \geqslant 0 & 1 \leqslant i \leqslant p \\
(2) & \sum_{i=1}^{p} \alpha_i = W_{\text{total}} & \\
(3) & \sum_{j=1}^{i} \alpha_j c_j + \alpha_i w_i \leqslant T_f & 1 \leqslant i \leqslant p \ (i\text{-th worker})
\end{array}\right.
\end{array}
$$

THEOREM 8.2. *The optimal solution is given by the solution of the linear program above.*

Theorem 8.2 is a direct consequence of the previous two lemmas. Note that inequalities (3) will be in fact equalities in the solution of the linear program, so that we can easily derive a closed-form expression for the α_i's (although it is far less elegant than for bus platforms). It is important to point out that this is linear programming with rational numbers, and hence of polynomial complexity.

As stated above, the variant where the master is capable of computing while communicating to one of its children can be solved by adding a fictitious worker P_{p+1} with $c_{p+1} = 0$ and $w_{p+1} = w_0$. The previous analysis shows that the master is kept busy all the time, as expected (otherwise more units of load could be processed). However, if the master is not fixed a priori but rather can be freely chosen among all processors, we should introduce new variables c_{ij} to denote the communication time of one unit of load from P_i to P_j. But we have no easy rule to decide which processor should be elected as master (we had such a rule for bus platforms). Instead, we can still solve $p+1$ linear programs (with P_i as master in the i-th program) and retain the best solution.

Exercise 8.1 explains how to extend the linear programming approach to general (multi-level) heterogeneous trees.

8.1.5 Going Further

The divisible load approach is more powerful than the classical approach with atomic tasks. Relaxing the integer constraint has led us to nice closed-form formulas and linear programs. However, there are some limitations to this approach.

The first limitation appears when we include return messages. After all, if the master M sends data to the workers, they might need to return some results to it after they finish computing. For simplicity, let the size of the initial and return messages be equal, so that P_i needs $\alpha_i c_i$ time units to return its message to M. Note that, in general, we may have different message sizes in both directions, as workers may, say, either filter data (shorter return messages) or generate large cryptographic keys (longer return messages). Even with this simplifying assumption, none of our results can be extended.

With return messages we need to specify whether the 1-port model used is unidirectional or bidirectional, because the master can possibly overlap communication of initial and return messages. We adopt a bidirectional setting here, but the results are similar for the unidirectional variant. For a given choice of participating workers, a given ordering of initial messages, and a given ordering of return messages, we can still write a linear program to compute the optimal assignment of the load to the workers. But there is a wide combinatorial space to explore! Figure 8.7 shows an example in which the optimal solution utilizes only two workers out of three. Intuitively, we could expect the order of return messages to be the same as that of input messages (FIFO strategy) or the converse (LIFO strategy). As shown in Figure 8.8, there are cases where the optimal solution is neither FIFO nor LIFO. In fact, the complexity of the problem with return messages is open. Note that in both these figures we simply indicate how many tasks are performed per time unit with the various schedules.

The second limitation of the divisible load model is related to the assumption that the master communicates a single message to each worker. In other words, there is only one round to serve the workers. As illustrated in Figure 8.9, this induces long idle times, especially for the last served workers. A way to circumvent this problem is to use a multi-round approach, by which the master sends several messages to the workers in a round-robin fashion. If things go smoothly, everything is nicely pipelined, communications and computations overlap well, and the idle time is greatly reduced. But if we allow multi-round algorithms with our simple linear cost model (where the cost to send/receive a message is proportional to its length), we are readily convinced that the optimal solution is to use an infinite number of arbitrary small-size rounds. To be realistic, we have to add start-up overheads in the communication time. This affine cost model is mandatory to study multi-round strategies. Unfortunately, the divisible scheduling problem becomes NP-hard with the affine model [79], even if we restrict to single-round algorithms. We need yet another relaxation! In the next section, we relax the objective function:

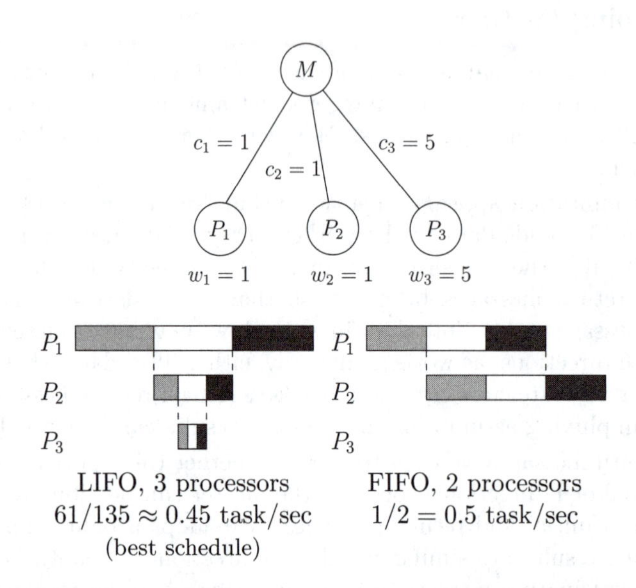

FIGURE 8.7: With return messages, all processors do not always participate in the computation.

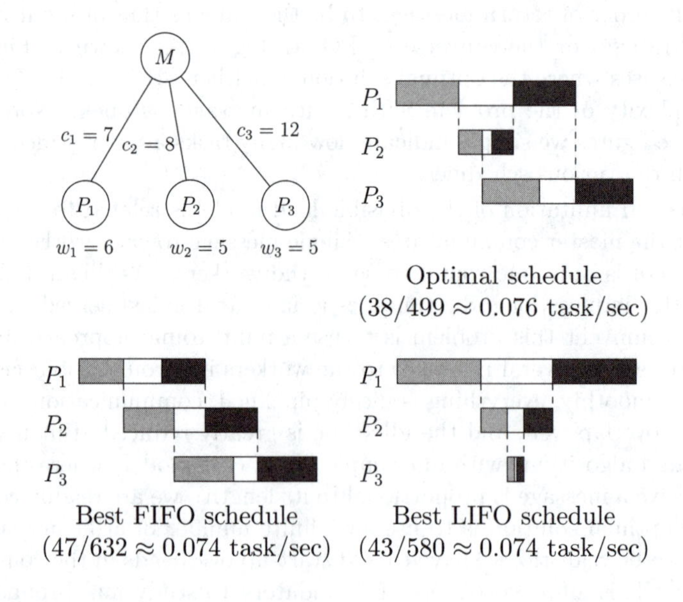

FIGURE 8.8: With return messages, the optimal schedule may be neither LIFO nor FIFO.

We forget about makespan minimization and instead we move to throughput optimization.

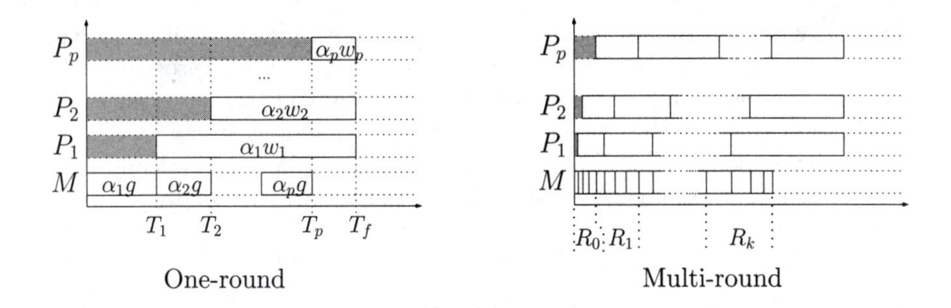

FIGURE 8.9: One-round versus multi-round.

8.2 Steady-State Scheduling

In this section, we revisit once more the master-worker paradigm on heterogeneous platforms, star- or tree-shaped. We somewhat blend the traditional approach and the divisible load approach: We consider atomic independent tasks, but we compute a steady-state schedule where we allocate rational numbers of tasks to processors before reconstructing the actual final schedule with a regular (integer) task assignment. A crucial point is that the goal is no longer to minimize makespan, but rather to maximize throughput.

8.2.1 Motivation

An idea to circumvent the difficulty of makespan minimization is to lower the ambition of the scheduling objective. Instead of aiming at the absolute minimization of the execution time, why not consider asymptotic optimality? Often, the motivation for deploying an application on a parallel platform is that the number of tasks is very large. In this case, the optimal execution time with the optimal schedule may be very large and a small deviation from it is likely acceptable. To state this informally: If there is a nice (e.g., polynomial) way to derive, say, a schedule whose length is two hours and three minutes, as opposed to an optimal schedule that would run for only two hours, we would be satisfied.

This approach has been pioneered by Bertsimas and Gamarnik [29]. Steady-state scheduling allows one to relax the scheduling problem in many ways. The costs of the initialization and clean-up phases are neglected. The initial integer formulation is replaced by a continuous or rational formulation. The precise scheduling of computations and communications is not required, or at least not before the optimal schedule is outlined. The main idea is to characterize the activity of each resource during each time unit: which (rational) fraction of time is spent computing, which is spent receiving or sending to which neighbor. Such activity variables are gathered into a linear program, which includes conservation laws that characterize the global behavior of the system. The actual schedule then arises naturally from these quantities.

We first work out an example with a tree platform before analytically solving the problem for heterogeneous star-shaped platforms, and finally for general tree platforms.

8.2.2 Working Out an Example

Consider the tree-shaped platform in Figure 8.10(a). As in Section 8.1, the master initially holds a collection of independent tasks to be processed. It can either decide to process these tasks itself or to delegate some to its children, which in turn face the same dilemma: which tasks to process in place, and which to send to which child?

An abstract view of the platform is given in Figure 8.10(b). For convenience we name the processors A, B, C, and D, rather than with indices. Each processor is a node in the tree. Each node is labeled by the time needed by the processor to compute a task. Each edge is labeled by a number that represents the time to send a task along this edge. For instance, it takes 6 time units for processor C to compute a task, and 1 time unit for processor C to send a task to processor D.

In the example, the master A decides to serve its fast-communicating child C first. C delegates the task to its own child D, which executes it. We see all these events in Figure 8.10(c). Because D requires 2 time units to execute a task, this scheme can be reproduced every two time units. This is illustrated in Figure 8.10(d), together with the fact that A can overlap its own computations with the sending of messages.

Next, the master realizes that although D is saturated, it can send more tasks to C, which can process them itself. C is slow-computing but fast-communicating, and A sends a new task to C every 6 time units so as to saturate it (Figure 8.10(e)). There remain idle times on the network: Exactly 2 time units out of every 6. Because B needs 2 time units to receive a task, A shifts ahead by one unit its third message intended to D (Figure 8.10(f)), and then every third communication to D, so that B can receive and execute a task once every 6 time units (Figure 8.10(g)). The network is now saturated, and we have derived the final schedule. As outlined in Figure 8.10(h), this schedule is cyclic. After a short initialization period (two time units), a steady-state is reached: The whole pattern of computations and communications repeats itself every 6 time units. Note that all nodes are saturated except B, which is fast-computing but slow-communicating.

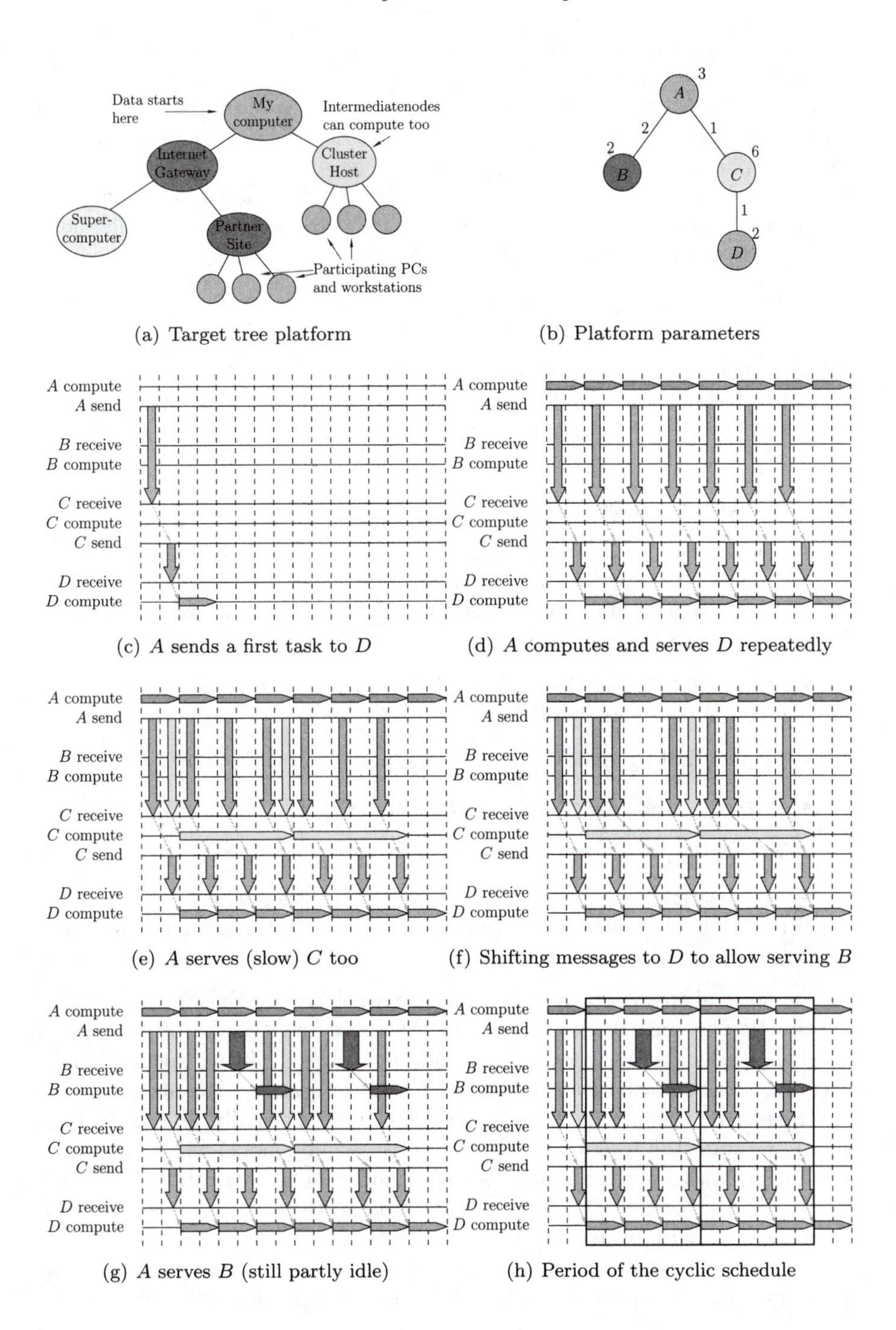

(a) Target tree platform (b) Platform parameters

(c) A sends a first task to D (d) A computes and serves D repeatedly

(e) A serves (slow) C too (f) Shifting messages to D to allow serving B

(g) A serves B (still partly idle) (h) Period of the cyclic schedule

FIGURE 8.10: Reaching optimal steady-state on an example tree platform.

We say that we have a *period* of length $T_{\mathsf{period}} = 6$. During each period seven tasks are computed, two by A, one by B, one by C, and three by D. The *throughput* is $\varrho = \frac{7}{6}$. If the execution lasts T time units, the schedule will execute almost ϱT tasks: a few tasks are "lost" during the initialization and clean-up phases. To state this more formally, let $\mathrm{nb}(T)$ be the actual number of tasks executed by the cyclic schedule within T time units, and let $\mathrm{nb}_{\mathrm{opt}}(T)$ be the optimal number of tasks that can be executed by any schedule, be it periodic or not. We show in Section 8.2.5 that

$$\lim_{T\to\infty} \frac{\mathrm{nb}(T)}{\mathrm{nb}_{\mathrm{opt}}(T)} = 1 \, .$$

We say that the schedule is *asymptotically optimal.* In the next section we provide a systematic way to derive a cyclic schedule whose throughput is maximum, and we establish its asymptotic optimality.

8.2.3 Star-Shaped Platforms

In this section, we target heterogeneous star-shaped platforms. We use exactly the same notations as in Section 8.1.4, but the master M initially holds a large collection of atomic tasks instead of a divisible load application. Refer to Figure 8.6 for notations:

- The master M (or P_0) sends tasks to workers *sequentially*, and without preemption (1-port model).
- There is full computation/communication overlap at each worker.
- Worker P_i receives a task in c_i time units.
- Worker P_i processes a task in w_i time units.
- The master M does not compute any task (but a master with computation time w_0 can be simulated as a worker with the same computation time and zero communication time).

The optimal steady-state is defined as follows: For each worker, determine the fraction of time spent computing tasks, and the fraction of time spent receiving tasks; for the master, determine the fraction of time spent communicating along each communication link. The objective is to maximize the (average) number of tasks processed per time unit. Formally, after a start-up phase, we want the resources to operate in a periodic mode, where worker P_i executes α_i tasks per time unit. We point out that α_i is a rational number, not an integer, so that there will remain some work to reconstruct a feasible schedule, i.e., with an integer number of tasks.

First we express the constraints for computations: P_i takes time $\alpha_i w_i$ to compute α_i tasks, and it computes these tasks within one time unit, so necessarily

$$\alpha_i w_i \leqslant 1 \, . \tag{8.1}$$

As for communications, the master M sends tasks sequentially to the workers, and it takes $\alpha_i c_i$ time units to send α_i tasks along the link to P_i, so necessarily

$$\sum_{i=1}^{p} \alpha_i c_i \leqslant 1 \,. \tag{8.2}$$

Finally, the objective is to maximize throughput, namely,

$$\varrho = \sum_i \alpha_i \,.$$

Altogether, we have a linear programming problem with rational unknowns:

$$\begin{array}{l} \text{MAXIMIZE } \varrho, \\ \text{SUBJECT TO} \\ \left\{ \begin{array}{ll} & \varrho = \sum_{i=1}^{p} \alpha_i \\ & \sum_{i=1}^{p} \alpha_i c_i \leqslant 1 \\ \forall i, & \alpha_i \geqslant 0 \\ \forall i, & \alpha_i w_i \leqslant 1 \end{array} \right. \end{array}$$

It turns out that the linear program is so simple that it can be solved analytically. Indeed it is a fractional knapsack problem [44] with value-to-cost ratio $\frac{1}{c_i}$. We should start with the "item" (worker) of the largest ratio, i.e., the smallest c_i, and take (assign) as many tasks as we can, i.e., $\min\left(\frac{1}{c_i}, \frac{1}{w_i}\right)$. Here is the detailed procedure:

1. Sort the workers by increasing communication times. Re-number them so that $c_1 \leqslant c_2 \ldots \leqslant c_k$.

2. Let q be the largest index so that $\sum_{i=1}^{q} \frac{c_i}{w_i} \leqslant 1$. Workers P_1 to P_q will be fully active (and each of them will execute $\frac{1}{w_i}$ tasks per time unit). If $q < p$, let $\varepsilon = 1 - \sum_{i=1}^{q} \frac{c_i}{w_i}$, otherwise let $\varepsilon = 0$. Worker P_{q+1} (if it exists) will be only partially active, and will execute $\min(\frac{\varepsilon}{c_{q+1}}, \frac{1}{w_{q+1}})$ tasks per time unit.

3. Workers P_{q+2} to P_p (if they exist) are discarded; they will not participate in the computation.

4. The optimal throughput is then

$$\varrho = \sum_{i=1}^{q} \frac{1}{w_i} + \min(\frac{\varepsilon}{c_{q+1}}, \frac{1}{w_{q+1}}) \,.$$

When $q = p$ the result is expected: It basically says that workers can be fed with tasks fast enough so that they are all kept computing steadily. However, if $q < p$, the result is surprising. Indeed, if the communication bandwidth is limited, some workers will partially starve. In the optimal solution these partially starved workers are those with slow communication rates, *regardless* of their processing speeds. In other words, a slow processor with a fast communication link is to be preferred to a fast processor with a slow communication link. This optimal strategy is often called *bandwidth-centric* because it delegates work to the fastest communicating workers, regardless of their computing speeds. Of course, slow workers will not contribute much to the overall throughput.

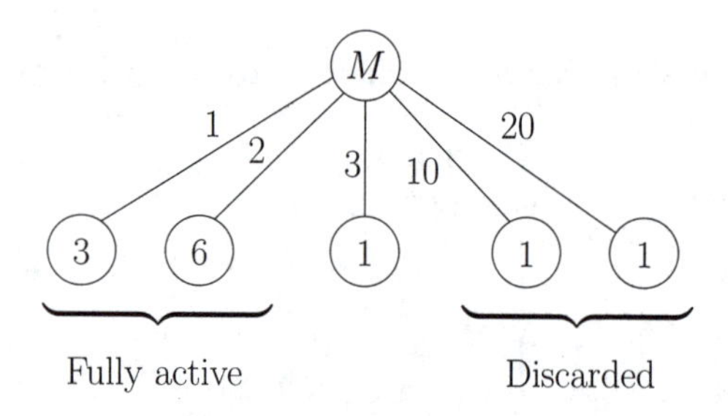

FIGURE 8.11: An example tree platform illustrating the bandwidth-centric strategy.

TABLE 8.1: Achieved throughput for the bandwidth-centric strategy on the example tree platform.

Tasks	Communication	Computation
6 tasks to P_1	$6c_1 = 6$	$6w_1 = 18$
3 tasks to P_2	$3c_2 = 6$	$3w_2 = 18$
2 tasks to P_3	$2c_3 = 6$	$2w_3 = \mathbf{2}$

Consider the example shown in Figure 8.11. Workers are sorted by non-decreasing c_i. We see that $\frac{c_1}{w_1} + \frac{c_2}{w_2} = \frac{2}{3} < 1$ and that $\frac{c_1}{w_1} + \frac{c_2}{w_2} + \frac{c_3}{w_3} = \frac{2}{3} + 3 \geqslant 1$, so that $q = 2$ and $\varepsilon = \frac{1}{3}$ in the previous formula. Therefore, P_1 and P_2 will be fully active, contributing $\alpha_1 + \alpha_2 = \frac{1}{w_1} + \frac{1}{w_2} = \frac{1}{3} + \frac{1}{6}$ tasks per time unit. P_3 will only be partially active, contributing $\alpha_3 = \min(\frac{\varepsilon}{c_{q+1}}, \frac{1}{w_{q+1}} = \min(\frac{1}{9}, 1) = \frac{1}{9}$. P_4 and P_5 will be discarded. The optimal throughput is $\varrho = \frac{1}{3} + \frac{1}{6} + \frac{1}{9} = \frac{11}{18} \approx 0.6$. Table 8.1 shows that 11 tasks are computed every $T_{\mathsf{period}} = 18$ time units.

It is important to point out that if we had used a purely greedy (demand-driven) strategy, we would have reached a much lower throughput. Indeed, the master would serve the workers in round-robin fashion, and we would execute only 5 tasks every 36 time units, therefore achieving a throughput of only $\varrho = 5/36 \approx 0.14$. The conclusion is that even when resources are cheap and abundant, resource selection is key to performance.

Once one has obtained the solution to the linear program defined above, say (α, ϱ), one needs to reconstruct a (periodic) schedule. In other terms, one needs to decide in which specific activities each computation and communication resource is involved during each period. More precisely, we need to define T_{period} such that during a period (i) an integral number of tasks is processed by each processor, and (ii) an integral number of messages goes through each link.

We express all the rational numbers α_i, $1 \leqslant i \leqslant q$, as $\alpha_i = \frac{u_i}{v_i}$, where the u_i and the v_i are relatively prime integers. We also write $\alpha_{q+1} = \min(\frac{\varepsilon}{c_{q+1}}, \frac{1}{w_{q+1}}) = \frac{u_{q+1}}{v_{q+1}}$. The period of the schedule is set to $T_{\mathsf{period}} = \operatorname{lcm}(v_1, \ldots, v_q, v_{q+1})$, the least common multiple of the denominators. In the example in Figure 8.11, we had $\alpha_1 = \frac{1}{3}$, $\alpha_2 = \frac{1}{6}$, $\alpha_3 = \frac{1}{9}$, and $\operatorname{lcm}(3, 6, 9) = 18$. This is how we found the period used in Table 8.1.

In steady-state, during each period of duration T_{period}:

- Master M sends $\alpha_i T_{\mathsf{period}}$ tasks to P_i, $1 \leqslant i \leqslant q+1$. These $q+1$ messages are sent in any order. A total of $\varrho T_{\mathsf{period}}$ tasks are sent during each period. All these communications share the network, but Equation (8.2) ensures that link bandwidths are not exceeded.

- Worker P_i executes the $\alpha_i T_{\mathsf{period}}$ tasks that it has received during the last period. Equation (8.1) ensures that P_i has enough time to execute these tasks.

Obviously, the first and last periods are different: No computation takes place during the first period, and no communication during the last one. Note that the steady-state regime is reached no later than the beginning of the second period.

Altogether, we have a periodic schedule, which is described in compact form: Because it arises from the linear program, $\log(T_{\mathsf{period}})$ is indeed a number polynomial in the problem size, but T_{period} itself is not. Hence, describing what happens at every time unit during the period would be exponential in the problem size. Instead, we have a more "compact" description of the schedule: We only need the duration of the p time intervals during which the master sends tasks to each worker (some possibly zero), and the duration of the p time intervals during which each worker computes (again, some possibly zero).

We conclude this section by explaining how to modify the linear program to use the bounded multi-port model instead of the 1-port model. Here is the

1-port linear program again:

$$\begin{array}{l}\text{MAXIMIZE } \varrho,\\ \text{SUBJECT TO}\\ \left\{\begin{array}{lll} & \varrho = \sum_{i=1}^{p} \alpha_i & \text{(i)}\\ & \sum_{i=1}^{p} \alpha_i c_i \leqslant 1 & \text{(ii)}\\ \forall i, & \alpha_i \geqslant 0 & \text{(iii)}\\ \forall i, & \alpha_i w_i \leqslant 1 & \text{(iv)}\end{array}\right.\end{array}$$

Because messages can now be sent in parallel, we replace Equation (ii) by

$$\forall i, \quad \alpha_i c_i \leqslant 1 ,$$

which states that the bandwidth of the link from M to P_i is not exceeded. Let δ be the volume of data (in bytes) that needs to be sent for one task. We can rewrite the last equation as

$$\forall i, \quad \frac{\alpha_i \delta}{B_i} \leqslant 1 , \tag{ii-a}$$

where B_i is the bandwidth of the link (thus $c_i = \frac{\delta}{B_i}$). We also have to enforce a global bound related to the bandwidth B of the master's network card:

$$\frac{(\sum_{i=1}^{p} \alpha_i)\delta}{B} \leqslant 1 . \tag{ii-b}$$

Replacing Equation (ii) by both Equations (ii-a) and (ii-b) is all that is needed to change to the bounded multi-port model!

8.2.4 Tree-Shaped Platforms

The linear programming approach extends nicely to general tree platforms under the bidirectional 1-port model. For a node P_i in the tree, let w_i denote the time it needs to compute a task, and α_i the (fractional) number of tasks that it executes every time unit. For any edge $P_i \to P_j$ in the tree, let $c_{i,j}$ denote the time needed to send a task from P_i to P_j, and sent$(P_i \to P_j)$ the (fractional) number of tasks that are sent by P_i to P_j every time unit.

The master $M = P_0$ is the root of the tree. If $P_{0_1}, P_{0_2}, \ldots, P_{0_k}$ denote its children in the tree, we have an equation quite similar to Equation (8.1):

$$\sum_{j=1}^{k} \text{sent}(P_0 \to P_{0_j}) c_{0,0_j} \leqslant 1 .$$

If the master computes tasks as well, then we have

$$\alpha_0 w_0 \leqslant 1 .$$

Consider now an internal node P_i. Let P_{i_0} denote its parent in the tree, and let $P_{i_1}, P_{i_2}, \ldots, P_{i_k}$ denote its children in the tree. As before, there are equations to constrain the computations and communications of P_i:

$$\alpha_i w_i \leqslant 1 ,$$

$$\sum_{j=1}^{k} \text{sent}(P_i \to P_{i_j}) c_{i,i_j} \leqslant 1 .$$

But there is another constraint to write, namely a *conservation law*: The number of tasks received by P_i is equal to the sum of the number of tasks that it executes itself, plus the number of tasks that it sends to its children:

$$\text{sent}(P_{i_0} \to P_i) = \alpha_i + \sum_{j=1}^{k} \text{sent}(P_i \to P_{i_j}) .$$

It is important to understand that the conservation law applies to the steady-state operation: Overall, the flow of incoming tasks equals the flow of tasks consumed in place plus the flow of outgoing tasks. The law links tasks that arrive during different periods: Incoming tasks arrive during the current period, while consumed and outgoing tasks had arrived during the previous period. We can check the conservation law for node C in the example shown in Figure 8.10. We have $\text{sent}(A, C) = \frac{2}{3}$, $\alpha_C = \frac{1}{6}$, and $\text{sent}(C, D) = \frac{1}{2}$. We see in Figure 8.10 that these quantities pertain to tasks from different periods.

Finally, if P_i is a tree leaf whose parent is P_{i_0}, we easily derive that

$$\alpha_i w_i \leqslant 1 ,$$

$$\text{sent}(P_{i_0} \to P_i) c_{i_0,i_j} \leqslant 1 ,$$

$$\text{sent}(P_{i_0} \to P_i) = \alpha_i .$$

The total throughput to optimize is given by

$$\varrho = \sum_{\text{all} P_i} \alpha_i .$$

We have built a linear program, and we can follow the same steps as for star-shaped platforms to define the period and reconstruct the final schedule.

8.2.5 Asymptotic Optimality

We keep the notations of the previous section and we now prove that steady-state scheduling is asymptotically optimal. Given a tree platform and a time bound T, let $\text{nb}_{\text{opt}}(T)$ be the optimal number of tasks that can be computed by any schedule, be it periodic or not, within T time units. We have the following result:

PROPOSITION 8.1. $nb_{opt}(T) \leqslant \varrho \times T$.

Proof. Consider an optimal schedule. Consider any node P_i, with parent P_{i_0} and children $P_{i_1}, P_{i_2}, \ldots, P_{i_k}$. Let $t_i(T)$ be the total number of tasks that are executed by P_i within the T time units. Similarly, for each edge $e_{i,i_j} : P_i \to P_{i_j}$ in the tree, let $t_{i,i_j}(T)$ be the total number of tasks that have been forwarded by P_i to P_{i_j} within the T time units. The following equations hold true:

- $t_i(T) \cdot w_i \leqslant T$ (time for P_i to process its tasks);
- $\sum_{j=1}^{k} t_{i,i_j}(T) \cdot c_{i,i_j} \leqslant T$ (time for P_i to forward outgoing tasks in the 1-port model);
- $t_{i_0,i}(T) \cdot c_{i_0,i} \leqslant T$ (time for P_i to receive incoming tasks in the 1-port model);
- $t_{i_0,i}(T) = t_i(T) + \sum_{j=1}^{k} t_{i,i_j}(T)$ (conservation equation).

Let $\alpha_i = \frac{w_i t_i(T)}{T}$, $\text{sent}(P_i \to P_{i_j}) = \frac{c_{i,i_j} t_{i,i_j}(T)}{T}$, and $\text{sent}(P_{i_0} \to P_i) = \frac{c_{i_0,i} t_{i_0,i}(T)}{T}$. All the equations of the previous linear program hold, hence $\sum_i \frac{\alpha_i}{w_i} \leqslant \varrho$, the optimal value. Going back to the original variables, we derive

$$\text{nb}_{\text{opt}}(T) = \sum_i t_i(T) \leqslant \varrho \times T \,. \qquad \square$$

Essentially Proposition 8.1 says that no schedule can execute more tasks than the optimal steady-state. There remains to bound the potential loss due to the initialization and the clean-up phase. Consider the following algorithm (assume T is large enough):

- Solve the linear program: Compute the maximal throughput ϱ, and compute all the values for α_i, $\text{sent}(P_i \to P_j)$, and r_{ij} to determine T_{period}. For each processor P_i, determine per_i, the total number of tasks that it receives per period. Note that all these quantities are independent of T: They depend only upon w_i and c_{ij}, characteristics of the platform.
- Initialization: The master sends per_i tasks to each processor P_i. This requires I units of time, where I is a constant independent of T.
- Let J be the maximum time for each processor to consume per_i tasks ($J = \max_i\{per_i.w_i\}$). Again, J is a constant independent of T.
- Let $r = \lfloor \frac{T-I-J}{T_{\text{period}}} \rfloor$.
- Steady-state scheduling: During r periods of duration T_{period}, operate the platform in steady-state.

- Clean-up: Do not forward any task but consume in place all remaining tasks. This requires at most J time units. Do nothing during the very last units ($T - I - J$ may not be evenly divisible by T).

The number of tasks processed by the above algorithm within T time units is equal to $\text{nb}(T) = (r + 1) \times T_{\text{period}} \times \varrho$.

PROPOSITION 8.2. *The previous scheduling algorithm based upon steady-state operation is asymptotically optimal:*

$$\lim_{T \to +\infty} \frac{nb(T)}{nb_{opt}(T)} = 1 \, .$$

Proof. Using Lemma 8.1, $\text{nb}_{\text{opt}}(T) \leqslant \varrho T$. From the description of the algorithm, we have $\text{nb}(T) = ((r + 1)T_{\text{period}}).\varrho \geqslant (T - I - J).\varrho$. This proves the result because I, J, T_{period}, and ϱ are constants independent of T. □

8.2.6 Summary

In addition to its simplicity, which makes it possible to tackle more complex problems, steady-state scheduling has two main advantages over traditional scheduling:

- *Efficiency:* Steady-state scheduling provides, by definition, a periodic schedule, which is described in compact form and is thus possible to implement efficiently in practice. This is to be contrasted with the need to list the starting dates and target processors for each task in the application.

- *Adaptability:* Because the schedule is periodic, it is possible to dynamically record the observed performance during the current period, and to inject this information into the algorithm that will compute the optimal schedule for the next period. This makes it possible to react on-the-fly to resource availability variations, which is the common case on non-dedicated federated platforms such as Grid platforms [54, 25].

With steady-state scheduling one can give efficient solutions to complex scheduling problems. We sketch two examples below, but we do not detail them because their difficulty goes beyond the scope of this book. We refer the reader to the bibliographical notes at the end of the chapter instead.

- Our first example is the problem of throughput optimization in the case of several master-worker applications that run simultaneously and compete for resources. This is an obvious generalization of the material presented in this section to concurrent applications. A natural objective would be to maximize the minimum throughput achieved over all applications (or a minimum weighted throughput if the applications have different priority levels). This is the classical Max-Min fairness approach used in networks [28, 88]. It turns out that casting all constraints into a linear program, in a way that is akin to what we have done in this section, still works well for this kind of objective function.

- Our second example is for the broadcasting of long messages on heterogeneous platforms (whose interconnection graph is arbitrary and may include cycles and redundant paths). Slicing the message into many packets and organizing the pipelined broadcast of these packets in a steady-state regimen makes it possible to derive asymptotically optimal algorithms for solving this NP-hard problem.

8.3 Workflow Scheduling

In this section, we study the scheduling of simple but fundamental applications that belong to the broad class of so-called *scientific workflows.* While scientific workflows in general are typically structured as arbitrary cDAGs, in this section we consider "chains," i.e., applications structured as a sequence of stages. Each stage corresponds to a different computational task. In a way typical of the execution of workflows in practice, we assume that the application must process a (large) number of data sets, each of which must go through all the stages. As a result, it should be possible to achieve an efficient pipelined execution. Each stage has its own communication and computation requirements: It reads an input from the previous stage, processes the data, and outputs a result to the next stage. Initial data are input to the first stage and final results are obtained as the output from the last stage. The pipeline operates in synchronous mode: After some initialization delay, a new task is completed every period. The period is defined as the longest "cycle-time" to operate a stage, and it is the inverse of the throughput that can be achieved.

Each pipeline stage is a sequential task that may write some global data structure, to disk or to memory, for each processed data set. In this case tasks must always be processed in a sequential order within a stage. Moreover, due to possible local updates, each stage must be mapped onto a single processor. For a given stage, we cannot process half of the tasks on one processor and the remaining half on another without maintaining global information, which might be costly and difficult to implement. In other words, a processor that is assigned a stage will execute the operations required by this stage (input, computation, and output) for all the tasks fed into the pipeline. Other assumptions are possible and we refer the reader to the bibliographical notes at the end of the chapter.

What is the objective function? As pointed out in Section 8.2, an important metric for parallel applications that consists of many individual computations

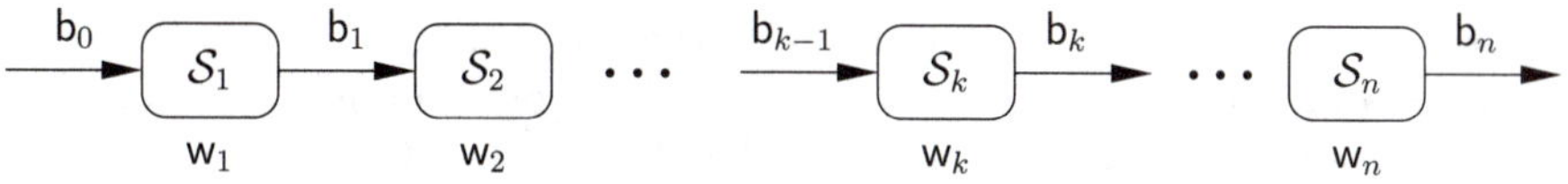

$b_0 \rightarrow \mathcal{S}_1 \xrightarrow{b_1} \mathcal{S}_2 \cdots \xrightarrow{b_{k-1}} \mathcal{S}_k \xrightarrow{b_k} \cdots \mathcal{S}_n \xrightarrow{b_n}$ (weights w_1, w_2, w_k, w_n)

FIGURE 8.12: The application pipeline.

is the throughput. The throughput measures the aggregate rate of data processing; it is the rate at which data sets can enter the system. Equivalently, as pointed out before, the inverse of the throughput, defined as the period, is the time interval required between the beginning of the execution of two consecutive data sets. The period minimization problem can be stated informally as follows: Which stage to assign to which processor so that the largest period of a processor is kept minimal? We consider several variants, in which we require the mapping to be one-to-one (a processor is assigned at most one stage), or interval-based (a processor is assigned an interval of consecutive stages). In addition to these two mapping categories, we target three different platform types: First, *Fully Homogeneous* platforms with homogeneous network links and processors; second, *Communication Homogeneous* platforms with homogeneous network links but heterogeneous processors; third, *Fully Heterogeneous* platforms with heterogeneous network links and processors.

We start by defining the application and architectural framework more precisely in Section 8.3.1. In Section 8.3.2, we study the complexity of the period minimization problem for all combinations: one-to-one or interval mappings; and *Fully Homogeneous*, *Communication Homogeneous* or *Fully Heterogeneous* platforms. We also touch upon an interesting extension of the well-known chains-to-chains problem: Given an array of n elements $a_1, a_2, \ldots, a_n$, the original chains-to-chains problem is to partition the array into p intervals whose element sums are well balanced (technically, the aim is to minimize the largest sum of the elements of any interval). This problem has been extensively studied in the literature (see the survey [95]). It amounts to load balancing n computations whose ordering must be preserved (hence the restriction to intervals) onto p identical processors. The advent of heterogeneous clusters naturally leads to the following generalization: Can we partition the n elements into p intervals whose element sums match p prescribed values (the processor speeds) as closely as possible? Theorem 8.6 shows the NP-hardness of this natural generalization of the chains-to-chains problem.

We conclude in Section 8.3.3 with brief mentions of other possible objective functions, such as response time minimization. The response time is the time elapsed between the beginning and the end of the execution of a single data set, which is clearly an important metric.

8.3.1 Framework

We consider a pipeline with n stages $\mathcal{S}_k$, $1 \leqslant k \leqslant n$, as illustrated in Figure 8.12. Tasks are fed into the pipeline and processed from stage to stage, until they exit the pipeline after the last stage. The k-th stage $\mathcal{S}_k$ receives an input from the previous stage, of size b_{k-1}, performs a number of w_k operations, and outputs data of size b_k to the next stage. The first stage $\mathcal{S}_1$ receives an initial input of size b_0, while the last stage $\mathcal{S}_n$ returns a final result of size b_n.

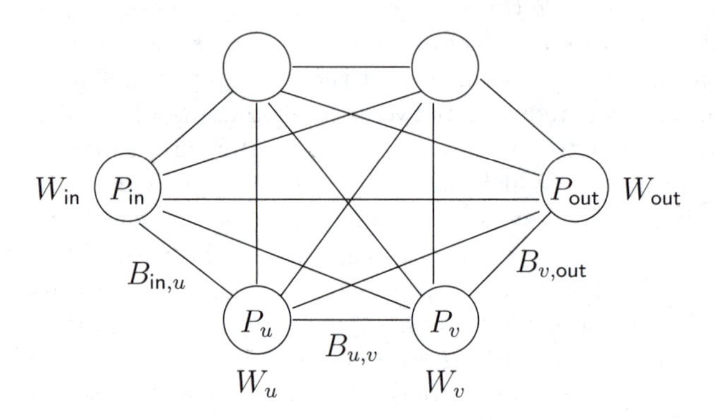

FIGURE 8.13: The target platform.

We target a platform with p processors P_u, $1 \leqslant u \leqslant p$, that are fully interconnected (see Figure 8.13). There is a bidirectional link $\mathsf{link}_{u,v} : P_u \leftrightarrow P_v$ with bandwidth $B_{u,v}$ between each processor P_u and P_v, For the sake of simplicity, we enforce the unidirectional variant of the 1-port model: A given processor can be involved in a single communication at any time unit, either a send or a receive. Note that independent communications between distinct processor pairs can take place simultaneously (this was not possible in star-shaped platforms). However, remember that in the unidirectional variant of the 1-port model a given processor cannot send and receive data in parallel (while this was allowed for tree-shaped platforms in Section 8.2.4).

In the most general case, we have fully heterogeneous platforms, with different processor speeds and link bandwidths. The speed of processor P_u is denoted as W_u, and it takes X/W_u time units for P_u to execute X operations. We also enforce a linear cost model for communications, hence it takes $X/B_{u,v}$ time units to send (resp. receive) a message of size X to (resp. from) P_v. We classify below particular cases that are important, both from a theoretical and practical perspective:

Fully Homogeneous– homogeneous processors ($W_u = W$) and links ($B_{u,v} = B$), which corresponds to typical parallel platforms.

Communication Homogeneous– heterogeneous processors ($W_u \neq W_v$) and homogeneous links ($B_{u,v} = B$), which corresponds to workstations on a LAN.

Fully Heterogeneous– heterogeneous processors ($W_u \neq W_v$) and links ($B_{u,v} \neq B_{u',v'}$), which is the most general model and corresponds, for instance, to hierarchical platforms comprising several clusters interconnected by backbone links.

Finally, we assume the existence of two special additional processors P_{in} and P_{out}. The initial input data for each task resides on P_{in}, while all final results must be returned to P_{out}.

The mapping problem consists in assigning application stages to processors. If we restrict the search to *one-to-one mappings*, we require that each stage $\mathcal{S}_k$ of the application pipeline be mapped onto a distinct processor $P_{\mathsf{alloc}(k)}$ (which is possible only if $n \leqslant p$). The function alloc associates a processor index to each stage index. For convenience, we create two fictitious stages, $\mathcal{S}_0$ and $\mathcal{S}_{n+1}$, and we assign $\mathcal{S}_0$ to P_{in} and $\mathcal{S}_{n+1}$ to P_{out}. What is the period of $P_{\mathsf{alloc}(k)}$, i.e., the minimum delay between the processing of two consecutive tasks? To answer this question, we need to know to which processors the previous and next stages are assigned. Let $t = \mathsf{alloc}(k-1)$, $u = \mathsf{alloc}(k)$, and $v = \mathsf{alloc}(k+1)$. P_u needs $\mathsf{b}_{k-1}/B_{t,u}$ time units to receive the input data from P_t, w_k/W_u time units to process it, and $\mathsf{b}_k/B_{u,v}$ time units to send the result to P_v, hence a cycle-time of $\mathsf{b}_{k-1}/B_{t,u} + \mathsf{w}_k/W_u + \mathsf{b}_k/B_{u,v}$ time units for P_u. Because of the 1-port communication model, these three steps are serialized (see Figure 8.14 for an illustration). The *period* achieved with the mapping is the maximum of the cycle-times of the processors, which corresponds to the rate at which the pipeline can be activated.

In this simple instance, the optimization problem can be stated as follows: Determine a one-to-one allocation function $\mathsf{alloc} : [1,n] \to [1,p]$ (augmented with $\mathsf{alloc}(0) = \mathsf{in}$ and $\mathsf{alloc}(n+1) = \mathsf{out}$) such that

$$T_{\mathsf{period}} = \max_{1 \leqslant k \leqslant n} \left\{ \frac{\mathsf{b}_{k-1}}{B_{\mathsf{alloc}(k-1),\mathsf{alloc}(k)}} + \frac{\mathsf{w}_k}{W_{\mathsf{alloc}(k)}} + \frac{\mathsf{b}_k}{B_{\mathsf{alloc}(k),\mathsf{alloc}(k+1)}} \right\} \quad (8.3)$$

is minimized. We denote this optimization problem ONE-TO-ONE MAPPING.

time unit	1	2	3	4	5	6	7	8	9	10	11	12	13	14	15	16	17	18	...
P_1	★	■	■	△	△	★	■	■	△	△	★	■	■	△	△	★	■	■	...
P_2				★	★	■	△	△	★	★	■	△	△	★	★	■	△	△	...
P_3							★	★	■	△		★	★	■	△		★	★	...

FIGURE 8.14: An example of one-to-one mapping with three stages and processors. Each processor periodically receives input data from its predecessor (★), performs some computation (■), and outputs data to its successor (△). Note that these operations are shifted in time from one processor to another. The cycle-time of P_1 and P_2 is 5 while that of P_3 is 4, hence $T_{\mathsf{period}} = 5$.

Unfortunately, one-to-one mappings may be needlessly restrictive. Natural extensions are *interval mappings*, in which each participating processor is assigned an interval of consecutive stages. Note that when $p < n$ interval mappings are mandatory. Intuitively, assigning several consecutive tasks to the same processor will increase its computational load, but will also decrease communication. The best interval mapping may turn out to be a one-to-one mapping, or instead may utilize only a very small number of fast computing processors interconnected by high-speed links.

Let us write the optimization problem associated to interval mappings formally. We need to express that the intervals achieve a partition of the original set of stages $\mathcal{S}_1$ to $\mathcal{S}_n$. We search for a partition of $[1..n]$ into m intervals $I_j = [d_j, e_j]$ such that $d_j \leqslant e_j$ for $1 \leqslant j \leqslant m$, $d_1 = 1$, $d_{j+1} = e_j + 1$ for $1 \leqslant j \leqslant m-1$, and $e_m = n$. Recall that the function $\mathsf{alloc} : [1, n] \to [1, p]$ associates a processor index to each stage index. In a one-to-one mapping, this function was a one-to-one assignment. In an interval mapping, for $1 \leqslant j \leqslant m$, the whole interval I_j is mapped onto the same processor $P_{\mathsf{alloc}(d_j)}$, i.e., for $d_j \leqslant i \leqslant e_j$, $\mathsf{alloc}(i) = \mathsf{alloc}(d_j)$. Also, two intervals cannot be mapped to the same processor, i.e., for $1 \leqslant j, j' \leqslant m$, $j \neq j'$, $\mathsf{alloc}(d_j) \neq \mathsf{alloc}(d_{j'})$. The period is expressed as

$$T_{\mathsf{period}} = \max_{1 \leqslant j \leqslant m} \left\{ \frac{\mathsf{b}_{d_j - 1}}{B_{\mathsf{alloc}(d_j - 1), \mathsf{alloc}(d_j)}} + \frac{\sum_{i=d_j}^{e_j} \mathsf{w}_i}{W_{\mathsf{alloc}(d_j)}} + \frac{\mathsf{b}_{e_j}}{B_{\mathsf{alloc}(d_j), \mathsf{alloc}(e_j + 1)}} \right\}. \tag{8.4}$$

Note that $\mathsf{alloc}(d_j - 1) = \mathsf{alloc}(e_{j-1}) = \mathsf{alloc}(d_{j-1})$ for $j > 1$ and $d_1 - 1 = 0$. Also, $e_j + 1 = d_{j+1}$ for $j < m$, and $e_m + 1 = n + 1$. We still assume that $\mathsf{alloc}(0) = \mathsf{in}$ and $\mathsf{alloc}(n+1) = \mathsf{out}$. The optimization problem INTERVAL MAPPING is to determine the mapping that minimizes T_{period}, over all possible partitions into intervals, and over all mappings of these intervals to the processors.

With two mapping strategies (one-to-one and interval) and three platform types, we have six problems to solve. The next section assesses the complexity of these six problems when the objective is to minimize the period (or to maximize the throughput).

8.3.2 Period Minimization

Throughout this section, the objective is to determine a mapping that minimizes the period. We start with one-to-one mappings on simpler platforms:

THEOREM 8.3. *For Fully Homogeneous and Communication Homogeneous platforms, the optimal* ONE-TO-ONE MAPPING *can be determined in polynomial time.*

Proof. For *Fully Homogeneous* platforms the problem is straightforward: all mappings have the same period. For *Communication Homogeneous* platforms,

```
1  GREEDY ASSIGNMENT()
2    Work with the n fastest processors, numbered P_1 to P_n
3    where W_1 ⩽ W_2 ⩽ ... ⩽ W_n
4    Mark all stages S_1 to S_n as free
5    for u = 1 to n do
6      Pick any free stage S_k s.t. b_{k-1}/B + w_k/W_u + b_k/B ⩽ T_period
7      Assign S_k to P_u. Mark S_k as already assigned
8      If no stage is found return "failure"
9    Return "success"
```

ALGORITHM 8.1: Greedy assignment algorithm for a given period T_{period}.

we provide a constructive proof: we outline a binary-search algorithm that iterates until the optimal period is found.

The optimal period belongs to the set $\mathcal{T} = \{\frac{\mathsf{b}_{k-1}}{B} + \frac{\mathsf{w}_k}{W_u} + \frac{\mathsf{b}_k}{B},\ 1 \leqslant k \leqslant n, 1 \leqslant u \leqslant p\}$, because it is equal to the cycle-time of some processor P_u executing some stage $\mathcal{S}_k$. First we compute the set $\mathcal{T}$ and sort its elements into an array $\mathcal{T}_A$. Then, we binary search the array $\mathcal{T}_A$ for the optimal period, testing at each step whether the current element T_{period} is a feasible value. To do so, we use the greedy assignment procedure described hereafter. Initially, the current element T_{period} is the median of $\mathcal{T}_A$. If the greedy assignment procedure returns "failure" we increase the period by jumping to the median of the elements of $\mathcal{T}_A$ that are larger than T_{period}, and if it returns "success," we jump to the median of the elements of $\mathcal{T}_A$ that are smaller than T_{period}. The algorithm terminates in $\lceil \log(\mathcal{T}) \rceil$ iterations. Note that $|\mathcal{T}| \leqslant pn$, and hence the total computation time is $O((pn + \text{cost}_{GA}) \log(pn))$, where cost_{GA} is the cost of the greedy assignment procedure.

We now describe the greedy assignment algorithm for a given T_{period} value. Recall that there are n stages to map onto $p \geqslant n$ processors in a one-to-one fashion. We target *Communication Homogeneous* platforms with heterogeneous processors ($W_u \neq W_v$) but with homogeneous links ($B_{u,v} = B$). First, we retain only the fastest n processors, which we rename $P_1, P_2, \ldots, P_n$ such that $W_1 \leqslant W_2 \leqslant \ldots \leqslant W_n$. Then, we consider the processors in the order P_1 to P_n, i.e., from the slowest to the fastest, and greedily assign to them any free, that is, not already assigned, stage that they can process within the period. Algorithm 8.1 details the procedure.

Before providing the proof, we proceed with a small example. Consider an application with three stages with respective computation requirements 1, 1, and 100; three processors of speed 1, 1, and 100; and no communication costs whatsoever. Can we achieve $T_{\text{period}} = 1$? The algorithm starts by assigning stages to the slowest processors, i.e., to the first two processors. Only the first two stages can be assigned to these processors to fit into the period, so they are chosen, and the last stage then fits when assigned to the fastest processor. It cannot be assigned to one of the first two processors because the required period would be exceeded.

The proof that the greedy procedure returns a solution if and only if there exists a solution of period T_{period} is done via a simple exchange argument. Consider a valid one-to-one assignment of period T_{period}, denoted by A, and assume that in this assignment S_{k_1} is assigned to P_1. Note first that the greedy procedure will indeed find a stage to assign to P_1 and cannot fail, since S_{k_1} can be chosen. If the choice of the greedy procedure is actually S_{k_1}, we proceed by induction with P_2. If the greedy procedure has selected another stage S_{k_2} for P_1, we find which processor, say P_u, has been assigned this stage in the valid assignment A. Then, we exchange the assignments of P_1 and P_u in A. Because P_u is faster than P_1, which could process S_{k_1} in time in the assignment A, P_u can process S_{k_1} in time, too. Because S_{k_2} has been mapped on P_1 by the greedy procedure, P_1 can process S_{k_1} in time. So the exchange is valid. We can consider the new assignment A, which is valid and which assigns the same stage to P_1 as the greedy procedure. The proof proceeds by induction with P_2 as before.

The complexity of the greedy assignment procedure is $O(n^2)$, because of the two loops over processors and stages. Altogether, since $n \leqslant p$, the complexity of the whole algorithm is $O((pn\log(pn))$, which is indeed polynomial in the problem size. □

We now proceed to interval mappings. First we show that the complexity is polynomial for *Fully Homogeneous* platforms:

THEOREM 8.4. *For Fully Homogeneous platforms, the optimal* INTERVAL MAPPING *can be determined in polynomial time.*

Proof. Consider an application with n stages $\mathcal{S}_1$ to $\mathcal{S}_n$ to be mapped onto a fully homogeneous platform with p processors. Let W and B, respectively, denote the processor speed and the link bandwidth. We compute recursively the value $c(i,j,k)$, which is the optimal period that can be achieved by any interval-based mapping of stages S_i to S_j using exactly k processors. The goal is to determine $\min_{1\leqslant k\leqslant p} c(1,n,k)$. The recursion can be written, including two initializations, as

$$\begin{cases} c(i,j,k) = \min\limits_{\substack{q+r=k\\ 1\leqslant q,r\leqslant k-1}} \left\{ \min\limits_{i\leqslant \ell\leqslant j-1} \{\max\left(c(i,\ell,q), c(\ell+1,j,r)\right)\}\right\} \\ \begin{cases} c(i,j,1) = \dfrac{\mathsf{b}_{i-1}}{B} + \dfrac{\sum_{k=i}^{j} \mathsf{w}_k}{W} + \dfrac{\mathsf{b}_j}{B} \\ c(i,j,k) = +\infty \;\text{ if }\; k > j-i+1\,. \end{cases} \end{cases}$$

The first initialization gives the period obtained by mapping stages S_i to S_j to a single processor. The second initialization means that no mapping can be found with more processors than stages. The main recursion is easy to justify: To compute $c(i, j, k)$, we search over all possible partitionings into two subintervals $[i..\ell]$ and $[\ell+1..j]$. Furthermore, we try every possible processor number q for the first interval and every possible processor number r for the second interval, where $q + r = k$. We explore all possibilities with such a strategy. The complexity of this dynamic programming algorithm is bounded by $O(n^3p^2)$ (there are five nested loops to execute the algorithm, three on pipeline stages and two on processors). □

It is not possible to extend the previous dynamic programming algorithm to deal with *Communication Homogeneous* platforms. This is because the algorithm intrinsically relies on identical processors in the recursion. Heterogeneous processors would execute sub-intervals with different cycle-times. Because of this additional difficulty, the INTERVAL MAPPING problem for *Communication Homogeneous* platforms is very combinatorial and, indeed, we have the following result:

THEOREM 8.5. *For Communication Homogeneous platforms, the (decision problem associated to the)* INTERVAL MAPPING *optimization problem is NP-complete.*

Theorem 8.5 is a consequence of Theorem 8.6, which states the complexity of the heterogeneous one-dimensional (1-D) partitioning problem. We introduce this problem and prove Theorem 8.6 before returning to the proof of Theorem 8.5.

The chains-to-chains problem– Given an array of n elements $a_1, a_2, \ldots, a_n$, the 1-D partitioning problem, also known as the chains-to-chains problem, is to partition the array into p intervals whose element sums are almost identical. More precisely, we search for a partition of $[1..n]$ into p consecutive intervals $\mathcal{I}_1, \mathcal{I}_2, \ldots, \mathcal{I}_p$, where $\mathcal{I}_k = [d_k, e_k]$ and $d_k \leqslant e_k$ for $1 \leqslant k \leqslant p$, $d_1 = 1$, $d_{k+1} = e_k + 1$ for $1 \leqslant k \leqslant p-1$, and $e_p = n$. The objective is to minimize

$$\max_{1 \leqslant k \leqslant p} \sum_{i \in \mathcal{I}_k} a_i = \max_{1 \leqslant k \leqslant p} \sum_{i=d_k}^{e_k} a_i.$$

The chains-to-chains problem amounts to load balancing n computations whose ordering must be preserved (hence the restriction to intervals) onto p identical processors. Here a_i corresponds to the execution time of the i-th task, and the sum of the elements in interval $\mathcal{I}_k$ is the load of the processor to which $\mathcal{I}_k$ is assigned. Replacing a_i by w_i, we recognize a simplified version of our INTERVAL MAPPING optimization problem on *Fully Homogeneous* platforms, without any communication cost (and the dynamic programming algorithm of Theorem 8.4 can be simplified, as seen in Exercise 8.3).

With different-speed processors, the 1-D partitioning problem can be stated as follows: The goal is to partition the n elements into p intervals whose element sums match p prescribed values (the processor speeds) as closely as possible. Let $W_1, W_2, \ldots, W_p$ denote these values. We search for a partition of $[1..n]$ into p intervals $\mathcal{I}_k = [d_k, e_k]$ and for a permutation σ of $\{1, 2, \ldots, p\}$, with the objective to minimize

$$\max_{1 \leqslant k \leqslant p} \frac{\sum_{i \in \mathcal{I}_k} a_i}{W_{\sigma(k)}}.$$

Formally, we define:

DEFINITION 8.1 (HETERO-1D-PARTITION-DEC). Given n elements a_1, $a_2, \ldots, a_n$; p values $W_1, W_2, \ldots, W_p$; and a bound K, can we find a partition of $[1..n]$ into p intervals $\mathcal{I}_1, \mathcal{I}_2, \ldots, \mathcal{I}_p$, with $\mathcal{I}_k = [d_k, e_k]$ and $d_k \leqslant e_k$ for $1 \leqslant k \leqslant p$, $d_1 = 1$, $d_{k+1} = e_k + 1$ for $1 \leqslant k \leqslant p - 1$, and $e_p = n$, and a permutation σ of $\{1, 2, \ldots, p\}$, such that

$$\max_{1 \leqslant k \leqslant p} \frac{\sum_{i \in \mathcal{I}_k} a_i}{W_{\sigma(k)}} \leqslant K\ ?$$

THEOREM 8.6. *The* HETERO-1D-PARTITION-DEC *problem is NP-complete.*

Proof. The HETERO-1D-PARTITION-DEC problem clearly belongs to the class NP: Given a solution, it is easy to verify in polynomial time that the partition into p intervals is valid and that the maximum sum of the elements in a given interval divided by the corresponding W value does not exceed the bound K. To establish the completeness, we use a reduction from NUMERICAL MATCHING WITH TARGET SUMS (NMWTS), which is NP-complete in the strong sense [57]. We consider an instance $\mathfrak{I}_1$ of NMWTS. Given $3m$ numbers $x_1, x_2, \ldots, x_m$, $y_1, y_2, \ldots, y_m$, and $z_1, z_2, \ldots, z_m$, does there exist two permutations σ_1 and σ_2 of $\{1, 2, \ldots, m\}$, such that $x_i + y_{\sigma_1(i)} = z_{\sigma_2(i)}$ for $1 \leqslant i \leqslant m$? Because NMWTS is NP-complete in the strong sense, we can encode the $3m$ numbers in unary and assume that the size of $\mathfrak{I}_1$ is $O(m + M)$, where $M = \max_i\{x_i, y_i, z_i\}$. We also assume that $\sum_{i=1}^{m} x_i + \sum_{i=1}^{m} y_i = \sum_{i=1}^{m} z_i$, otherwise $\mathfrak{I}_1$ cannot have a solution.

We build the following instance $\mathfrak{I}_2$ of HETERO-1D-PARTITION-DEC (we use the formulation in terms of task weights and processor speeds, which is more intuitive):

- We define $n = (M + 3)m$ tasks, whose weights are shown below:

$$A_1 \underbrace{111...1}_{M} C\ D \mid A_2 \underbrace{111...1}_{M} C\ D \mid \ldots \mid A_m \underbrace{111...1}_{M} C\ D$$

  Here, $B = 2M$, $C = 5M$, $D = 7M$, and $A_i = B + x_i$ for $1 \leqslant i \leqslant m$. To define the a_i formally for $1 \leqslant i \leqslant n$, let $N = M + 3$. We have for

$1 \leqslant i \leqslant m$

$$\begin{cases} a_{(i-1)N+1} = A_i = B + x_i \\ a_{(i-1)N+j} = 1 \text{ for } 2 \leqslant j \leqslant M+1 \\ a_{iN-1} = C \\ a_{iN} = D \ . \end{cases}$$

- For the number of processors (and intervals), we choose $p = 3m$. As for the speeds, we let W_i be the speed of processor P_i where, for $1 \leqslant i \leqslant m$

$$W_i = B + z_i, \qquad W_{m+i} = C + M - y_i, \qquad W_{2m+i} = D \ .$$

Finally, we ask whether there exists a solution matching the bound $K = 1$. Clearly, the size of $\mathfrak{I}_2$ is polynomial in the size of $\mathfrak{I}_1$. We now show that instance $\mathfrak{I}_1$ has a solution if and only if instance $\mathfrak{I}_2$ does.

Suppose first that $\mathfrak{I}_1$ has a solution, with permutations σ_1 and σ_2 such that $x_i + y_{\sigma_1(i)} = z_{\sigma_2(i)}$. For $1 \leqslant i \leqslant m$:

- We map each task A_i and the following $y_{\sigma_1(i)}$ tasks of weight 1 onto processor $P_{\sigma_2(i)}$.
- We map the following $M - y_{\sigma_1(i)}$ tasks of weight 1 and the next task, of weight C, onto processor $P_{m+\sigma_1(i)}$.
- We map the next task, of weight D, onto the processor P_{2m+i}.

We do have a valid partition of all the tasks into $p = 3m$ intervals. For $1 \leqslant i \leqslant m$, the load and speed of the processors are indeed equal:

- The load of $P_{\sigma_2(i)}$ is $A_i + y_{\sigma_1(i)} = B + x_i + y_{\sigma_1(i)}$ and its speed is $B + z_{\sigma_2(i)}$.
- The load of $P_{m+\sigma_1(i)}$ is $M - y_{\sigma_1(i)} + C$, which is equal to its speed.
- The load and speed of P_{2m+i} are both equal to D.

The mapping does achieve the bound $K = 1$, and hence a solution to $\mathfrak{I}_2$.

Suppose now that $\mathfrak{I}_2$ has a solution, i.e., a mapping matching the bound $K = 1$. We first observe that $W_i < W_{m+j} < W_{2m+k} = D$ for $1 \leqslant i, j, k \leqslant m$. Indeed $W_i = B + z_i \leqslant B + M = 3M$, $5M \leqslant W_{m+j} = C + M - y_j \leqslant 6M$, and $D = 7M$. Hence, each of the m tasks of weight D must be assigned to a processor of speed D, and it is the only task assigned to this processor. These m singleton assignments divide the set of tasks into m intervals, namely, the set of tasks before the first task of weight D, and the $m - 1$ sets of tasks lying between two consecutive tasks of weight D. The total weight of each of these m intervals is $A_i + M + C > B + M + C = 10M$, while the largest speed of the $2m$ remaining processors is $6M$. Therefore, each of them must be assigned to at least two processors each. However, there remain only $2m$ available processors, hence each interval is assigned exactly two processors.

Consider such an interval A_i 111...1 C with M tasks of weight 1, and let P_{i_1} and P_{i_2} be the two processors assigned to this interval. Tasks A_i and C are not assigned to the same processor (otherwise the whole interval would be assigned to the same processor). So P_{i_1} receives task A_i and h_i tasks of weight 1 while P_{i_2} receives $M - h_i$ tasks of weight 1 and task C. The weight of P_{i_2} is $M - h + C \geqslant C = 5M$ while $W_i \leqslant 3M$ for $1 \leqslant i \leqslant m$. Hence, P_{i_1} must be some P_i, $1 \leqslant i \leqslant m$ while P_{i_2} must be some P_{m+j}, $1 \leqslant j \leqslant m$. Because this holds true on each interval, this defines two permutations, $\sigma_2(i)$ and $\sigma_1(i)$, such that $P_{i_1} = P_{\sigma_2(i)}$ and $P_{i_2} = P_{\sigma_1(i)}$. Because the bound $K = 1$ is achieved, we have

- $A_i + h_i = B + x_i + h_i \leqslant B + z_{\sigma_2(i)}$
- $M - h_i + C \leqslant C + M - y_{\sigma_1(i)}$.

Therefore, $y_{\sigma_1(i)} \leqslant h_i$ and $x_i + h_i \leqslant z_{\sigma_2(i)}$, and $\sum_{i=1}^m x_i + \sum_{i=1}^m y_i \leqslant \sum_{i=1}^m x_i + \sum_{i=1}^m h_i \leqslant \sum_{i=1}^m z_i$.

By hypothesis, $\sum_{i=1}^m x_i + \sum_{i=1}^m y_i = \sum_{i=1}^m z_i$, hence all inequalities are tight, and in particular $\sum_{i=1}^m x_i + \sum_{i=1}^m h_i = \sum_{i=1}^m z_i$. We can deduce that $\sum_{i=1}^m y_i = \sum_{i=1}^m h_i = \sum_{i=1}^m z_i - \sum_{i=1}^m x_i$, and since $y_{\sigma_1(i)} \leqslant h_i$ for all i, we have $y_{\sigma_1(i)} = h_i$ for all i. Similarly, we deduce that $x_i + h_i = z_{\sigma_2(i)}$ for all i, and therefore $x_i + y_{\sigma_1(i)} = z_{\sigma_2(i)}$. Altogether, we have found a solution for $\mathfrak{I}_1$, which concludes the proof. □

Back to the proof of Theorem 8.5. Obviously, the INTERVAL MAPPING optimization problem belongs to the class NP. Any instance of the HETERO-1D-PARTITION problem with n tasks a_i, p processor speeds W_i, and bound K can be converted into an instance of the INTERVAL MAPPING problem with n stages of weight $\mathsf{w}_i = a_i$, letting all communication costs $\mathsf{b}_i = 0$, targeting a *Communication Homogeneous* platform with p processor speeds W_i and homogeneous links of bandwidth $B = 1$, and trying to achieve a period that is no greater than K. This concludes the proof. □

As a consequence of Theorem 8.5, the interval mapping problem is also NP-hard for *Fully Heterogeneous* platforms. It turns out it is also the case of the one-to-one mapping problem on such platforms:

THEOREM 8.7. *For Fully Heterogeneous platforms, the (decision problems associated to the)* ONE-TO-ONE MAPPING *optimization problem is NP-complete.*

Proof. The problem clearly belongs to the class NP: Given a solution, it is easy to verify in polynomial time that all stages are processed, and that the maximum cycle-time of any processor does not exceed the bound on the period. To establish the completeness, we use a reduction from MINIMUM METRIC BOTTLENECK WANDERING SALESPERSON PROBLEM (MMBWSP) [9, 46]. We consider an instance $\mathcal{I}_1$ of MMBWSP. Given a set $C = \{c_1, c_2, \ldots, c_m\}$

of m cities, an initial city $s \in C$, a final city $f \in C$, distances $d(c_i, c_j) \in \mathbb{N}$ satisfying the triangle inequality, and a bound $K \geqslant 2$ on the largest distance, does there exist a simple path from the initial city s to the final city f passing through all cities in C, i.e., a permutation $\pi : [1..m] \to [1..m]$ such that $c_{\pi(1)} = s$, $c_{\pi(m)} = f$, and $d(c_{\pi(i)}, c_{\pi(i+1)}) \leqslant K$ for $1 \leqslant i \leqslant m-1$? To simplify notations, and without loss of generality, we renumber the cities so that $s = c_1$ and $f = c_m$ (i.e., $\pi(1) = 1$ and $\pi(m) = m$).

We build the following instance $\mathcal{I}_2$ of our mapping problem (note that the same instance will work out for the two variants: ONE-TO-ONE MAPPING and INTERVAL MAPPING):

- For the application: $n = 2m - 1$ stages that for convenience we denote as

$$\to \mathcal{S}_1 \to \mathcal{S}'_1 \to \mathcal{S}_2 \to \mathcal{S}'_2 \to \ldots \to \mathcal{S}_{m-1} \to \mathcal{S}'_{m-1} \to \mathcal{S}_m \to$$

  For each stage $\mathcal{S}_i$ or $\mathcal{S}'_i$ we set $\mathsf{b}_i = \mathsf{w}_i = 1$ (as well as $\mathsf{b}_0 = \mathsf{b}_n = 1$), so that the application is perfectly homogeneous.

- For the platform (see Figure 8.15):

  - We use $p = m + \binom{m}{2}$ processors that for convenience we denote as P_i, $1 \leqslant i \leqslant m$ and $P_{i,j}$, $1 \leqslant i < j \leqslant m$. We also use an input processor P_{in} and an output processor P_{out}. The speed of each processor P_i or P_{ij} has the same value $s = \frac{1}{2K}$ (note that we have a computation-homogeneous platform), except for P_{in} and P_{out}, whose speeds are 0.
  - The communication links shown in Figure 8.15 have a larger bandwidth than the others, and are referred to as *fast* links. More precisely, $b_{P_{\mathsf{in}}, P_1} = b_{P_m, P_{\mathsf{out}}} = 1$, and $b_{P_i, P_{ij}} = b_{P_{ij}, P_j} = \frac{2}{d(c_i, c_j)}$ for $1 \leqslant i < j \leqslant m$. All the other links have a very small bandwidth $B = \frac{1}{5K}$ and are referred to as *slow* links. The intuition here is that slow links are too slow to be used for the mapping.

Finally, we ask whether there exists a solution with period $3K$. Clearly, the size of $\mathcal{I}_2$ is polynomial (and even linear) in the size of $\mathcal{I}_1$. We now show that instance $\mathcal{I}_1$ has a solution if and only if instance $\mathcal{I}_2$ does.

Suppose first that $\mathcal{I}_1$ has a solution. We map stage $\mathcal{S}_i$ onto $P_{\pi(i)}$ for $1 \leqslant i \leqslant m$, and stage $\mathcal{S}'_i$ onto processor $P_{\pi(i),\pi(i+1)}$ for $1 \leqslant i \leqslant m-1$. The cycle-time of P_1 is $1 + 2K + \frac{d(c_{\pi(1)}, c_{\pi(2)})}{2} \leqslant 1 + 2K + \frac{K}{2} \leqslant 3K$. Quite similarly, the cycle-time of P_m is smaller than $3K$. For $2 \leqslant i \leqslant m-1$, the cycle-time of $P_{\pi(i)}$ is $\frac{d(c_{\pi(i-1)}, c_{\pi(i)})}{2} + 2K + \frac{d(c_{\pi(i)}, c_{\pi(i+1)})}{2} \leqslant 3K$. Finally, for $1 \leqslant i \leqslant m-1$, the cycle-time of $P_{\pi(i),\pi(i+1)}$ is $\frac{d(c_{\pi(i)}, c_{\pi(i+1)})}{2} + 2K + \frac{d(c_{\pi(i)}, c_{\pi(i+1)})}{2} \leqslant 3K$. The mapping does achieve a period that is no greater than $3K$, hence a solution to $\mathcal{I}_2$.

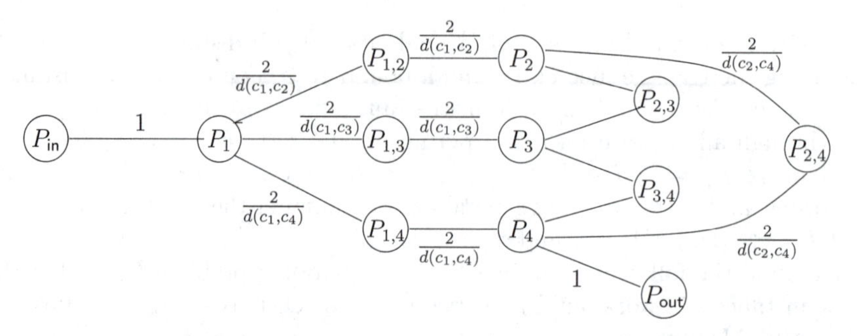

FIGURE 8.15: The platform used in the reduction for Theorem 8.7.

TABLE 8.2: Summary of complexity results for the different instances of the workflow mapping problem.

	Fully Hom.	*Comm. Hom.*	*Hetero.*
One-to-one	polynomial (bin.search)		NP-hard
Interval	polynomial (dyn. prog.)	NP-hard	NP-hard

Suppose now that $\mathcal{I}_2$ has a solution, i.e., a mapping of period lower than $3K$. We first observe that each processor is assigned at most one stage by the mapping, because executing two stages would require at least $2K + 2K$ units of time, which would be too large to match the period. Next, we observe that any slow link of bandwidth $\frac{1}{5K}$ cannot be used in the solution; otherwise, the period would exceed $5K$.

The input processor P_{in} has a single fast link to P_1, so necessarily P_1 is assigned stage $\mathcal{S}_1$ (i.e., $\pi(1) = 1$). As observed above, P_1 cannot execute any other stage. Because of fast links, stage $\mathcal{S}'_1$ must be assigned to some $P_{1,j}$; we let $j = \pi(2)$. Again, because of fast links and the one-to-one constraint, the only choice for stage $\mathcal{S}_2$ is $P_{\pi(2)}$. Necessarily $j = \pi(2) \neq \pi(1) = 1$, otherwise P_1 would execute two stages. We proceed similarly for stage S'_2, assigned to some $P_{2,k}$ (let $k = \pi(3)$) and stage S_3 assigned to $P_{\pi(3)}$. Owing to the one-to-one constraint, $k \neq 1$ and $k \neq j$, i.e., $\pi : [1..3] \to [1..m]$ is a one-to-one mapping. By induction, we build the full permutation $\pi : [1..m] \to [1..m]$. Because the output processor P_{out} has a single fast link to P_m, necessarily P_m is assigned stage $\mathcal{S}_m$, hence $\pi(m) = m$.

We have built the desired permutation. It now remains to show that for $1 \leqslant i \leqslant m-1$: $d(c_{\pi(i)}, c_{\pi(i+1)}) \leqslant K$. The cycle-time of processor $P_{\pi(i)}$ is $\frac{d(c_{\pi(i)},c_{\pi(i+1)})}{2} + 2K + \frac{d(c_{\pi(i)},c_{\pi(i+1)})}{2} \leqslant 3K$, hence $d(c_{\pi(i)}, c_{\pi(i+1)}) \leqslant K$. Altogether, we have found a solution for $\mathcal{I}_1$, which concludes the proof. □

Table 8.2 summarizes all previous complexity results. We see that one level of heterogeneity (in processor speed) is enough to make interval mapping NP-hard, while two levels of heterogeneity (adding different link bandwidths) are required to make one-to-one mapping NP-hard as well.

8.3.3 Response Time Minimization

Let us briefly discuss the response time minimization problem. Recall that an interval mapping is a partition of $[1..n]$ into m intervals $I_j = [d_j, e_j]$, and that the whole interval I_j is mapped onto the same processor $P_{\mathsf{alloc}(d_j)}$. As usual, assume that $\mathsf{alloc}(0) = \mathsf{in}$ and $\mathsf{alloc}(n+1) = \mathsf{out}$. Each data set traverses all stages, but only communications between two stages mapped on the same processors take zero time units. Overall, the response time is expressed as

$$T_{\mathsf{rt}} = \sum_{1 \leqslant j \leqslant m} \left\{ \frac{\mathsf{b}_{d_j-1}}{B_{\mathsf{alloc}(d_j-1),\mathsf{alloc}(d_j)}} + \frac{\sum_{i=d_j}^{e_j} \mathsf{w}_i}{W_{\mathsf{alloc}(d_j)}} \right\} + \frac{\mathsf{b}_n}{B_{\mathsf{alloc}(n),\mathsf{alloc}(n+1)}} . \quad (8.5)$$

The response time for a one-to-one mapping obeys the same formula (with the restriction that each interval has length 1).

Note that it may well be the case that different data sets have different response times (because they are mapped onto different processor sets), hence the response time is defined as the maximum response time over all data sets. Intuitively, the response time is small when all communications are zeroed out. An obvious candidate mapping would be to map all stages on the fastest processor, which will result in a large period. This is a general observation: Minimizing the response time is antagonistic to minimizing the period. The bibliographical notes at the end of the chapter point to works that tackle a bi-criteria approach, aiming at minimizing the response time under period constraints (or the converse).

We conclude this section with the following result:

PROPOSITION 8.3. *The response time minimization problem is polynomial on Fully Homogeneous and Communication Homogeneous platforms.*

Proof. We start with interval mappings, which are more natural to minimize the response time as many communications will be zeroed out. On *Fully Homogeneous* platforms, the optimal solution is to map all stages on a single (arbitrary) processor, because this zeroes out all communications except input/output ones. On *Communication Homogeneous* platforms, the optimal solution is to map all the stages on the fastest processor, for the same reason.

All one-to-one mappings on *Fully Homogeneous* platforms have the same response time. On *Communication Homogeneous* platforms, the optimal solution is to assign the most computationally expensive stages to the fastest processors, in a greedy manner (largest stage on fastest processor, second largest on second fastest, and so on). □

In Exercise 8.4 we see that the response time minimization problem is NP-hard for one-to-one mappings on *Fully Heterogeneous* platforms. To the best of our knowledge, the complexity is open for interval mappings on such platforms, at least at the time this book is being written.

8.4 Hyperplane Scheduling (or Scheduling at Compile-Time)

In this section, we focus on yet another topic, that of compiler transformations to schedule parallel loop nests. This section can be viewed as an introduction to automatic parallelization techniques, a major area of research for scheduling and mapping. Rather than writing a parallel program, every user would dream of a compiler that would do all the work:

1. find the parallelism in a sequential program (identify which statements are dependent and which are not);

2. transform the program to expose the parallel statements;

3. execute the program efficiently on a target parallel platform.

This dream has not come (fully) true yet, but in spite of many setbacks a lot of progress has been made in several directions. In this section, we focus solely on the automatic parallelization of so-called *uniform loop nests.* We explain simple but representative results, motivated by two seminal papers by Karp, Miller and Winograd [71], and by Lamport [77] (see the bibliographical notes at the end of the chapter for further information).

8.4.1 Uniform Loop Nests

Consider the following fragment of code:

$$
\begin{aligned}
S_1 &: a \leftarrow b+1 \\
S_2 &: b \leftarrow a-1 \\
S_3 &: a \leftarrow c-2 \\
S_4 &: d \leftarrow c
\end{aligned}
$$

Here a, b, c, and d are four un-aliased variables (pointing to distinct memory locations). They are initialized somewhere before the code fragment and possibly re-used later. The sequential execution order $<_{seq}$ is simply the textual order $<_{text}$ of the four statements, first S_1, then S_2, then S_3, and finally S_4. Dependence analysis is the name of the technique for determining which of the four statements are independent, and hence could be executed in arbitrary order, and which are not. We already encountered this problem in Section 7.1.1 with the example of the triangular system. Here we aim at automating the process.

Statement S_1 writes variable a while statement S_2 reads it, hence they are dependent. If we change their order, we may use a wrong value of a, and change the result of the program. Thus, S_1 and S_2 cannot be executed in parallel; they must remain executed one after the other. But which one first? The answer is obvious, it is S_1 because it precedes S_2 in the sequential execution order. What we have found here is a *flow* dependence from S_1 to S_2, i.e., a write-then-read dependence.

Focusing on variable a again, we see that S_2 reads a while S_3 writes it. Both statements are therefore dependent. But this time this is not a flow dependence from S_3 to S_2. It is of course impossible to have a dependence that violates the sequential execution order. Instead, what we have here is an *anti* dependence from S_2 to S_3, i.e., a read-then-write dependence: The value of a must be read by S_2 before being updated by S_3. Let us continue with a. We find yet another kind of dependence from S_1 to S_3, because both statements write to a (and of course the write of S_1 must precede that of S_3). This is an *output* dependence from S_1 to S_3, i.e., a write-then-write dependence. However, because there is a dependence (flow) from S_1 to S_2 and another dependence (anti) from S_2 to S_3, the latter dependence (output) from S_1 to S_3 is not an actual dependence between successors. It can be retrieved from the transitive closure of the dependence relation, just as explained in Section 7.1.1. We provide a more formal explanation below. Note that there is another dependence involving variable b, namely, an anti-dependence from S_1 to S_2. Finally, statement S_4 is independent of the others, even from S_3, which reads the same variable c (but there is no write of c, hence the order does not matter).

```
1 for i = 0 to N do
2     for j = 0 to N do
3         S1(i,j) : a(i,j) = b(i,j-6) + d(i-1,j+3)
4         S2(i,j) : b(i+1,j-1) = c(i+2,j+5) + 1
5         S3(i,j) : c(i+3,j-1) = a(i,j+2)
6         S4(i,j) : d(i,j-1) = a(i,j-1) - 1
```

ALGORITHM 8.2: An example of uniform loop nest.

The definitions of flow, anti, and output dependences can be extended to *uniform loop nests* as follows. Consider the example in Algorithm 8.2. In this example, there are four statements, S_1 to S_4, all surrounded by the loops on i and j. We say the loops are *perfectly nested* because all statements are surrounded by the same loops. Each loop is defined by a loop counter that takes, one after the other, integral values from the lower bound to the upper bound. The iterations of n perfectly nested loops are represented by a vector of size n, called the *iteration vector*. The set of all possible values of the iteration vector (defined by the lower and upper bounds of the loops) is called the *iteration domain*. In the example, the loops i and j define a two-dimensional iteration vector $I = (i, j)$ whose iteration domain *Dom* is the set of integral vectors in the square $1 \leqslant i, j \leqslant N$:

$$Dom = \{(i, j) \in \mathbb{Z}^2, 0 \leqslant i \leqslant N, 0 \leqslant j \leqslant N\} .$$

When running the program, each value of an iteration vector I corresponds to a given execution of the code surrounded by the loops; in particular, each statement S_k is executed for each value I of the iteration vector. Such an execution is called an *operation* and is denoted by $S(I)$. These operations are carried out in a predefined order, called the *sequential order*, which we denote $<_{seq}$, as previously. The sequential order is not difficult to characterize for uniform loop nests:

$$S(I) <_{seq} T(J) \Leftrightarrow (I <_{lex} J) \text{ or } (I = J \text{ and } S <_{text} T),$$

where $<_{lex}$ is the *lexicographic order* of the iteration vectors, and $<_{text}$ is the textual order of statements in the program's source code. In the example, we have $S_3(2, 5) <_{seq} S_1(3, 1)$, $S_3(2, 5) <_{seq} S_2(2, 6)$, and $S_3(2, 5) <_{seq} S_4(2, 5)$.

There is a dependence between operations $S_i(I)$ and iteration $S_j(J)$ if:

- $S_i(I)$ is executed before $S_j(J)$.
- $S_i(I)$ and $S_j(J)$ refer to a memory location M, and at least one of these references is a write.
- The memory location M is not written between iteration I and iteration J.

We retrieve flow, anti and output dependences, depending on whether there is a single write to M (flow if the read occurs after, anti if it comes before) or two writes (output). The last condition is to ensure that the dependence is indeed between successors in the dependence graph.

The dependence vector between iteration $S_i(I)$ and iteration $S_j(J)$ is defined as $d_{(i,I),(j,J)} = J - I$. The loop nest is said to be *uniform* if the dependence vectors $d_{(i,I),(j,J)}$ do not depend on either I or J, and we denote them simply $d_{i,j}$. We point out that each dependence vector $d_{i,j}$ is lexicographically positive (its first nonzero component is greater than 0), due to the semantics of the sequential execution: If there is a dependence from $S_i(I)$ to iteration $S_j(J)$, then I precedes J, i.e., $I \leqslant_{lex} J$ and $d_{i,j} = J - I \geqslant_{lex} 0$. We can then represent the dependences as an oriented graph with k nodes (the statements) linked by edges corresponding to the dependence vectors.

In the example, Algorithm 8.2, we see that variable $a(i, j)$ is produced (written) by statement $S_1(i, j)$ and consumed (read) by statement $S_4(i, j+1)$, hence a uniform dependence from S_1 to S_4 of vector $d_{1,4} = \begin{pmatrix} 0 \\ 1 \end{pmatrix}$. How did we find this? First, since $S_4(i, j)$ reads $a(i, j-1)$, then $S_4(i, j+1)$ reads $a(i, j)$, the memory location written by $S_1(i,j)$. We do have $I = (i,j) <_{seq} J = (i, j+1)$. No other operation writes into this location in between, since in this loop nest every memory location is written only once. We have "manually" checked the three conditions stated above.

How can we automate the process of finding dependences? For each candidate array and statement pair, we can write a system of equations, and try to solve it. Let us do this for array a and the pair (S_1, S_3). S_1 writes into a and S_3 reads it, so we search if there exists a flow dependence from $S_1(I)$ to $S_3(J)$, for some $I, J \in Dom$ with $I <_{seq} J$. Letting $I = (i,j)$ and $J = (i', j')$, we obtain:

- $i = i'$ and $j = j' + 2$: we write $a(i,j)$ in $S_1(i,j)$, and read $a(i', j'+2)$ in $S_3(i', j')$;

- $i < i'$, or $i = i'$ and $j < j'$: this is the condition $I <_{seq} J$.

We see that there is no solution, and hence no flow dependence from S_1 to S_3. But we see (because we are used to looking at loop nests!) that there is an anti-dependence from S_3 to S_1. With the same notations, looking for an anti dependence from $S_3(J)$ to $S_1(I)$, for some $I, J \in Dom$ with $J <_{seq} I$, and letting $J = (i', j')$, $I = (i,j)$, we obtain:

- $i' = i$ and $j' + 2 = j$: we read $a(i', j'+2)$ in $S_3(i,j)$, and write $a(i,j)$ in $S_1(i,j)$;

- $i' < i$ or $i' = i$ and $j' < j$.

We derive the solution $i = i'$ and $j = j' + 2$, hence a uniform anti dependence, with dependence vector $d_{3,1} = \begin{pmatrix} 0 \\ 2 \end{pmatrix}$.

The above illustrates how parallelizing compiler software works. In general, the situation is more complicated, because we have to check the third condition for a dependence to exist, namely that no other operation writes to the same variable in between. For the anti dependence from $S_3(J)$ to $S_1(I)$, if we had found many solutions, we should have kept only the last write, i.e., the lexicographic maximum over all index vectors J that lead to a solution for a given I. But as already mentioned, each memory location is written only once in the example, which simplifies the problem considerably.

Altogether, after checking all candidate arrays and statement pairs, we obtain the following list of dependences and dependence vectors:

$$S_1 \longrightarrow S_4 : \text{ array } a, \text{ flow }, d_{1,4} = \begin{pmatrix} 0 \\ 1 \end{pmatrix}$$

$$S_3 \longrightarrow S_1 : \text{ array } a, \text{ anti }, d_{3,1} = \begin{pmatrix} 0 \\ 2 \end{pmatrix}$$

$$S_2 \longrightarrow S_1 : \text{ array } b, \text{ flow }, d_{2,3} = \begin{pmatrix} 1 \\ 5 \end{pmatrix}$$

$$S_3 \longrightarrow S_2 : \text{ array } c, \text{ flow }, d_{3,2} = \begin{pmatrix} 1 \\ -6 \end{pmatrix}$$

$$S_4 \longrightarrow S_1 : \text{ array } d, \text{ flow }, d_{4,1} = \begin{pmatrix} 1 \\ -4 \end{pmatrix}$$

We now obtain the so-called dependence matrix:

$$D = (d_{1,4}, d_{3,1}, d_{2,3}, d_{3,2}, d_{4,1}) = \begin{pmatrix} 0 & 0 & 1 & 1 & 1 \\ 1 & 2 & 5 & -6 & -4 \end{pmatrix}.$$

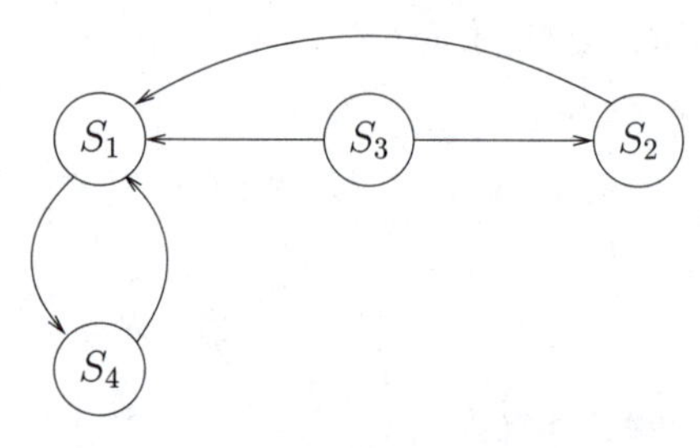

FIGURE 8.16: Dependence graph for the example in Algorithm 8.2.

The dependence graph is shown in Figure 8.16. It captures all the information extracted from the dependence analysis. Note that this information is not 100% accurate. Each uniform dependence seems to occur for each operation over the entire iteration domain, while it is not the case at the boundary. For instance, with the anti dependence from S_3 to S_1, the condition $j = j'+2$ cannot be satisfied if $j' = N-1$ or $j' = N$. However, the condition is satisfied on most points of the domain, so it makes sense to approximate dependences as we did in the graph. What is important is to always over-approximate, i.e., to record more dependences than there may actually exist. Over-approximations can reduce parallelism but will never lead to a violation of the semantics of the original program, while under-approximations are simply... dangerous.

8.4.2 Lamport's Hyperplane Method

In this section, we examine the problem of scheduling uniform loop nests with infinite processors. Consider a uniform loop nest whose computation domain *Dom* is the integral polyhedron

$$Dom = \{p \in \mathbb{Z}^n, Ap \leqslant b, \text{ where } A \in \mathbb{Z}^{a \times n}, \text{ and } b \in \mathbb{Z}^a\} ,$$

and whose dependence matrix D is:

$$D = \left(d_1 \; d_2 \; \ldots \; d_m \right) .$$

Here n is the nest depth, i.e., the dimension of index points, a is the number of constraints that define the shape of the domain, and m is the number of dependence vectors, so that the constraint matrix A is of dimension $a \times n$ and the dependence matrix D is of dimension $n \times m$.

Back to the example in Figure 8.16, we have $Dom = \{(i,j) \in \mathbb{Z}^2, 0 \leqslant i \leqslant N, 0 \leqslant j \leqslant N\}$, which translates into

$$A = \begin{pmatrix} -1 & 0 \\ 1 & 0 \\ 0 & -1 \\ 0 & 1 \end{pmatrix} \text{ and } b = \begin{pmatrix} 0 \\ N \\ 0 \\ N \end{pmatrix} = N \times b_0, \text{ where } b_0 = \begin{pmatrix} 0 \\ 1 \\ 0 \\ 1 \end{pmatrix},$$

so that $a = 4$ and $n = 2$. Recall that we have $m = 5$ dependence vectors.

Given a uniform loop nest, we write $p_1 \prec p_2$ if $p_2 = p_1 + d_i$ for some $d_i \in D$, i.e., when p_1 and p_2 are two points of *Dom* such that p_2 depends upon p_1. As defined in Section 7.2, a schedule is a function $\sigma : Dom \to \mathbb{Z}$ such that for any arbitrary index points $p_1, p_2 \in Dom$, $\sigma(p_1) < \sigma(p_2)$ if $p_1 \prec p_2$. In other words, a schedule is a mapping that assigns a time of execution to each operation of the nest in such a way that dependences are preserved, nothing new since Chapter 7!

Using the terminology of Section 7.2, each operation $p \in Dom$ is a unitary task that requires one time unit. The total execution time of a schedule σ is

$$T_\sigma = 1 + \max(\sigma(p), p \in Dom) - \min(\sigma(q), q \in Dom) .$$

With an unlimited number of processors, T_σ is minimum when σ schedules computations as soon as their operands are available, i.e., when σ is the free schedule σ_{free} of Section 7.3. But the free schedule requires that one traverse the dependence graph and that one generate in extension the scheduling date $\sigma_{free}(p)$ for each operation $p \in Dom$. Unfortunately, there are too many operations for this to be possible (there are $(N+1)^2$ operations in the example in Figure 8.16). Instead, compilers use *linear schedules.* A linear schedule is a mapping $\sigma_\pi : Dom \to \mathbb{Z}$ defined by

$$\sigma_\pi(p) = \lfloor \pi.p \rfloor \text{ for } p \in Dom ,$$

such that dependences are preserved. The vector $\pi \in Q^{1\times n}$ (π has rational coefficients) is called an *admissible scheduling vector*. Not any vector π can be chosen as an admissible scheduling vector, since dependences must be preserved.

The basic idea of linear schedules is to try to transform the original uniform loop nests into an equivalent nest in which all the internal loops except the first one are parallel, as in

```
1 for time=time_min to time_max do
2   forall p ∈ E(time) in parallel do
3     S_1(p)
4     ...
5     S_k(p)
```

The external loop corresponds to an iteration time and $E(time)$ is the set of all the points computed at the step *time*. Such points must not be linked by dependence vectors so that they can be executed simultaneously. With a linear schedule σ_π, we have $E(time) = \{p \in Dom, \lfloor \pi.p \rfloor = time\}$. The scheduling vector π defines a family of affine hyperplanes $H(t)$ orthogonal to π such that the set of points executed at a time t is equal to the intersection of the domain with the hyperplane $H(t)$. The flow of computation goes from $H(t)$ to $H(t+1)$ by a translation; hence the name of the method that we describe in this section: the computations progress like a wavefront parallel to the family of hyperplanes $H(t)$.

Admissible scheduling vectors π are easy to characterize for uniform loop nests:

$$\pi d \geqslant 1 \text{ for each dependence vector } d \in D .$$

Let us explain this: If $p_1, p_2 \in Dom$ are two points such that $p_1 \prec p_2$, i.e., $p_2 = p_1 + d_i$ for some $d_i \in D$, we must have $\sigma_\pi(p_1) < \sigma_\pi(p_2)$, i.e., $\lfloor \pi p_1 \rfloor + 1 \leqslant \lfloor \pi(p_1 + d_i) \rfloor$. This inequality is satisfied if $\pi d_i \geqslant 1$. We retrieve Lamport's condition:

$$\pi D \geqslant 1 ,$$

which means that $\pi d_i \geqslant 1$ for all $d_i \in D$. Note that this condition is sufficient for π to be an admissible scheduling vector, and it is necessary unless the domain *Dom* is very small, a situation very unlikely to occur in practice.

How can we find an admissible scheduling vector? It turns out to be straightfoward, owing to the fact that all dependence vectors are lexicographically positive. Assume that the columns of matrix $D = (d_1, \ldots, d_m)$, whose dimension is $n \times m$, have been sorted lexicographically. Let k_1 be the index of the first nonzero component of d_1, the "smallest" vector of D. Since d_1 is positive, $d_{1,k_1} > 0$. Take $\pi_{k_1} = 1$ and $\pi_k = 0$ for $k_1 < k \leqslant n$. Before defining the first components of π, we remark that whatever their values, $\pi d_1 > 0$. Now, let k_2 be the index of the first nonzero component of d_2. Because d_1

is lexicographically smaller than d_2, $k_2 \leqslant k_1$. Let $\pi_k = 0$ for $k_2 < k < k_1$ and take for π_{k_2} the smallest positive integer such that $\pi d_2 > 0$, modifying maybe π_{k_1} if $k_2 = k_1$. Continuing the process, we obtain an admissible vector. Let us execute this procedure for the example in Figure 8.16 to find an admissible vector $\pi = \begin{pmatrix} \pi_1 \\ \pi_2 \end{pmatrix}$. After sorting the dependence matrix, we have $D = \begin{pmatrix} 0 & 0 & 1 & 1 & 1 \\ 1 & 2 & -6 & -4 & 5 \end{pmatrix}$. We have $n = 2$, $k_1 = 2$, and we take $\pi_2 = 1$ so that $\pi d_1 \geqslant 1$, whatever the value of π_1. We have $k_2 = 2$ and $\pi d_2 \geqslant 1$, so we do not need to change the value of π_2. Next $k_3 = 1$ and the condition $\pi d_3 \geqslant 1$ writes $\pi_1 - 6\pi_2 \geqslant 1$; so we take $\pi_1 = 7$. We can keep the value $\pi = \begin{pmatrix} 7 \\ 1 \end{pmatrix}$ because we already have $\pi d_4 \geqslant 1$ and $\pi d_5 \geqslant 1$.

We know how to find an admissible scheduling vector. But how do we find the optimal one? We first start with the example in Figure 8.16, and then move to the general procedure. As discussed above, the scheduling vector $\pi = \begin{pmatrix} \pi_1 \\ \pi_2 \end{pmatrix}$ is admissible if and only if $\pi_2 \geqslant 1$ and $\pi_1 \geqslant 1 + 6\pi_2$, which implies that both π_1 and π_2 are positive. The total execution time of σ_π is

$$T_{\sigma_\pi} = 1 + \max(\lfloor \pi p \rfloor,\, p \in Dom) - \min(\lfloor \pi q \rfloor,\, q \in Dom) \ .$$

Here Dom is the square $\{0 \leqslant i, j \leqslant N\}$ and π has positive components, so $max(\lfloor \pi p \rfloor,\, p \in Dom) = \lfloor (\pi_1 + \pi_2) N \rfloor$ and $\min(\lfloor \pi q \rfloor,\, q \in Dom) = 0$. We see that the optimal scheduling vector is the one that we found already, $\pi = \begin{pmatrix} 7 \\ 1 \end{pmatrix}$, and the optimal execution time is $T_{\sigma_\pi} = 1 + 8N$.

For the general case, consider a uniform loop nest. The best linear schedule is the one that minimizes $T_{\sigma_\pi} = 1 + \max(\lfloor \pi p \rfloor,\, p \in Dom) - \min(\lfloor \pi q \rfloor,\, q \in Dom)$ over all rational vectors π such that $\pi D \geqslant 1$. We write

$$T_{linear} = \min(T_{\sigma_\pi},\, \pi \in \mathbb{Q}^n,\, \pi D \geqslant 1) \ . \tag{8.6}$$

Consider the usual continuous relaxation of the integer programming problem of Equation (8.6). In other words, we move from Dom, the computation domain with integer points, to $\overline{Dom}$, the "extended" computation domain with rational points:

$$\overline{Dom} = \{p \in \mathbb{Q}^n,\ Ap \leqslant b\} \ ,$$

$$T^*_{\sigma_\pi} = \max_{p,q \in \overline{Dom}} \pi(p - q) \ ,$$

$$T^*_{linear} = \min(T^*_{\sigma_\pi},\, \pi \in \mathbb{Q}^n,\, \pi D \geqslant 1) \ .$$

We claim that T_{linear}, the time of the best linear schedule, verifies

$$T_{linear} \leqslant 2 + T^*_{linear} \ .$$

To see why we may lose at most two time units when moving from Dom to $\overline{Dom}$, we simply write the following:

- For $p \in Dom$, let $\pi p = R + r$, where $R = \lfloor \pi p \rfloor$ and $0 \leqslant r < 1$. Similarly, for $q \in Dom$, $\pi q = S + s$, where $S = \lfloor \pi q \rfloor$ and $0 \leqslant s < 1$.
- We have $\sigma_\pi(p) - \sigma_\pi(q) = R - S$ and $\pi(p-q) = R - S + (r-s)$, where $-1 < r - s < 1$, hence $\sigma_\pi(p) - \sigma_\pi(q) \leqslant \pi(p-q)+1$. We get $T_{\sigma_\pi} \leqslant 2 + T^*_{\sigma_\pi}$.
- Because $T_{linear} \leqslant T_{\sigma_\pi}$ for all scheduling vectors π, we have $T_{linear} \leqslant 2 + T^*_{\sigma_\pi}$. This is valid for all π, hence $T_{linear} \leqslant 2 + T^*_{linear}$.

Finding the optimal scheduling vector in T^*_{linear} is equivalent to solving the problem

$$\min_{\pi D \geqslant 1} \max_{Ap \leqslant b, Aq \leqslant b} \pi(p-q) \, .$$

Consider first that π is fixed. By the duality theorem of linear programming [106, p. 90],

$$\begin{cases} Ap \leqslant b \\ Aq \leqslant b \\ \max \pi(p-q) \end{cases} \quad \text{is equivalent to:} \quad \begin{cases} X_1 A = \pi \\ X_2 A = -\pi \\ X_1 \geqslant 0 \\ X_2 \geqslant 0 \\ \min (X_1 + X_2) b \, , \end{cases}$$

the existence of the maximum implying the existence of the minimum. Thus, the search for the optimal scheduling vector can be done by solving the following linear problem:

$$\begin{cases} \pi D \geqslant 1 \\ X_1 A = \pi \\ X_2 A = -\pi \\ X_1 \geqslant 0 \\ X_2 \geqslant 0 \\ \min (X_1 + X_2) b \, . \end{cases}$$

We remark that this problem is linear in b. The search for the best scheduling vector over the family of domains $\{Ax \leqslant Nb\}$ is reduced to the search on the domain $\{Ax \leqslant b\}$, which can be done without knowing the parameter N, thus at compile-time.

For the example in Figure 8.16, we let $\pi = (\pi_1, \pi_2)$, $X_1 = (a_1, b_1, c_1, d_1)$, and $X_2 = (a_2, b_2, c_2, d_2)$. We solve the following optimization problem:

$$\begin{cases} XD \geqslant 1 : \pi_2 \geqslant 1, 2\pi_2 \geqslant 1, \pi_1 - 6\pi_2 \geqslant 1, \pi_1 - 4\pi_2 \geqslant 1, \pi_1 + 5\pi_2 \geqslant 1 \\ X_1 A = X : -a_1 + b_1 = \pi_1, -c_1 + d_1 = \pi_2 \\ X_2 A = -X : -a_2 + b_2 = -\pi_1, -c_2 + d_2 = -\pi_2 \\ X_1 \geqslant 0 : a_1 \geqslant 0, b_1 \geqslant 0, c_1 \geqslant 0, d_1 \geqslant 0 \\ X_2 \geqslant 0 : a_2 \geqslant 0, b_2 \geqslant 0, c_2 \geqslant 0, d_2 \geqslant 0 \\ N \times \min (X_1 + X_2) b : N \times (b_1 + b_2 + d_1 + d_2) \, . \end{cases}$$

To solve this problem, we can use the simplex method provided by packages such as GLPK [62] or Maple [39]. We obtain the solution: $\pi = (7, 1)$, $X_1 =$

$(0, 7, 0, 1)$ and $X_2 = (7, 0, 1, 0)$. The total execution time is $T^*_{linear} = N \times (7 + 0 + 1 + 0) = 8N$.

The last thing to do is to rewrite the loop nest to expose the parallelism. We consider again the example of Figure 8.16. We need to transform the original loop nest into

```
1 for time=time_min to time_max do
2 |  forall p ∈ E(time) in parallel do
3 |  |  ...
```

We already have $time = 7i + j$, which we extend into

$$\begin{pmatrix} time \\ proc \end{pmatrix} = \begin{pmatrix} 7 & 1 \\ 1 & 0 \end{pmatrix} \begin{pmatrix} i \\ j \end{pmatrix}.$$

The trick is that the 2×2 matrix is unimodular, i.e., has an integral inverse:

$$\begin{pmatrix} i \\ j \end{pmatrix} = \begin{pmatrix} 0 & 1 \\ 1 & -7 \end{pmatrix} \begin{pmatrix} time \\ proc \end{pmatrix}.$$

Because $0 \leqslant i, j \leqslant N$, we can compute the new loop bounds for *time* and *proc*:

- $time = 7i + j$, hence $time_{min} = 0$ and $time_{max} = 8N$.
- $i = proc$, hence $0 \leqslant proc \leqslant N$. Also, $j = time - 7proc$, hence $0 \leqslant time - 7proc \leqslant N$. We derive $\lfloor \frac{time-N}{7} \rfloor \leqslant proc \leqslant \lfloor \frac{time}{7} \rfloor$.

Altogether, we can rewrite the loop nest as shown in Algorithm 8.3.

```
1 for time = 0 to 8N do
2 |  for proc = max(0, ⌊(time−N)/7⌋) to min(N, ⌊time/7⌋) do
3 |  |  a(proc, time − 7proc) ←
   |  |        b(proc, time − 7proc − 6) + d(proc − 1, time − 7proc + 3)
4 |  |  b(proc + 1, time − 7proc − 1) ← c(proc + 2, time − 7proc + 5) + 1
5 |  |  c(proc + 3, time − 7proc − 1) ← a(proc, time − 7proc + 2)
6 |  |  d(proc, time − 7proc − 1) ← a(proc, time − 7proc − 1) − 1
```

ALGORITHM 8.3: Rewriting the loop nest to expose parallelism.

At each instant *time*, all nodes $p = \begin{pmatrix} proc \\ time - 7proc \end{pmatrix} \in Dom$ belong to $E(time)$ and can be executed in parallel. If we have N available processors, we can execute all iterations of the inner loop simultaneously on distinct processors (hence the name "*proc*" for the first component). Again, it is possible to automate the process of rewriting the loop nest in the general case. This relies on sophisticated mathematical tools such as Hermite normal forms and Fourier-Motzkin elimination, and we point the reader to references [49, 5] for more details.

Bibliographical Notes

The divisible load model has been widely studied in the last several years, after having been popularized by the landmark book by Bharadwaj, Ghose, Mani, and Robertazzi [32]. See also the introductory papers [33, 102]. Recent results on star and tree networks are surveyed in [19].

Steady-state scheduling was pioneered by Bertsimas and Gamarnik [29]. See [13, 20, 21, 18] for recent applications of the technique.

Workflow scheduling is a hot topic. A few papers related to the material covered in this chapter are [113, 31, 30, 110, 24]. Bi-criteria optimization (response time with period constraints, or the converse) is discussed in [114, 22, 23, 117].

Finally, loop nest scheduling dates back to the seminal papers of Karp, Miller, and Winograd [71] and Lamport [77]. Many loop transformation algorithms are described in the books by Banerjee [10], by Wolfe [119], and by Allen and Kennedy [5]. An introduction to the topic of uniform recursions, loop parallelization, and software pipelining is provided in the book by Darte, Robert, and Vivien [49].

8.5 Exercises

We offer quite a comprehensive set of exercises. Exercise 8.1 is devoted to divisible load scheduling on tree platforms. Exercise 8.2 deals with the steady-state approach for multiple applications. The next two exercises are on the topic of workflow scheduling: Exercise 8.3 tackles the classic chains-to-chains problem, while Exercise 8.4 explores the complexity of response time minimization. Finally, the last two exercises study dependences in loop nests. Exercise 8.5 applies basic techniques from Section 8.4. Exercise 8.6 is more involved and investigates the problem of removing dependence cycles.

◇ Exercise 8.1 : Divisible Load Scheduling on Heterogeneous Trees

In this exercise we extend the linear programming approach to schedule divisible load applications on general multi-level trees in which each node is associated to a processor. The master P_0 serially sends a single message to each of its children (one round). The master has no processing power (same reason as in Section 8.1.4) but every other node is capable of computing and communicating to one of its children simultaneously. However, because of the one-round hypothesis, no overlap can occur with the incoming communication from the node's parent.

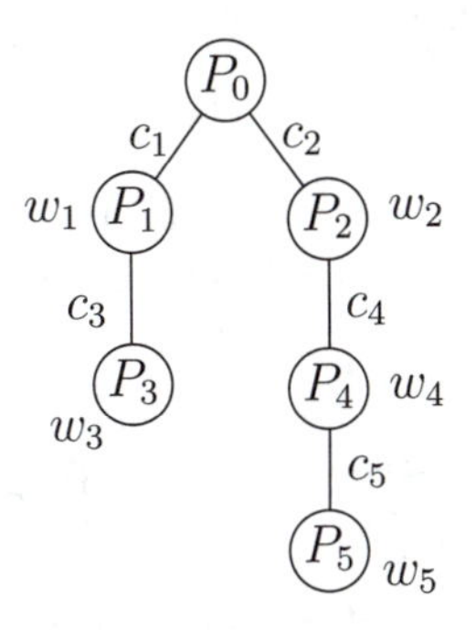

FIGURE 8.17: Heterogeneous tree of processors.

1. Consider the tree shown in Figure 8.17, with a master P_0 and five workers, P_1 to P_5. The cycle-time of worker P_i is w_i. Let W_{total} be the total amount of work, and let α_i be the fraction executed by P_i for $1 \leqslant i \leqslant 5$. Assume that $c_1 \leqslant c_2$, so that the master P_0 will serve P_1 before P_2 in the optimal solution (same proof as for Lemma 8.3). Write the linear program that characterizes the optimal makespan T for the tree.

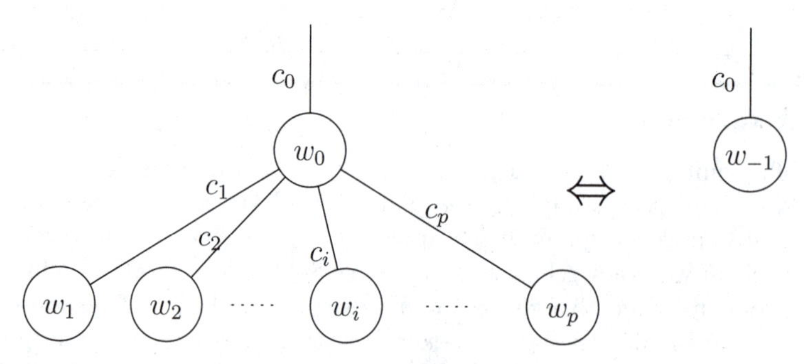

FIGURE 8.18: Replacing a single-level tree by an equivalent node.

2. Now we investigate another method to solve the problem, based on a node reduction technique illustrated in Figure 8.18. Explain how a single-level tree network with parent P_0 (with input link of communication-time c_0 and cycle-time w_0) and p children P_i, $1 \leqslant i \leqslant p$ (with input link of communication-time c_i and cycle-time w_i), where $c_1 \leqslant c_2 \leqslant \ldots \leqslant c_p$, is equivalent to a single node with the same communication-time c_0 and cycle-time $w_{-1} = 1/W$, where W is the solution to the linear program:

$$\begin{array}{l}\text{MAXIMIZE } W,\\ \text{SUBJECT TO}\\ \left\{\begin{array}{ll}(1)\ \alpha_i \geqslant 0 & 0 \leqslant i \leqslant p\\ (2)\ \sum_{i=0}^{p} \alpha_i = W & \\ (3)\ W c_0 + \alpha_0 w_0 = 1 & \\ (4)\ W c_0 + \sum_{j=1}^{i} \alpha_j c_j + \alpha_i w_i = 1 & 1 \leqslant i \leqslant p\end{array}\right.\end{array}$$

3. Explain how the previous reduction can be used to compute the optimal solution for a general multi-level tree. Can you compare this approach to that in Question 1?

◇ Exercise 8.2 : Steady-State Scheduling of Multiple Applications

In this exercise we consider multiple applications that execute concurrently on heterogeneous tree platforms and compete for processors and network links. Each application consists of a large number of independent and identical tasks that originate at the tree's root. Each application is also given a weight that quantifies its relative value (one can think of this weight as a priority). The goal is to maximize throughput while executing tasks from each application in the same ratio as their weights.

Here is a summary of the notations:

- P_0 is the root processor and $P_{p(u)}$ is the parent of node P_u for $u \neq 0$.

- $\Gamma(u)$ is the set of indices of the children of node P_u.
- Node P_u can perform c_u operations per time unit, and, if $u \neq 0$, can receive b_u bytes from its parent $P_{p(u)}$.
- Application k has weight $w^{(k)}$, and each task of type k involves $b^{(k)}$ bytes and $c^{(k)}$ operations.

1. Let $\alpha^{(k)}$ be the throughput achieved for application k, i.e., the number of tasks of type k that are executed on the platform every time unit, in steady-state mode. Explain why maximizing

$$\min_k \left\{ \frac{\alpha^{(k)}}{w^{(k)}} \right\}$$

is a better objective function than maximizing

$$\sum_k \left\{ \frac{\alpha^{(k)}}{w^{(k)}} \right\}.$$

2. Write a linear program to maximize the former objective function

$$\min_k \left\{ \frac{\alpha^{(k)}}{w^{(k)}} \right\}$$

.

◇ Exercise 8.3 : Chains-to-Chains

Given an array of n elements $a_1, a_2, \ldots, a_n$, the chains-to-chains problem consists in partitioning the array into p consecutive intervals $\mathcal{I}_1, \mathcal{I}_2, \ldots, \mathcal{I}_p$, where $\mathcal{I}_k = [d_k, e_k]$ and $d_k \leqslant e_k$ for $1 \leqslant k \leqslant p$, $d_1 = 1$, $d_{k+1} = e_k + 1$ for $1 \leqslant k \leqslant p - 1$, and $e_p = n$. The objective is to minimize

$$\max_{1 \leqslant k \leqslant p} \sum_{i \in \mathcal{I}_k} a_i = \max_{1 \leqslant k \leqslant p} \sum_{i=d_k}^{e_k} a_i.$$

1. Give a dynamic programming algorithm to solve the problem. What is its complexity?

2. Give a binary search algorithm to solve the problem. What is its complexity?

◇ Exercise 8.4 : Response Time of One-to-One Mappings

Show that the response time minimization problem is NP-hard for one-to-one mappings on *Fully Heterogeneous* platforms.

⋄ Exercise 8.5 : Dependence Analysis and Scheduling Vectors

1. Perform a dependence analysis of the following program fragment:

S_1: $A \leftarrow B + C$
S_2: $D \leftarrow E + F$
S_3: $C \leftarrow A + C$
S_4: $E \leftarrow C + D$
S_5: $E \leftarrow 1$

2. Perform a dependence analysis of the following loop nest:

for $i = 1$ **to** N **do**
  for $j = 1$ **to** N **do**
    S_1: $a(i+1, j-1) \leftarrow b(i, j+4) + c(i-2, j-3) + 1$
    S_2: $b(i-1, j) \leftarrow a(i, j) - 1$
    S_3: $c(i, j) \leftarrow a(i, j-2) + b(i, j)$

3. What are the admissible scheduling vectors for the previous loop nest? Determine the one that minimizes the execution time. Use this vector to re-write the loop nest so that the innermost loop is parallel.

4. Write a two-dimensional loop nest with three statements, S_1, S_2, and S_3, whose dependences are as follows:

- From S_1 to S_2: an anti dependence of distance vector $(1, -2)$
- From S_1 to S_3: a flow dependence of distance vector $(1, -1)$
- From S_2 to S_1: an anti dependence of distance vector $(1, 2)$
- From S_2 to S_3: a flow dependence of distance vector $(1, 4)$
- From S_3 to S_1: an anti dependence of distance vector $(0, 1)$
- From S_3 to S_2: a flow dependence of distance vector $(0, 3)$

⋄ Exercise 8.6 : Dependence Removal

For this exercise we define the reduced dependence graph (RDG) of a loop nest as the graph whose nodes are the statements of the nest, and whose edges are the dependences between the statements. But here we are only interested in the *type* of the dependences, not in their distance vector. Thus, edges will be labeled by *f*, *a*, or *o*, to identify flow, anti, and output dependences.

1. Compute the RDG of the following loop nest:

for $i = 1$ **to** N **do**
  S_1: $a(i) \leftarrow b(i) + c(i)$
  S_2: $a(i+1) \leftarrow a(i) + 2d(i)$

Try to break the dependence cycle by introducing a temporary variable.

2. In the general case, the node splitting technique proceeds as shown below:

before splitting	after splitting
for $i = 1$ **to** N **do**	**for** $i = 1$ **to** N **do**
$\ldots$	$\ldots$
S_k: $lhs(f(i)) \leftarrow rhs(\ldots)$	S'_k: $temp(f(i)) \leftarrow rhs(\ldots)$
	S_k: $lhs(f(i)) \leftarrow temp(f(i))$
$\ldots$	
$\ldots$	$\ldots$
S_i: $\cdots \quad \leftarrow lhs(g(i))$	S_i: $\cdots \quad \leftarrow temp(g(i))$

Assume that the access function lhs (standing for *left-hand side*) is one-to-one. Show the effect of the node splitting technique on the six possible dependence types: dependences incoming to S_k of type flow, anti and output, and dependences outgoing from S_k of type flow, anti and output.

3. Let G be the RDG of a loop nest, and let G' be the graph obtained after splitting all the nodes. Show that a cycle in G' is either uniquely composed of flow dependences, or uniquely composed of output dependences. In addition, show that any cycle in G' corresponds to an existing cycle in G.

4. Apply the node splitting technique to the following loop nest:

for $i = 4$ **to** N **do**
  S_1: $a(i+5) \leftarrow c(i-3) + b(2i+2)$
  S_2: $b(2i) \leftarrow a(i-1) + 1$
  S_3: $a(i) \leftarrow c(i+5) + 1$
  S_4: $c(i) \leftarrow b(2i-4)$

8.6 Answers

Exercise 8.1 (Divisible Load Scheduling on Heterogeneous Trees)

▷ **Question 1.** We give the linear program and comment on the equations afterwards:

MINIMIZE T,

SUBJECT TO

$$\begin{cases}
(i)\ \alpha_i \geqslant 0 & 1 \leqslant i \leqslant 5\\
(ii)\ \sum_{i=1}^{5} \alpha_i = W_{\text{total}} & \\
(0)\ (\alpha_1+\alpha_3)c_1 + (\alpha_2+\alpha_4+\alpha_5)c_2 \leqslant T & \\
(1)\ (\alpha_1+\alpha_3)c_1 + \alpha_1 w_1 \leqslant T & \\
(1')\ (\alpha_1+\alpha_3)c_1 + \alpha_3 c_3 \leqslant T & \\
(2)\ (\alpha_2+\alpha_4+\alpha_5)c_2 + \alpha_2 w_2 \leqslant T & \\
(2')\ (\alpha_2+\alpha_4+\alpha_5)c_2 + \alpha_4 c_4 \leqslant T & \\
(3)\ (\alpha_1+\alpha_3)c_1 + \alpha_3 c_3 + \alpha_3 w_3 \leqslant T & \\
(4)\ (\alpha_2+\alpha_4+\alpha_5)c_2 + \alpha_4 c_4 + \alpha_4 w_4 \leqslant T & \\
(4')\ (\alpha_2+\alpha_4+\alpha_5)c_2 + \alpha_4 c_4 + \alpha_5 c_5 \leqslant T & \\
(5)\ (\alpha_2+\alpha_4+\alpha_5)c_2 + \alpha_4 c_4 + \alpha_5 c_5 + \alpha_5 w_5 \leqslant T &
\end{cases}$$

Equation (0) expresses the total communication time for the master P_0. Equation (1) says that after the end of its incoming communication, internal node P_1 should be constantly computing. Equation (1′) says that after the end of its incoming communication, P_1 should should be constantly sending data to P_3 while computing. Note that Equation (1′) is useless, due to Equation (3). Equation (3) states that P_3 keeps computing after receiving its own data, which come from the master together with the data for P_1 (hence the term $(\alpha_1 + \alpha_3)c_1$), and which is then forwarded by P_1 (hence the term $\alpha_3 c_3$). Finally, Equations (2), (4) and (5) are the counterparts for the right side of the tree (equations (2′) and (4′) are useless as well).

▷ **Question 2.** Here, instead of minimizing the time T_f required to execute load W, we aim at determining the maximum amount of work W that can be processed in one time unit. Obviously, after the end of the incoming communication, the parent P_0 should be constantly computing (Equation (3)). We know that all children (i) participate in the computation and (ii) terminate execution at the same time. Finally, the optimal ordering for the children is given by Lemma 8.3. Equation (4) expresses the communication and computation time for each child, using the optimal ordering. This completes the proof. Note that because Equations (3) and (4) are equalities, we could derive a closed-form expression for $w_{-1} = 1/W$.

▷ **Question 3.** First we traverse the tree from bottom to top, replacing each single-level tree by the equivalent node. We do this until there remains a single star. We solve the problem for the star, using the results of Section 8.1.4. Then, we traverse the tree from top to bottom, and undo each transformation in the reverse order. Going back to a reduced node, we know for how long it is supposed to be working. We know the optimal ordering of its children and we know for which amount of time each of the children is supposed to work. If one of these children is a leaf node, we have computed its load. If it is instead a reduced node, we apply the transformation recursively.

Instead of this pair of tree traversals, we could write down the linear program for a general tree, exactly as we did for the example in Figure 8.17. Briefly, when it receives data, a given node knows exactly what to do: Compute itself all the remaining time, and forward data to its children in decreasing bandwidth order. The problem is that the size of the linear program would grow proportionally to the size of the tree. Consequently, the recursive solution is preferred, at least for large platforms.

Exercise 8.2 (Steady-State Scheduling of Multiple Applications)

▷ **Question 1.** Assume first that all application weights are identical (say equal to 1). Maximizing the minimum throughput ensures a fair processing of each application. By contrast, maximizing the sum of the throughputs would unduly favor applications with a large computation-to-communication ratio; executing more tasks of such applications requires less time and fewer resources than for other applications.

Adding weights in the picture simply allows for setting priorities among applications. They receive a fair treatment, as the number of tasks that are executed for a given application is no less than the respective share that is due to this application.

▷ **Question 2.** We use the following variables:

- $\widetilde{\alpha}_u^{(k)}$, the number of tasks of type k executed by P_u per time unit.
- $rec_v^{(k)}$, the number of tasks of type k received by P_v from $P_{p(v)}$ per time unit.

We derive the following equations:

$$
\text{Maximize } \min_k \left\{ \frac{\alpha^{(k)}}{w^{(k)}} \right\} \text{ under the constraints}
$$

$$
\begin{cases}
\forall k, \quad \sum_u \widetilde{\alpha}_u^{(k)} = \alpha^{(k)} & \text{definition of } \alpha^{(k)} \\
\forall k, \forall u \neq 0, \quad rec_u^{(k)} = \widetilde{\alpha}_u^{(k)} + \sum_{v \in \Gamma(u)} rec_v^{(k)} & \text{data movement conservation} \\
\forall u, \quad \sum_k \widetilde{\alpha}_u^{(k)} \cdot c^{(k)} \leqslant c_u & \text{computation limit at node } P_u \\
\forall u, \quad \sum_{v \in \Gamma(u)} \frac{\sum_k rec_v^{(k)} . b^{(k)}}{b_v} \leqslant 1 & \text{communication limit out of } P_u \\
\forall k, u \quad \widetilde{\alpha}_u^{(k)} \geqslant 0 \text{ and } \quad rec_v^{(k)} \geqslant 0 & \text{non-negativity}
\end{cases}
$$

The first equation sums the contribution $\widetilde{\alpha}_u^{(k)}$ of each processor P_u into $\alpha^{(k)}$ for each task type k. The second equation is the conservation law: For each task type k, the number of tasks that a processor receives is equal to what it consumes plus what it sends to all its children. The third equation states that P_u computes all of its load within one time unit. The fourth equation is the counterpart of the third equation, but for communications with the 1-port model: everything is serialized, hence the summation. Note that it would be easy to use the bounded multi-port model instead of the one-port model. We would write

$$\forall u, \quad \sum_{v \in \Gamma(u)} \sum_{k} rec_v^{(k)} \cdot b^{(k)} \leqslant B_u \ ,$$

where B_u is the total bandwidth of the communication card of P_u, and for each incoming link

$$\forall u, \quad \sum_{k} \widetilde{\alpha}_u^{(k)} \cdot b^{(k)} \leqslant b_u \ .$$

Exercise 8.3 (Chains-to-Chains)

▷ **Question 1.** Define the cost of a partitioning as the maximum interval sum. For $1 \leqslant i \leqslant n$ and $1 \leqslant k \leqslant p$, let $g(i,k)$ be the optimal cost when partitioning $[a_1, \ldots, a_i]$ into k intervals. We want to find $g(n,p)$ with initial values $g(i,1) = \sum_{j=1}^{i} a_j$ for $1 \leqslant i \leqslant n$. For $k \geqslant 2$ the recursion is

$$g(i,k) = \min_{1 \leqslant s \leqslant i-1} \left(\max \left(g(s,k-1), \sum_{j=s+1}^{i} a_j \right) \right) .$$

With the above equation we simply try all possible splits into $[1,s]$ (with $k-1$ intervals) and the single interval $[s+1,i]$.

We can speed up computations by pre-computing all values $f(s,i) = \sum_{j=s+1}^{i} a_j$ in time $O(n)$. The complexity becomes $O(np)$.

▷ **Question 2.** Given a bound M, can we partition $[a_1, a_2, \ldots, a_n]$ into p intervals $\mathcal{I}_1, \mathcal{I}_2, \ldots, \mathcal{I}_p$ whose sums are not greater than M? We answer this question with a simple greedy algorithm. For $\mathcal{I}_1$ we start with a_1 and include the next elements until the sum exceeds M. Formally, e_1 is defined by $\sum_{i=1}^{e_1} a_i \leqslant M$ and $\sum_{i=1}^{e_1+1} a_i > M$. For $\mathcal{I}_2$ we start from a_{e_1+1} and repeat the procedure. We have a solution if and only if we reach the last element in p steps. Checking for a solution with M has a cost $O(n)$.

Obviously, a lower bound for M is $L = \max\left(\max_i(a_i), \frac{1}{p}\sum_{i=1}^{n} a_i\right)$. An upper bound is $U = \sum_{i=1}^{n} a_i$. The number of iterations in the binary search for M is bounded by $\log(U-L)$, which is polynomial in the problem size.

Exercise 8.4 (Response Time of One-to-One Mappings)

The problem clearly belongs to NP. We use a reduction from the TRAVELING SALESMAN PROBLEM (TSP), which is NP-complete [57]. Consider an arbitrary instance Inst_1 of TSP, i.e., a complete graph $G = (V, E, c)$, where $c(e)$ is the cost of edge e, a source vertex $s \in V$, a sink vertex $t \in V$, and a bound K. Is there a Hamiltonian path in G from s to t with cost no greater than K?

We build an instance Inst_2 of the one-to-one response time minimization problem. We consider an application with $n = |V|$ identical stages. All application costs are unit costs: $w_i = \delta_i = 1$ for all i. In terms of processors, in addition to P_{in} and P_{out}, we use $p = n = |V|$ identical processors of unit speed: $s_i = 1$ for all i. We simply write i for the processor P_i that corresponds to vertex $v_i \in V$.

We only tinker with the link bandwidths: We interconnect P_{in} and s, P_{out} and t with links of bandwidth 1. We interconnect i and j with a link of bandwidth $\frac{1}{c(e_{i,j})}$. All the other links are very slow (say their bandwidth is smaller than $\frac{1}{K+n+3}$). We ask whether we can achieve a response time $T_{\text{rt}} \leqslant K'$, where $K' = K + n + 2$. Clearly, the size of Inst_2 is linear in the size of Inst_1.

Because we have as many processors as stages, a solution to Inst_2 uses all processors. We need to map the first stage on s and the last one on t, otherwise the input/output cost already exceeds K'. We spend 2 time units for input/output, and n time units for computing (one unit per stage/processor). There remain exactly K time units for inter-processor communications, i.e., for the total cost of the Hamiltonian path that goes from s to t. We cannot use any slow link either. Hence, we have a solution for Inst_2 if and only if we have one for Inst_1.

Exercise 8.5 (Dependence Analysis and Scheduling Vectors)

▷ **Question 1.** Consider the first statement, which writes to A. The only other statement that uses A is S_3. Since S_3 is a read, and since it comes after S_1 in the sequential order, we have found a flow dependence. We proceed similarly for all the instructions, one after the other, and we obtain:

- Flow $S_1 \to S_3$: variable A
- Flow $S_2 \to S_4$: variable D
- Anti $S_1 \to S_3$: variable C
- Flow $S_3 \to S_4$: variable C
- Anti $S_2 \to S_4$: variable E
- Anti $S_2 \to S_5$: variable E (transitive)

- Output $S_4 \to S_5$: variable E

▷ **Question 2.** S_1 writes to $a(i+1, j-1)$ and S_2 reads $a(i,j)$, hence there is a dependence between both statements. We use the method described in Section 8.4.1. If we look for a dependence from $S_1(i,j)$ to $S_2(i',j')$ (it will be a flow dependence), we must have $(i,j) <_{seq} (i',j')$ with $i+1 = i'$ and $j-1 = j'$. If we find such a dependence, its distance vector is $(1,-1)$. Note that if we look for a dependence from $S_2(i,j)$ to $S_1(i',j')$ (hence an anti dependence), we have the equations $i = i'+1$ and $j = j'-1$ with $(i,j) <_{seq} (i',j')$, i.e., $i < i'$ or $i = i'$ and $j < j'$, and hence no solution. With a little care we can compute the difference of the index pairs and orient the dependence in the right direction (that of the sequential order). But watch out for possible mistakes! Consider the case of S_2, which writes to $b(i-1,j)$, and of S_1, which reads $b(i, j+4)$. The first component of the distance vector must be nonnegative, hence we obtain an anti dependence from S_1 to S_2. Finally, we have the following dependences:

- Flow $S_1 \to S_2$: variable a, distance vector $(1,-1)$
- Flow $S_1 \to S_3$: variable a, distance vector $(1,1)$
- Flow $S_3 \to S_1$: variable c, distance vector $(2,3)$
- Anti $S_1 \to S_2$: variable b, distance vector $(1,4)$
- Anti $S_3 \to S_2$: variable b, distance vector $(1,0)$

▷ **Question 3.** The dependence matrix is

$$D = \begin{pmatrix} 1 & 1 & 1 & 1 & 2 \\ -1 & 0 & 1 & 4 & 3 \end{pmatrix}.$$

We look for scheduling vectors $\pi = \begin{pmatrix} a \\ b \end{pmatrix}$ such that $\pi d_i \geqslant 1$ for all $d_i \in D$. We derive the constraints

$$\begin{cases} a - b \geqslant 1 \\ a \geqslant 1 \\ a + b \geqslant 1 \\ a + 4b \geqslant 1 \\ 2a + 3b \geqslant 1 \end{cases}$$

These equations characterize all admissible vectors. Note that we have several choices, including $\begin{pmatrix} 1 \\ 0 \end{pmatrix}$, $\begin{pmatrix} 2 \\ 1 \end{pmatrix}$, $\begin{pmatrix} 5 \\ -1 \end{pmatrix}$, and many others.

It is easy to determine the best vector because the domain Dom is a square. Indeed, with $\pi = \begin{pmatrix} a \\ b \end{pmatrix}$, operation $p = (0,0) \in Dom$ is scheduled at time 0 and operation $p = (N,0) \in Dom$ is scheduled at time aN. Hence, the makespan is at least $|a|N$. If we look for an admissible vector with $a = 1$, the only solution is $b = 0$ (because of the first two constraints). But with $\pi = \begin{pmatrix} 1 \\ 0 \end{pmatrix}$, the makespan is indeed N, so it is the optimal scheduling vector.

To re-write the loop nest, we do nothing, because we can use the identity matrix as the unimodular transformation matrix. We merely mark the second loop as parallel (to show that we have worked hard):

```
for i = 1 to N do
  for j = 1 to N in parallel do
    S1: a(i+1, j−1) ← b(i, j+4)+c(i−2, j−3)+1
    S2: b(i − 1, j) ← a(i, j) − 1
    S3: c(i, j) ← a(i, j − 2) + b(i, j)
```

▷ **Question 4.** We prepare a loop nest of the form

```
for i = 1 to N do
  for j = 1 to N do
    S1: a(i, j) ← ...
    S2: b(i, j) ← ...
    S3: c(i, j) ← ...
```

We need an anti dependence $(1, -2)$ from S_1 to S_2, so we let S_1 read $b(i+1, j-2)$ since S_2 writes to $b(i, j)$. For the flow dependence $(1, -1)$ from S_1 to S_3, we let S_3 read $a(i-1, j+1)$. Proceeding likewise for the other dependences, we obtain a solution:

```
for i = 1 to N do
  for j = 1 to N do
    S1: a(i, j) ← b(i + 1, j − 2)
    S2: b(i, j) ← a(i + 1, j + 2) + c(i, j − 3)
    S3: c(i, j) ← a(i − 1, j + 1) + a(i, j + 1) + b(i − 1, j − 4)
```

Exercise 8.6 (Dependence Removal)

▷ **Question 1.** We have an output dependence from S_2 to S_1, because $a(i+1)$ is written by S_2 at iteration i before being re-written by S_1 at iteration $i+1$. There is also a flow dependence from S_1 to S_2 because of $a(i)$. We get the following RDG:

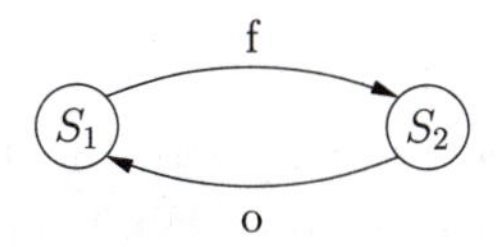

Introducing a temporary variable enables us to "split" statement S_1 and to remove the cycle:

for $i = 1$ **to** N **do**
S_1': $temp(i) \leftarrow b(i) + c(i)$
S_1: $a(i) \leftarrow temp(i)$
S_2: $a(i+1) \leftarrow temp(i) + 2d(i)$

The RDG of the new loop nest is as follows:

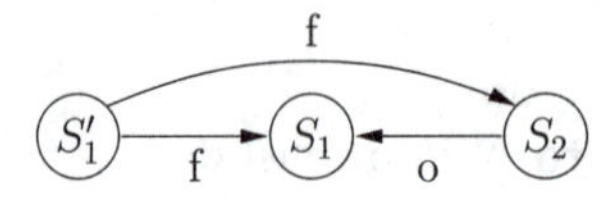

The cycle has been broken. We can split (or "distribute") the loop to exhibit the parallelism:

for $i = 1$ **to** N **do**
S_1': $temp(i) \leftarrow b(i) + c(i)$
$a(1) \leftarrow temp(1)$
for $i = 1$ **to** N **do**
S_2: $a(i+1) \leftarrow temp(i) + 2d(i)$

▷ **Question 2.** We examine the six possibilities. We obtain the following transformations:

- **Flow dependence incoming to** S_k (Figure 8.19). Assume that one of the variables that are read in the right-hand side of S_k has been produced (written) by the left-hand side of S_i. After the split this variable is read in the right-hand side of S_k'; the flow dependence now points to S_k'. Also, do not forget the new flow dependence induced by *temp* from S_k' to S_k.

- **Anti dependence incoming to** S_k (Figure 8.20). Assume that statement S_i reads $lhs(f(i))$ before S_k writes to it. After the split, $lhs(f(i))$ is still read in S_k, and the anti dependence from S_i to S_k is unchanged.

- **Output dependence incoming to** S_k (Figure 8.21). Assume that statement S_i writes to $lhs(f(i))$ before S_k does. After the split this output dependence from S_i to S_k is unchanged.

- **Flow dependence outgoing from** S_k (Figure 8.22). Assume that statement S_i reads the value of $lhs(f(i))$ produced by S_k. After the split the access to $lhs(f(i))$ in S_i has been replaced by an access to $temp(f(i))$, whose value is produced by S_k'. There is a flow dependence from S_k' to S_i.

- **Anti dependence outgoing from** S_k (Figure 8.23). Assume that a variable read in the right-hand side of S_k is later written in the left-hand side of S_i. After the split this variable is read in the right-hand side of S_k', and the anti dependence now has its source in S_k' (but still points to S_i).

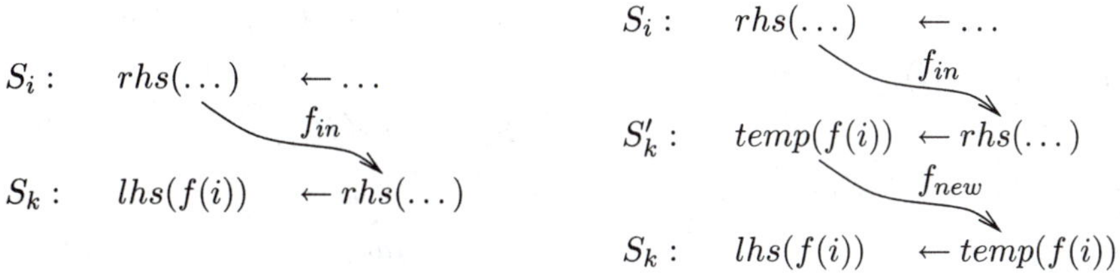

$$\begin{array}{lll} S_i: & rhs(\ldots) & \leftarrow \ldots \\ & \quad f_{in} & \\ S_k: & lhs(f(i)) & \leftarrow rhs(\ldots) \end{array}
\qquad
\begin{array}{lll} S_i: & rhs(\ldots) & \leftarrow \ldots \\ & \quad f_{in} & \\ S_k': & temp(f(i)) & \leftarrow rhs(\ldots) \\ & \quad f_{new} & \\ S_k: & lhs(f(i)) & \leftarrow temp(f(i)) \end{array}$$

FIGURE 8.19: Flow dependence incoming to S_k, before and after node splitting.

$$\begin{array}{lll} S_i: & & \leftarrow lhs(\ldots) \\ & \quad a_{in} & \\ S_k: & lhs(f(i)) & \leftarrow rhs(\ldots) \end{array}
\qquad
\begin{array}{lll} S_i: & & \leftarrow lhs(\ldots) \\ S_k': & temp(f(i)) & \leftarrow rhs(\ldots) \\ & a_{in} \quad f_{new} & \\ S_k: & lhs(f(i)) & \leftarrow temp(f(i)) \end{array}$$

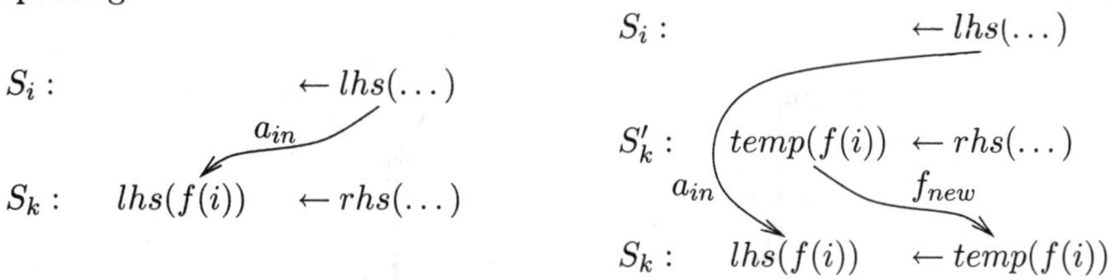

FIGURE 8.20: Anti dependence incoming to S_k, before and after node splitting.

$$\begin{array}{lll} S_i: & lhs(\ldots) & \leftarrow \ldots \\ & \quad o_{in} & \\ S_k: & lhs(f(i)) & \leftarrow rhs(\ldots) \end{array}
\qquad
\begin{array}{lll} S_i: & lhs(\ldots) & \leftarrow \ldots \\ S_k': & temp(f(i)) & \leftarrow rhs(\ldots) \\ & o_{in} \quad f_{new} & \\ S_k: & lhs(f(i)) & \leftarrow temp(f(i)) \end{array}$$

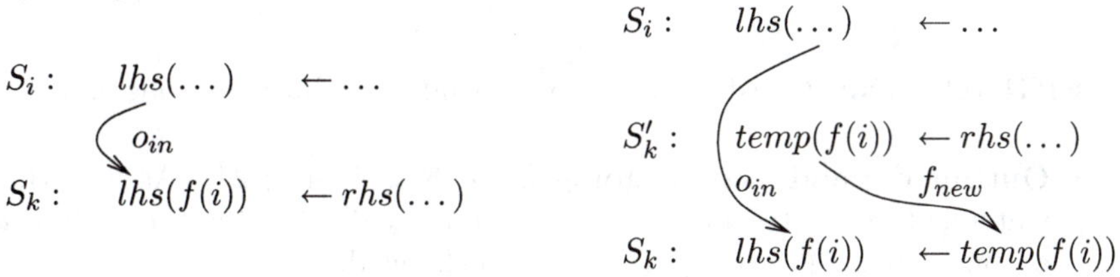

FIGURE 8.21: Output dependence incoming to S_k, before and after node splitting.

$$\begin{array}{lll} S_k: & lhs(f(i)) & \leftarrow rhs(\ldots) \\ & \quad f_{out} & \\ S_i: & & \leftarrow lhs(\ldots) \end{array}
\qquad
\begin{array}{lll} S_k': & temp(f(i)) & \leftarrow rhs(\ldots) \\ & \quad f_{new} & \\ S_k: & lhs(f(i)) & \leftarrow temp(f(i)) \\ & \quad f_{out} & \\ S_i: & & \leftarrow temp(\ldots) \end{array}$$

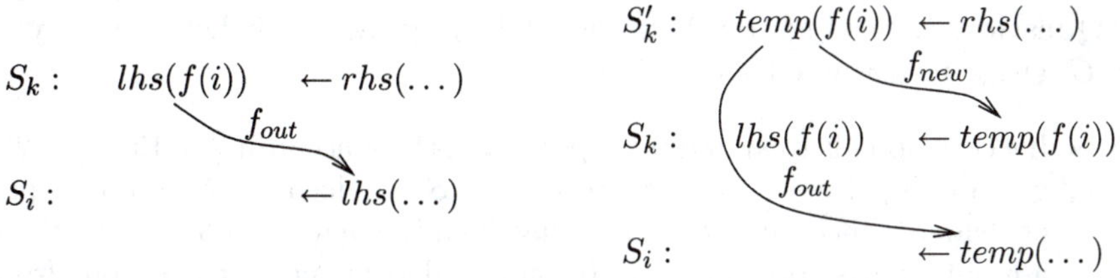

FIGURE 8.22: Flow dependence outgoing from S_k, before and after node splitting.

$$\begin{array}{lll} S_k: & lhs(f(i)) & \leftarrow rhs(\ldots) \\ & \quad a_{out} & \\ S_i: & rhs(\ldots) & \leftarrow \end{array}
\qquad
\begin{array}{lll} S_k': & temp(f(i)) & \leftarrow rhs(\ldots) \\ & \quad f_{new} & \\ S_k: & lhs(f(i)) & \leftarrow temp(f(i)) \\ & & \quad a_{out} \\ S_i: & rhs(\ldots) & \leftarrow \end{array}$$

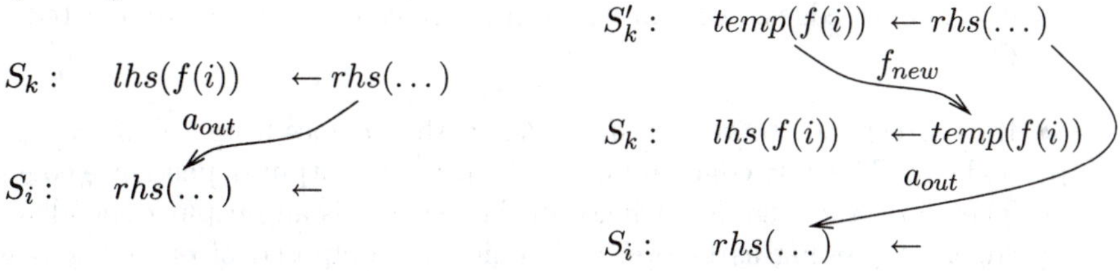

FIGURE 8.23: Anti dependence outgoing from S_k, before and after node splitting.

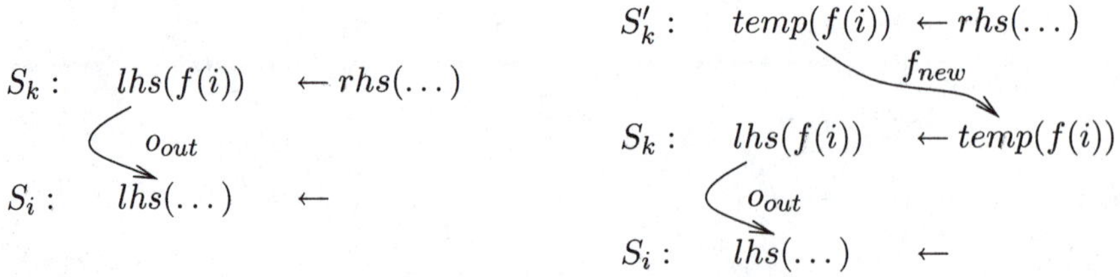

$S_k: \quad lhs(f(i)) \quad \leftarrow rhs(\ldots)$

$\quad o_{out}$

$S_i: \quad lhs(\ldots) \quad \leftarrow$

$S'_k: \quad temp(f(i)) \leftarrow rhs(\ldots)$

$\quad f_{new}$

$S_k: \quad lhs(f(i)) \quad \leftarrow temp(f(i))$

$\quad o_{out}$

$S_i: \quad lhs(\ldots) \quad \leftarrow$

FIGURE 8.24: Output dependence outgoing from S_k, before and after node splitting.

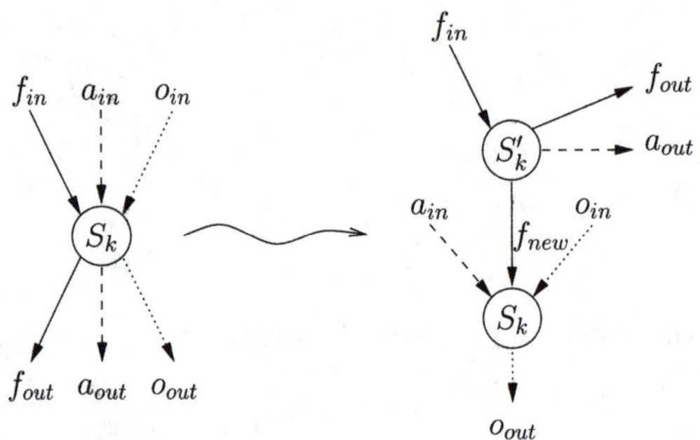

FIGURE 8.25: Transformations of dependences during a node split.

- **Output dependence outgoing from** S_k (Figure 8.24). Assume that statement S_i writes into $lhs(f(i))$ after S_k did. After the split this output dependence from S_k to S_i is unchanged.

All these transformations are synthesized in Figure 8.25.

▷ **Question 3.** See Figure 8.25 for help following the proof. Let C be a cycle in G' and consider an edge e of C:

- If e corresponds to an *output dependence*, then, according to Figure 8.25, e is an edge from a node S_k to a node S_i. Indeed, there is no output dependence incoming to or outgoing from the new node S'_k. Next, there are only edges corresponding to output dependences that go out from S_i. Therefore, the edge following e in the cycle C also corresponds to an output dependence. In conclusion, C is uniquely composed of output dependence edges. And all edges in C are edges that already existed in G.

- If e corresponds to an *anti dependence*, then e goes from a node S'_k to a node S_i. The only edges going out from S_i are output dependence edges. Therefore, the edge following e in the cycle C is an output dependence edge. Reasoning as before, C is uniquely composed of output dependence edges, which contradicts our assumption regarding e. Therefore, no cycle in G' may include an anti dependence edge.

- If e corresponds to a *flow dependence*, then either e has been created when splitting a node S_k (and then goes from S'_k to S_k), or e corresponds to an existing edge going from S_k to S_i in G (and then goes from S'_k to S'_i). We study both cases:
  - $e : S'_k \xrightarrow{f_{new}} S_k$: All edges going out from S_k correspond to output dependences, and the edge following e in the cycle C is an output dependence edge. Reasoning as before, C is uniquely composed of output dependence edges, a contradiction.
  - $e : S'_k \xrightarrow{f} S'_i$: There may exist flow or anti dependence edges going out from S'_i. But we have proved that there is no anti dependence edge in a cycle of G', hence the edge following e in the cycle C corresponds to a flow dependence. Therefore, C is uniquely composed of flow dependence edges. Furthermore, edges in C do not correspond to the new dependences introduced by the splits, thus they all were already present in G.

In summary, the splitting technique has not introduced any new dependence cycle. It has made it possible to break all cycles, except those uniquely composed of flow dependences and those uniquely composed of output dependences.

▷ **Question 4.** The RDG is shown in Figure 8.26(a).

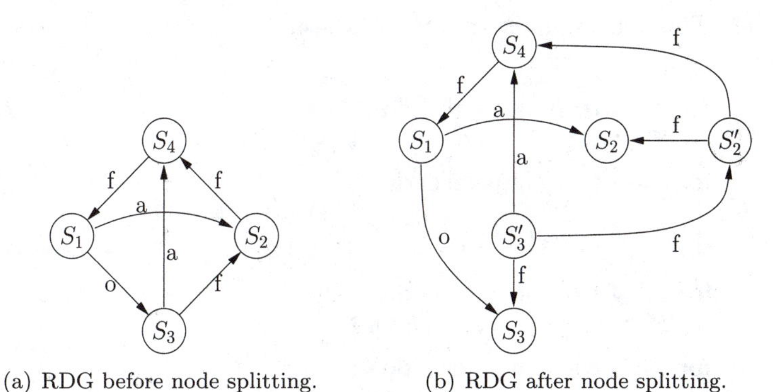

(a) RDG before node splitting. (b) RDG after node splitting.

FIGURE 8.26: RDG before and after node splitting.

There are six dependences in the nest:

- three flow dependences (from S_3 to S_2 because of variable a, from S_2 to S_4 because of b, and from S_4 to S_1 because of c);
- two anti dependences (from S_1 to S_2 because of b, and from S_3 to S_4 because of c);

- one output dependence (from S_1 to S_3 because of a).

The RDG is strongly connected. Splitting nodes S_2 and S_3 leads to the RDG shown in Figure 8.26(b).

There is no cycle in this new graph. We re-write the loop nest using temporary variables a_{temp} (for the split of S_3) and b_{temp} (for the split of S_2):

for $i = 4$ **to** N **do**

$$
\begin{aligned}
&S_1\text{: } a(i+5) \leftarrow c(i-3) + b(2i+2) \\
&S_2'\text{: } b_{\text{temp}}(2i) \leftarrow \begin{cases} a_{\text{temp}}(i-1)+1 \text{ if } i \geqslant 5 \\ a(i-1)+1 \text{ if } i = 4 \end{cases} \\
&S_2\text{: } b(2i) \leftarrow b_{\text{temp}}(2i) \\
&S_3'\text{: } a_{\text{temp}}(i) \leftarrow c(i+5)+1 \\
&S_3\text{: } a(i) \leftarrow a_{\text{temp}}(i) \\
&S_4\text{: } c(i) \leftarrow \begin{cases} b_{\text{temp}}(2i-4) \text{ if } i \geqslant 6 \\ b(2i-4) \text{ if } i \leqslant 5 \end{cases}
\end{aligned}
$$

Here, conditional assignments are necessary to take into account dependences originating from outside the loop nest.

What can we do now that we have broken all cycles? We can topologically sort the nodes, and get the ordering S_3', S_2', S_4, S_1, S_2, and S_3. Now, all dependences flow from a statement to statements following it in the list; there is no feedback edge. Hence, we can process the statements one after the other, because by doing so we only delay the operations of the following statements, which is safe. Because there are no self-dependences, all loops can be made parallel. This technique is known as *loop distribution*. For the example, we obtain

for $i = 4$ **to** N in parallel **do**

$\quad S_3'\text{: } a_{\text{temp}}(i) \leftarrow c(i+5)+1$

for $i = 4$ **to** N in parallel **do**

$$S_2'\text{: } b_{\text{temp}}(2i) \leftarrow \begin{cases} a_{\text{temp}}(i-1)+1 \text{ if } i \geqslant 5 \\ a(i-1)+1 \text{ if } i = 4 \end{cases}$$

for $i = 4$ **to** N in parallel **do**

$\quad S_3'\text{: } a_{\text{temp}}(i) \leftarrow c(i+5)+1$

for $i = 4$ **to** N in parallel **do**

$$S_4\text{: } c(i) \leftarrow \begin{cases} b_{\text{temp}}(2i-4) \text{ if } i \geqslant 6 \\ b(2i-4) \text{ if } i \leqslant 5 \end{cases}$$

for $i = 4$ **to** N in parallel **do**

$\quad S_1\text{: } a(i+5) \leftarrow c(i-3) + b(2i+2)$

for $i = 4$ **to** N in parallel **do**

$\quad S_2\text{: } b(2i) \leftarrow b_{\text{temp}}(2i)$

for $i = 4$ **to** N in parallel **do**

$\quad S_3\text{: } a(i) \leftarrow a_{\text{temp}}(i)$

Bibliography

[1] R. Agarwal, F. Gustavson, and M. Zubair. A High Performance Matrix Multiplication Algorithm on a Distributed-Memory Parallel Computer, Using Overlapped Communication. *IBM J. Research and Development*, 38(6):673–681, 1994.

[2] M. Ajtai, J. Komlos, and E. Szemeredi. An $O(n \log n)$ Sorting Network. In *Proc. of the 15th ACM Symposium on the Theory of Computing*, pages 1–19. ACM Press, 1983.

[3] S. G. Akl. *The Design and Analysis of Parallel Algorithms.* Prentice Hall, 1989.

[4] A. Alexandrov, M. Ionescu, K. Schauser, and C. Scheiman. LogGP: Incorporating Long Messages into the LogP model for Parallel Computation. *J. Parallel and Distributed Computing*, 44(1):71–79, 1997.

[5] R. Allen and K. Kennedy. *Optimizing Compilers for Modern Architectures.* Academic Press, 2002.

[6] G. Amdahl. The Validity of the Single Processor Approach to Achieving Large Scale Computing Capabilities. In *Proc. of the AFIPS Conference*, volume 30, pages 483–485. AFIPS Press, 1967.

[7] D. P. Anderson, J. Cobb, E. Korpela, M. Lebofsky, and D. Werthimer. SETI@home: An Experiment in Public-Resource Computing. *Communications of the ACM*, 45(11):56–61, November 2002.

[8] E. Anderson, Z. Bai, C. Bischof, J. Demmel, J. Dongarra, J. D. Croz, A. Greenbaum, S. Hammarling, A. McKenney, S. Ostrouchov, and D. Sorensen. *LAPACK Users' Guide, Second Edition.* SIAM, 1995. See also http://www.netlib.org/lapack.

[9] G. Ausiello, P. Crescenzi, G. Gambosi, V. Kann, A. Marchetti-Spaccamela, and M. Protasi. *Complexity and Approximation.* Springer-Verlag, 1999.

[10] U. Banerjee. *Loop Parallelization.* Kluwer Academic Publishers, 1994.

[11] M. Banikazemi, V. Moorthy, and D. K. Panda. Efficient Collective Communication on Heterogeneous Networks of Workstations. In *Proc. of the 27th International Conference on Parallel Processing*, pages 460–467. IEEE Computer Society Press, 1998.

[12] M. Banikazemi, J. Sampathkumar, S. Prabhu, D. Panda, and P. Sadayappan. Communication Modeling of Heterogeneous Networks of Workstations for Performance Characterization of Collective Operations. In *Proc. of the 8th Heterogeneous Computing Workshop*, pages 125–133. IEEE Computer Society Press, 1999.

[13] C. Banino, O. Beaumont, L. Carter, J. Ferrante, A. Legrand, and Y. Robert. Scheduling Strategies for Master-Slave Tasking on Heterogeneous Processor Platforms. *IEEE Trans. Parallel Distributed Systems*, 15(4):319–330, 2004.

[14] A. Bar-Noy, S. Guha, J. S. Naor, and B. Schieber. Message Multicasting in Heterogeneous Networks. *SIAM J. Computing*, 30(2):347–358, 2000.

[15] K. Batcher. Sorting Networks and their Applications. In *Proc. of the AFIP Spring Joint Computing Conference*, volume 32, pages 307–314. AFIPS Press, 1968.

[16] O. Beaumont, V. Boudet, A. Petitet, F. Rastello, and Y. Robert. A Proposal for a Heterogeneous Cluster ScaLAPACK (Dense Linear Solvers). *IEEE Trans. Computers*, 50(10):1052–1070, 2001.

[17] O. Beaumont, V. Boudet, F. Rastello, and Y. Robert. Matrix Multiplication on Heterogeneous Platforms. *IEEE Trans. Parallel Distributed Systems*, 12(10):1033–1051, 2001.

[18] O. Beaumont, L. Carter, J. Ferrante, A. Legrand, L. Marchal, and Y. Robert. Centralized versus Distributed Schedulers for Multiple Bag-of-Task Applications. In *Proc. of the International Parallel and Distributed Processing Symposium*. IEEE Computer Society Press, 2006.

[19] O. Beaumont, H. Casanova, A. Legrand, Y. Robert, and Y. Yang. Scheduling Divisible Loads on Star and Tree Networks: Results and Open Problems. *IEEE Trans. Parallel Distributed Systems*, 16(3):207–218, 2005.

[20] O. Beaumont, A. Legrand, L. Marchal, and Y. Robert. Pipelining Broadcasts on Heterogeneous Platforms. *IEEE Trans. Parallel Distributed Systems*, 16(4):300–313, 2005.

[21] O. Beaumont, A. Legrand, L. Marchal, and Y. Robert. Steady-State Scheduling on Heterogeneous Clusters. *Int. J. Foundations of Computer Science*, 16(2):163–194, 2005.

[22] A. Benoit, V. Rehn, and Y. Robert. Complexity Results for Throughput and Latency Optimization of Replicated and Data-Parallel Workflows. In *Proc. of the International Conference on Heterogeneous Computing (HeteroPar)*. IEEE Computer Society Press, 2007.

[23] A. Benoit, V. Rehn-Sonigo, and Y. Robert. Multi-Criteria Scheduling of Pipeline Workflows. In *Proc. of the International Conference on*

Heterogeneous Computing (HeteroPar), jointly published with the IEEE Cluster Conference. IEEE Computer Society Press, 2007.

[24] A. Benoit and Y. Robert. Mapping pipeline skeletons onto heterogeneous platforms. *J. Parallel and Distributed Computing*, 68(6):790–808, 2008.

[25] F. Berman, G. Fox, and T. Hey, editors. *Grid Computing: Making the Global Infrastructure a Reality*. John Wiley & Sons, 2003.

[26] F. Berman, R. Wolski, H. Casanova, W. Cirne, H. Dail, M. Faerman, S. Figueira, J. Hayes, G. Obertelli, J. Schopf, G. Shao, S. Smallen, N. Spring, A. Su, and D. Zagorodnov. Adaptive Computing on the Grid Using AppLeS. *IEEE Trans. Parallel Distributed Systems*, 14(4):369–382, 2003.

[27] A. J. Bernstein. Analysis of Programs for Parallel Processing. *IEEE Trans. Electronic Computers*, 15(5):757–762, 1966.

[28] D. Bertsekas and R. Gallager. *Data Networks*. Prentice Hall, 1987.

[29] D. Bertsimas and D. Gamarnik. Asymptotically Optimal Algorithms for Job Shop Scheduling and Packet Routing. *J. Algorithms*, 33(2):296–318, 1999.

[30] M. Beynon, T. Kurc, A. Sussman, and J. Saltz. Optimizing Execution of Component-Based Applications Using Group Instances. *Future Generation Computer Systems*, 18(4):435–448, 2002.

[31] M. Beynon, A. Sussman, U. Catalyurek, T. Kurc, and J. Saltz. Performance Optimization for Data Intensive Grid Applications. In *Proc. of the Third Annual International Workshop on Active Middleware Services*, pages 97–105. IEEE Computer Society Press, 2001.

[32] V. Bharadwaj, D. Ghose, V. Mani, and T. Robertazzi. *Scheduling Divisible Loads in Parallel and Distributed Systems*. IEEE Computer Society Press, 1996.

[33] V. Bharadwaj, D. Ghose, and T. Robertazzi. Divisible Load Theory: a New Paradigm for Load Scheduling in Distributed Systems. *Cluster Computing*, 6(1):7–17, 2003.

[34] P. Bhat, C. Raghavendra, and V. Prasanna. Efficient Collective Communication in Distributed Heterogeneous Systems. In *Proc. of the 19th International Conference on Distributed Computing Systems*, pages 15–24. IEEE Computer Society Press, 1999.

[35] P. Bhat, C. Raghavendra, and V. Prasanna. Efficient Collective Communication in Distributed Heterogeneous Systems. *J. Parallel and Distributed Computing*, 63:251–263, 2003.

[36] L. S. Blackford, J. Choi, A. Cleary, E. D'Azevedo, J. Demmel, I. Dhillon, J. Dongarra, S. Hammarling, G. Henry, A. Petitet, K. Stanley, D. Walker, and R. C. Whaley. *ScaLAPACK Users' Guide.* SIAM, 1997.

[37] P. Brucker. *Scheduling Algorithms.* Springer-Verlag, 2004.

[38] L. E. Cannon. *A Cellular Computer to Implement the Kalman Filter Algorithm.* PhD thesis, Montana State University, 1969.

[39] B. W. Char, K. O. Geddes, G. H. Gonnet, M. B. Monagan, and S. M. Watt. *Maple Reference Manual.* Watcom Publications, 1988.

[40] P. Chrétienne. Task Scheduling with Interprocessor Communication Delays. *European J. Operational Research*, 57(3):348–354, 1992.

[41] P. Chrétienne and C. Picouleau. Scheduling with Communication Delays: A Survey. In P. Chrétienne, E. G. C. Jr., J. K. Lenstra, and Z. Liu, editors, *Scheduling Theory and its Applications*, pages 65–89. John Wiley & Sons, 1995.

[42] E. G. Coffman. *Computer and Job-Shop Scheduling Theory.* John Wiley & Sons, 1976.

[43] R. Cole. Parallel Merge Sort. In *Proc. of the 27th Annual Symposium on Foundations of Computer Science*, pages 511–516. IEEE Computer Society Press, 1986.

[44] T. H. Cormen, C. E. Leiserson, and R. L. Rivest. *Introduction to Algorithms.* MIT Press, 1990.

[45] M. Cosnard and D. Trystram. *Algorithmes et Architectures Parallèles.* Interéditions, 1993.

[46] P. Crescenzi and V. Kann. A Compendium of NP Optimization Problems. World Wide Web document: http://www.nada.kth.se/~viggo/wwwcompendium/wwwcompendium.html.

[47] D. Culler, R. Karp, D. Patterson, A. Sahay, E. Santos, K. Schauser, R. Subramonian, and T. von Eicken. LogP: A Practical Model of Parallel Computation. *Communication of the ACM*, 39(11):78–85, 1996.

[48] D. E. Culler and J. P. Singh. *Parallel Computer Architecture: A Hardware/Software Approach.* Morgan Kaufmann, 1999.

[49] A. Darte, Y. Robert, and F. Vivien. *Scheduling and Automatic Parallelization.* Birkhaüser, 2000.

[50] F. Desprez. *Procédures de base pour le calcul scientifique sur machines parallèles à mémoire distribuée.* PhD thesis, École Normale Supérieure de Lyon, Jan. 1994.

[51] H. El-Rewini and M. Abd-el barr. *Advanced Computer Architecture and Parallel Processing.* Wiley-Interscience, 2005.

[52] H. El-Rewini and T. G. Lewis. *Distributed and Parallel Computing.* Manning, 1998.

[53] H. El-Rewini, T. G. Lewis, and H. H. Ali. *Task Scheduling in Parallel and Distributed Systems.* Prentice Hall, 1994.

[54] I. Foster and C. Kesselman, editors. *The Grid 2: Blueprint for a New Computing Infrastructure.* Morgan Kaufman, 2nd edition, 2003.

[55] G. Fox, S. Otto, and A. Hey. Matrix Algorithms on a Hypercube I: Matrix Multiplication. *Parallel Computing*, 4(1):17–31, 1987.

[56] P. Fraigniaud and P. Gauron. D2B: A de Bruijn Based Content-Addressable Network. *Theoretical Computer Science*, 355(1):65–79, 2006.

[57] M. R. Garey and D. S. Johnson. *Computers and Intractability, a Guide to the Theory of NP-Completeness.* W. H. Freeman and Company, 1991.

[58] A. Geist, A. Beguelin, J. Dongarra, W. Jiang, R. Manchek, and V. Sunderam. *PVM Parallel Virtual Machine: A Users'Guide and Tutorial for Networked Parallel Computing.* MIT Press, 1996.

[59] M. Gengler, S. Ubéda, and F. Desprez. *Initiation au Parallélisme.* Masson, 1996.

[60] A. Gerasoulis and T. Yang. A Comparison of Clustering Heuristics for Scheduling DAGs on Multiprocessors. *J. Parallel and Distributed Computing*, 16(4):276–291, 1992.

[61] A. Gibbons and W. Rytter. *Efficient Parallel Algorithms.* Cambridge University Press, 1988.

[62] GLPK: GNU Linear Programming Kit. http://www.gnu.org/software/glpk/.

[63] G. H. Golub and C. F. V. Loan. *Matrix Computations.* Johns Hopkins University Press, 1989.

[64] R. L. Graham. Bounds on Multiprocessing Timing Anomalies. *SIAM J. Applied Mathematics*, 17(2):416–429, 1969.

[65] J. L. Gustafson. Reevaluating Amdahl's Law. *IBM Systems J.*, 31(5):532–533, 1988.

[66] C. Hanen and A. Munier. An Approximation Algorithm for Scheduling Dependent Tasks on m Processors with Small Communication Delays. In *Proc. of the INRIA/IEEE Symposium on Emerging Technology and Factory Animation*, pages 167–189. IEEE Computer Science Press, 1995.

[67] R. W. Hockney. The Communication Challenge for MPP: Intel Paragon and Meiko CS-2. *Parallel Computing*, 20(3):389–398, 1994.

[68] B. Hong and V. Prasanna. Distributed Adaptive Task Allocation in Heterogeneous Computing Environments to Maximize Throughput. In *Proc. of the International Parallel and Distributed Processing Symposium*. IEEE Computer Society Press, 2004.

[69] T. C. Hu. Parallel Sequencing and Assembly Line Problems. *Operations Research*, 9(6):841–848, 1961.

[70] M. F. Kaashoek and D. R. Karger. Koorde: A Simple Degree-Optimal Distributed Hash Table. In *Proc. of the 2nd International Workshop on Peer-to-Peer Systems*, volume 2735 of *Lecture Notes in Computer Science*, pages 98–107. Springer-Verlag, 2003.

[71] R. M. Karp, R. E. Miller, and S. Winograd. The Organization of Computations for Uniform Recurrence Equations. *J. of the ACM*, 14(3):563–590, 1967.

[72] S. Khuller and Y. A. Kim. On Broadcasting in Heterogenous Networks. In *Proc. of the fifteenth Annual ACM-SIAM Symposium on Discrete Algorithms*, pages 1011–1020. SIAM, 2004.

[73] T. Kielmann, H. E. Bal, and K. Verstoep. Fast Measurement of LogP Parameters for Message Passing Platforms. In *Proc. of the 15th IPDPS. Workshops on Parallel and Distributed Processing*, volume 1800 of *Lecture Notes in Computer Science*, pages 1176–1183. Springer-Verlag, 2000.

[74] D. E. Knuth. *The Art of Computer Programming, volume 3: Sorting and Searching.* Addison-Wesley, 1973.

[75] C. H. Koelbel, D. B. Loveman, R. S. Schreiber, G. L. S. Jr., and M. E. Zosel. *The High Performance Fortran Handbook.* MIT Press, 1994.

[76] V. Kumar, A. Grama, A. Gupta, and G. Karypis. *Introduction to Parallel Computing.* Benjamin Cummings, 1994.

[77] L. Lamport. The Parallel Execution of DO Loops. *Communications of the ACM*, 17(2):83–93, 1974.

[78] A. Legrand, H. Renard, Y. Robert, and F. Vivien. Mapping and Load-Balancing Iterative Computations. *IEEE Trans. Parallel and Distributed Systems*, 15(6):546–558, 2004.

[79] A. Legrand, Y. Yang, and H. Casanova. NP-Completeness of the Divisible Load Scheduling Problem on Heterogeneous Star Platforms with Affine Costs. Research Report CS2005-0818, Computer Science and Engineering Dept., University of California, San Diego, March 2005.

[80] F. T. Leighton. *Introduction to Parallel Algorithms and Architectures: Arrays, Trees, Hypercubes.* Morgan Kaufmann, 1992.

[81] C. E. Leiserson. Fat-Trees: Universal Networks for Hardware-Efficient Supercomputing. *IEEE Trans. Computers*, 34(10):892–901, 1985.

[82] T. G. Lewis and H. El-Rewini. *Introduction to Parallel Computing.* Prentice Hall, 1992.

[83] C. Lin and L. Snyder. A Matrix Product Algorithm and its Comparative Performance on Hypercubes. In *Proc. of the Scalable High Performance Computing Conference*, pages 190–193. IEEE Computer Society Press, 1992.

[84] P. Liu. Broadcast Scheduling Optimization for Heterogeneous Cluster Systems. *J. Algorithms*, 42(1):135–152, 2002.

[85] S. H. Low. A Duality Model of TCP and Queue Management Algorithms. *IEEE/ACM Trans. Networking*, 4(11):525–536, 2003.

[86] D. Malkhi, M. Naor, and D. Ratajczak. Viceroy: A Scalable and Dynamic Emulation of the Butterfly. In *Proc. of the 21st annual ACM Symposium on Principles of Distributed Computing.* ACM Press, 2002.

[87] F. Manne and T. Sørevik. Partitioning an Array onto a Mesh of Processors. In *Proc. of the Workshop on Applied Parallel Computing in Industrial Problems and Optimization*, volume 1184 of *Lecture Notes in Computer Science*, pages 467–477. Springer-Verlag, 1996.

[88] L. Massoulié and J. Roberts. Bandwidth Sharing: Objectives and Algorithms. *IEEE/ACM Trans. Networking*, 10(3):320–328, 2002.

[89] J. A. M. McHugh. Hu's Precedence Tree Scheduling Algorithm: A Simple Proof. *Naval Research Logistics Quaterly*, 31:409–411, 1984.

[90] S. Miguet and Y. Robert. Path Planning on a Ring of Processors. *Int. J. Computer Mathematics*, 32(1-2):61–74, 1990.

[91] S. Miguet and Y. Robert. Parallélisation d'algorithmes de balayage d'image sur un anneau de processeurs. *Technique et Science Informatiques*, 10(4):287–296, 1991.

[92] D. A. Padua and M. J. Wolfe. Advanced Compiler Optimizations for Supercomputers. *Communications of the ACM*, 29(12):1184–1201, 1986.

[93] C. Picouleau. Two New NP-Complete Scheduling Problems with Communication Delays and Unlimited Number of Processors. Technical Report 91-24, IBP, Université Pierre et Marie Curie, France, Apr. 1991.

[94] C. Picouleau. Task Scheduling with Interprocessor Communication Delays. *Discrete Applied Mathematics*, 60(1-3):331–342, 1995.

[95] A. Pinar and C. Aykanat. Fast Optimal Load Balancing Algorithms for 1D Partitioning. *J. Parallel Distributed Computing*, 64(8):974–996, 2004.

[96] C. G. Plaxton, R. Rajaraman, and A. W. Richa. Accessing Nearby Copies of Replicated Objects in a Distributed Environment. In *Proc. of the ACM Symposium on Parallel Algorithms and Architectures*, pages 311–320. ACM Press, 1997.

[97] W. Press, S. Teukolsky, W. Vetterling, and B. Flannery. *Numerical Recipes: The Art of Scientific Computing.* Cambridge University Press, third edition, 2007.

[98] F. Rastello. *Partitionnement: optimisations de compilation et algorithmique hétérogène.* PhD thesis, École Normale Supérieure de Lyon, Sept. 2000.

[99] S. Ratnasamy, P. Francis, M. Handley, R. Karp, and S. Shenker. A Scalable Content-Addressable Network. In *Proc. of the ACM SIGCOMM Conference*, pages 161–172. ACM Press, 2001.

[100] J. Reif. *Synthesis of Parallel Algorithms.* Morgan Kaufmann, 1993.

[101] Y. Robert. *The Impact of Vector and Parallel Architectures on the Gaussian Elimination Algorithm.* Manchester University Press and John Wiley & Sons, 1991.

[102] T. Robertazzi. Ten Reasons to Use Divisible Load Theory. *IEEE Computer*, 36(5):63–68, 2003.

[103] A. Rowstron and P. Druschel. Pastry: Scalable, Decentralized Object Location, and Routing for Large-Scale Peer-to-Peer Systems. In *Proc. of the IFIP/ACM International Conference on Distributed Systems Platforms*, volume 2218 of *Lecture Notes in Computer Science*, pages 329–350. Springer-Verlag, 2001.

[104] A. I. T. Rowstron, A.-M. Kermarrec, M. Castro, and P. Druschel. SCRIBE: The Design of a Large-Scale Event Notification Infrastructure. In *Proc. of the Third International COST264 Workshop on Networked Group Communication*, volume 2233 of *Lecture Notes in Computer Science*, pages 30–43, 2001.

[105] T. Saif and M. Parashar. Understanding the Behavior and Performance of Non-Blocking Communications in MPI. In *Proc. of the 10th International Euro-Par Conference*, volume 3149 of *Lecture Notes in Computer Science*, pages 173–182. Springer-Verlag, 2004.

[106] A. Schrijver. *Theory of Linear and Integer Programming.* John Wiley & Sons, 1986.

[107] B. A. Shirazi, A. R. Hurson, and K. M. Kavi. *Scheduling and Load Balancing in Parallel and Distributed Systems.* IEEE Computer Science Press, 1995.

[108] O. Sinnen. *Task Scheduling for Parallel Systems.* John Wiley & Sons, 2007.

[109] M. Snir, S. W. Otto, S. Huss-Lederman, D. W. Walker, and J. Dongarra. *MPI: The Complete Reference.* MIT Press, 1996.

[110] M. Spencer, R. Ferreira, M. Beynon, T. Kurc, U. Catalyurek, A. Sussman, and J. Saltz. Executing Multiple Pipelined Data Analysis Operations in the Grid. In *Proc. of the ACM/IEEE Supercomputing Conference.* ACM Press, 2002.

[111] R. Steinmetz and K. Wehrle, editors. *Peer-to-Peer Systems and Applications*, volume 3485 of *Lecture Notes in Computer Science.* Springer-Verlag, 2005.

[112] I. Stoica, R. Morris, D. Karger, F. Kaashoek, and H. Balakrishnan. Chord: A Scalable Peer-To-Peer Lookup Service for Internet Applications. In *Proc. of the ACM SIGCOMM Conference*, pages 149–160. ACM Press, 2001.

[113] J. Subhlok and G. Vondran. Optimal Mapping of Sequences of Data Parallel Tasks. In *Proc. of the 5th ACM SIGPLAN Symposium on Principles and Practice of Parallel Programming*, pages 134–143. ACM Press, 1995.

[114] J. Subhlok and G. Vondran. Optimal Latency-Throughput Tradeoffs for Data Parallel Pipelines. In *Proc. of the ACM Symposium on Parallel Algorithms and Architectures*, pages 62–71. ACM Press, 1996.

[115] H. Topcuoglu, S. Hariri, and M. Y. Wu. Performance-Effective and Low-Complexity Task Scheduling for Heterogeneous Computing. *IEEE Trans. Parallel Distributed Systems*, 13(3):260–274, 2002.

[116] R. van de Geijn. *Using PLAPACK: Parallel Linear Algebra Package.* MIT Press, 1997.

[117] N. Vydyanathan, U. Catalyurek, T. Kurc, P. Saddayappan, and J. Saltz. An Approach for Optimizing Latency under Throughput Constraints for Application Workflows on Clusters. In *Proc. of the 13th International Euro-Par Conference*, volume 4641 of *Lecture Notes in Computer Science*, pages 173–183. Springer-Verlag, 2007.

[118] M. Waterman and T. Smith. Identification of Common Molecular Subsequences. *J. Molecular Biology*, 147:195–197, 1981.

[119] M. Wolfe. *High Performance Compilers For Parallel Computing.* Addison-Wesley, 1996.

[120] B. Zhao, L. Huang, J. Stribling, S. Rhea, A. Joseph, and J. Kubiatowicz. Tapestry: A Resilient Global-Scale Overlay for Service Deployment. *IEEE J. Selected Areas in Communications*, 22(1):41–53, 2004.

Index